DICTIONARY OF ENVIRONMENTAL HEALTH

OTHER BOOKS IN CLAY'S LIBRARY OF HEALTH AND THE ENVIRONMENT:

E Coli: Environmental Health Issues of VTEC 0157
Sharon Parry and Stephen Palmer 0415235952

Environmental Health and Housing
Jill Stewart 041525129X

Air Quality Assessment and Management
Owen Harrop 0415234115

Air Pollution, Second Edition
Jeremy Colls 0415255643

Environmental Health Procedures 6th edition
W.H. Bassett 0415257190

Also available from Spon Press:

Clay's Handbook of Environmental Health 18th Edition
Edited by W.H. Bassett 0419229604

Decision-making in Environmental Health
Edited by C. Corvalán, D. Briggs and G. Zielhuis
0419259406 HB, 0419259503 PB

Groundwater Quality Monitoring
S. Foster, P. Chilton and R. Helmer
0419258809 HB, 0419258906 PB

Legal Competence in Environmental Health
Terence Moran 0419230009

Monitoring Bathing Waters
Edited by Jamie Bartram and Gareth Rees
0419243704 HB, 0419243801 PB

Statistical Methods in Environmental Health
J. Pearson and A. Turton 0412484501

Toxic Cyanobacteria in Water
Edited by Ingrid Chorus and Jamie Bartram
0419239308

Upgrading Water Treatment Plants
Glen Wagner and Renato Pinheiro
0419260404 HB, 0419260501 PB

Water Pollution Control
Richard Helmer and Ivanildo Hespanhol
0419229108

Water Quality Assessments
Deborah Chapman
0419215905 HB, 0419216006 PB

Water Quality Monitoring
Jamie Bartram and Richard Balance
0419223207 HB, 0419217304 PB

Urban Traffic Pollution
D. Schwela and O. Zali 0419237208

DICTIONARY OF ENVIRONMENTAL HEALTH

David Worthington

Routledge
Taylor & Francis Group
LONDON AND NEW YORK

First published 2003
by Spon Press

Published 2021 by Routledge
2 Park Square, Milton Park, Abingdon, Oxon OX14 4RN
605 Third Avenue, New York, NY 10017

Routledge is an imprint of the Taylor & Francis Group, an informa business

Typeset in Times by Taylor & Francis Books Ltd

British Library Cataloguing in Publication Data
A catalogue record for this book is available
from the British Library

Library of Congress Cataloging in Publication Data
Worthington, David, 1953–
Dictionary of environmental health / David Worthington.
(Clay's library of health and the environment)
Includes bibliographical references and index.
1. Environmental health–Dictionaries. I. Title. II. Series.
RA566.W68 2002
616.9′8′03–dc21 2002029208

ISBN 13: 978-0-415-26724-3 (hbk)

For Product Safety Concerns and Information please contact our
EU representative: GPSR@taylorandfrancis.com
Taylor & Francis Verlag GmbH, Kaufingerstraße 24,
80331 München, Germany.

Clay's Library of Health and the Environment

An increasing breadth and depth of knowledge is required to tackle the health threats of the environment in the 21st century, and to accommodate the increasing sophistication and globalisation of policies and practices.

Clay's Library of Health and the Environment provides a focus for the publication of leading-edge knowledge in this field, tackling broad and detailed issues. The flagship publication *Clay's Handbook of Environmental Health*, now in its 18th edition, continues to serve environmental health officers and other professionals in over thirty countries.

Contents

Foreword

There is a particular satisfaction in introducing this new work into the *Clay's Library of Health and the Environment* series since it is believed to be the first work of this nature in the environmental health field and is, perhaps, long overdue.

As well as being an extremely comprehensive dictionary it also includes an extensive cross-referencing system, which enables the reader to research particular areas of interest. It should prove a valuable source of information not only to those studying or practising environmental and public health but also to those involved in the wide-ranging concerns about the environment and its effects on health. I believe that the *Dictionary* will have a central place in the literature of these areas of interest for many years to come.

W.H. Bassett
Series Editor
Exeter
2002

Preface

As far as I can ascertain this is the first dictionary dedicated specifically to the field of environmental health, a fact that I find both surprising and puzzling. Our relationship with the environment in all its many facets is an inescapable and, to me, intensely interesting one. Perhaps one reason for such an omission to date has been that the scope for such a publication is potentially so wide that it is difficult to know where to begin. It was certainly a daunting prospect to sit down in front of a blank screen in the recognition that I had agreed to attempt the task and that the publishers were, not unreasonably, expecting me to keep to an agreed deadline.

Following the line of reasoning that environmental health touches on almost everything sooner or later it is inescapable to conclude that this book is only one view of a diverse and complex world. That there will be omissions is inevitable since, by trying to capture the breadth of what we understand by the term, I will doubtless fail to satisfy every expectation regarding the degree to which any one facet could possibly be addressed. I would like to be able to add to this volume in the future in both breadth and depth but the opportunity to do so will depend predominantly on how you, the user, view such a prospect. At this time I can only submit this book for your consideration and hope it meets at least much of your expectation.

Getting any book to the marketplace is an arduous task and I would be remiss if I did not thank *in absentia* the many people who I will probably never meet but whose work was essential to the task. In particular I wish to record my gratitude to Alice Hudson from Spon Press for helping me through the process and to Bill Bassett for thinking of me and commissioning this work in the first place. Finally and most certainly not least I wish to record a heartfelt thanks to my wife Fev who tolerated my prolonged absences from normal life during its production – I hope you all feel it has been worthwhile.

David Worthington

A

à l'anglaise Refers to food that has been plainly cooked, such as that which has been boiled in water or in stock.

à la carte In a restaurant, refers to food that is specially prepared to order and not therefore part of a set menu.

à la lyonnaise Refers to a foodstuff, a principal ingredient of which is shredded or chopped onion.

à la reine Refers to a food dish principally containing or based upon chicken.

a priori A Latin phrase applied to mean that which refers to what went before. It is used especially in logical reasoning to deduce effect from cause or to describe prior knowledge, especially that which has been used unethically or illegally.

abomasum The fourth or 'true' stomach of a RUMINANT. The others are the RUMEN (first stomach), RETICULUM (second stomach) and the OMASUM (third stomach). The abomasum is concerned with introducing the digestive juices into the stomach contents for digestion further along the digestive tract. The abomasum is also sometimes known as the 'RENNET' stomach.

See also: RUMINANT DIGESTION

abortion The expulsion from the womb, within (in humans) the first 24 weeks of pregnancy, of a foetus that exhibits no life signs.

See also: SPONTANEOUS ABORTION

abscess A localised accumulation of PUS. Abscesses are caused by the body's defensive reaction to bacterial invasion, usually, but not exclusively, in response to infection by *Staphylococcus* or *Streptococcus* organisms. An abscess may be ACUTE or CHRONIC.

absolute humidity The totality of water vapour contained within a given volume of air.

See also: RELATIVE HUMIDITY

absolute temperature That measured from the starting point of ABSOLUTE ZERO. If measured in degrees Celsius (or centigrade) the scale is known as the Kelvin scale. If measured in degrees Fahrenheit the scale is known as the Rankine scale.

absolute zero The theoretical lowest possible temperature. It is calculated to be at −273.16°C, equivalent to −459.67°F.

See also: ABSOLUTE TEMPERATURE

absorbed dose A description of the quantity of radioactive energy that a dose of IONISING RADIATION imparts to a unit mass (usually, but not exclusively, of living tissue). It is measured in units known as 'grays' (symbol Gy) – 1 gray is equivalent to 1 joule per kilogram.

See also: EQUIVALENT DOSE

absorptance A measure of the capacity of a surface to absorb radiation (including light). The absorptance is calculated as the ratio between the radiation that falls upon a surface to that leaving (or being transmitted through) the same surface. Absorptance is sometimes known as the 'absorption factor'.

absorption factor *see* ABSORPTANCE

abstract A concise summary of a piece of work intended to convey to the reader the essential content of a main text but (usually necessarily) excluding much of the detail.

abstraction The removal of water from ponds, lakes, rivers or underground sources for domestic, recreational, commercial or industrial purposes.

abutment A prop or support that, by means principally of its mass, is able to resist the forces of movement in another mass, such as the wall of a building or similar.

acaricide A chemical substance that kills ticks and mites (*Acarina*).

ACC *see* AEROBIC COLONY COUNT

acceleration A speeding up in the rate of movement of an object, the rate of acceleration being the rate of speeding up of the object. The acceleration of a body in free fall in a vacuum in the earth's gravitational field is 9.806 m per second (i.e. 32.174 feet per second).

accelerator A chemical agent that is used to increase the rate of progress of a chemical reaction.

acclimatisation The process of adaptation to a variation in climate sufficient upon completion to allow normal functioning of the body.

acetabulum Variously, the socket of the hip-bone, the round sucker of leeches and flatworms or the socket whereby the leg of an insect is attached to the thorax.

acetic acid An acid commonly used in the food industry as a PRESERVATIVE.

acetonaemia A condition of metabolic disturbance found in (usually older) cattle and sheep due to an abnormal increase in the amount of ketones in the blood. The condition is characterised by a distinct smell of ACETONE in the urine, milk or on the breath. The ketone component in the blood rises as a result of low levels of usable glucose in the blood. The condition may arise as a temporary phenomenon such as in cattle kept indoors during the winter or after calving, or it might arise as a result of a more serious condition. Affected animals can exhibit malaise, depression and constipation, and can become delirious, generally only for relatively short periods. Acetonaemia is also known as ketosis.

acetone Acetone (CH_3COCH_3) is a colourless ketone with a distinctive odour, which is used as a solvent. Acetone is also known as propanone. It is found normally in small quantities in the blood and the urine. Levels become elevated during periods of starvation, in conditions such as diabetes mellitus or sometimes cancer, during prolonged vomiting and sometimes during fever – the latter especially in children. The detection of the odour of acetone in food animals during postmortem inspection is indicative of disease or poor condition.

See also: ACETONAEMIA; ACETONE BREATH

acetone breath A condition in which the odour of ACETONE can be detected on the breath. It is suggestive of a condition of diabetes mellitus as diabetics evacuate acetone from their bodies through their lungs during exhalation.

acetylsalicylic acid The chemical name for ASPIRIN.

acid A liquid that reacts with metallic oxides to form SALTS and water; reacts with (most) metals to produce salt and hydrogen; or reacts with carbonates to produce salts, CARBON DIOXIDE and water. Although they are not ELECTROLYTES in pure form, they become so when dissolved in water.

See also: ALKALI; BASE

acid barrier of the stomach The term used to describe the acid conditions in the stomach that can kill certain bacteria (see BACTERIUM), thereby effectively producing a barrier to infection by preventing viable but susceptible pathogenic bacteria from gaining access to the lower portions of the digestive tract.

acid-fast bacteria Bacteria (see BACTERIUM) that, when stained with a dye during microbiological examination, are not easily discoloured by the action of an acid or similar agent.

acid rain Rainwater that has absorbed gaseous pollutants such as CARBON DIOXIDE, nitrogen oxides or sulphur dioxide sufficient to lower the PH to become distinctly acidic; this is usually taken as being lower than around pH 5.0. The gases are commonly derived from ANTHROPOGENIC activity, such as from the combustion of HYDROCARBON fuel, although the phenomenon can occur naturally as a result of emissions of aerosols following volcanic activity. The deposition of acid rain over a period of time will lower the pH of the soil and of run-off waters, sufficiently in extreme cases to cause damage to forests, vegetation and aquatic environments.

acid soil An imprecise term but generally taken as any soil with a PH (variously, dependent on source) of less than 7.2 or 6.3.

acidophile A micro-organism that grows better in acid conditions. These are typically within the PH range 2 to 4.5.

See also: HIGH-ACID FOOD; LOW-ACID FOOD

acne A widespread inflammation of the skin of (especially) the face, neck and chest, and is most commonly associated with adolescents. It is caused by the proliferation of bacteria (see BACTERIUM) in the SEBACEOUS GLANDS, which then leads to infection. Acne rosacea (sometimes simply known as ROSACEA) is a chronic condition of flushing of the skin.

See also: BLACKHEAD; ECZEMA

Acquired Immune Deficiency Syndrome *see* AIDS

acre A measurement of area. It is equivalent to 4,840 square yards or 4,046.86 m^2 (i.e. approximately 0.4047 of a HECTARE).

acrodynia A disease of infants characterised by restless behaviour, weakness and reddening of the face or extremities. It was caused by chronic mercury poisoning due to the inclusion of mercury in teething preparations. It virtually disappeared following withdrawal of these preparations in the 1950s. The condition was also known as Pink's disease.

Act In the legal sense, a main piece of statute law passed by a legislature. It is generally known as PRIMARY LEGISLATION. An Act is usually a stand-alone piece of legislation that is not subservient to other legislation unless such other has precedence by being more specifically applicable to the case or situation in question.

See also: BILL; REGULATION

actinomycetes Filamentous bacteria (see BACTERIUM).

actuary A statistician. The term is especially applied to one who is employed in the insurance industry to weigh risks and to calculate premiums and dividends.

acuity Refers to a particular acuteness or keenness in capability. The term is especially applied to describe visual acuity, which is an ability to be able to discern or to distinguish highly detailed characteristics.

acupuncture A traditional method of healing practised originally by the ancient Chinese. It involves the insertion and subsequent rotation of needles in specific areas of the body dependent on the condition being treated. It is a proven anaesthetic and operations can be performed whilst the patient is under its influence. It has been suggested that a plausible explanation for its effectiveness is that it stimulates the production of ENDORPHINS.

acute A term commonly synonymous with sharp. When applied to DISEASE or adverse health event it means of short onset or duration, usually of a severe disease or in relation to poisoning. Acute also applies to describe a sharp pain. Acute is generally taken as the opposite of CHRONIC.

acute toxicity That in which the effects brought about as a result of the toxin occur within a few hours (or at most a few days) following exposure.

adduce To state or cite examples or to produce reasoning as evidence in support of a contention, such as in a court of law.

adduct A chemical group that is attached to a large molecule such as a protein or DEOXYRIBONUCLEIC ACID (DNA) by means of a COVALENT BOND. When applied to a muscle the term is used to describe the process of pulling, such as of a leg or arm.

adenitis An inflammation of a lymph gland or lymph node.

adenovirus A type of VIRUS. They have a hexon shape, no outer membrane and are stable at low pH, and they are consequently stable in the gut. Adenoviruses have been isolated from every species of mammal, bird and amphibian that has been studied. There are at least forty-seven serotypes that have been identified in man. The serotypes are divided into six sub-groups (A–F), based on a number of criteria, including the severity of the disease that they cause. Infections occur throughout the year and are usually regarded as endemic rather than epidemic. They are slow growing in cell culture, making laboratory isolation difficult, a factor that may result in many infections going undiagnosed.

Various serotypes cause a range of diseases, including infectious infantile respiratory disease, ocular disease and kidney disease. Person-to-person spread is significant. There is no evidence that they are involved in food-borne or water-borne outbreaks. Adenoviruses can be isolated from the stools of infected patients for up to 2 years following infection. Viruses are excreted in great numbers during the acute stage. Incubation is typically 8–10 days. Symptoms are typically vomiting and fever for 2 days followed by diarrhoea for 9–12 days. This latter may lead to problems of dehydration. Peak virus shedding occurs for 3–13 days. Symptoms are generally mild but the infection accounts for between 4–15 per cent of children hospitalised with gastroenteritis.

adiabatic The term used to describe a thermodynamic process that occurs without the loss or gain of heat.

adipose tissue The FAT storage tissue in the bodies of animals. It has three main uses. It is primarily a food storage mechanism but also provides thermal insulation and protects internal organs from external physical damage.

adobe A structure consisting of sun-dried but unburnt clay bricks. This is a very early construction material although it is ideally suited to the areas in which it was (and in some areas still is) used.

See also: COB

adrenaline A hormone secreted by the adrenal medulla (adrenal glands/suprarenal glands) situated above the kidneys in mammals. It is also known as the 'fight or flight' hormone because of its function to stimulate the sympathetic nervous system when the animal is stressed. The hormone raises blood pressure, increases the amount of blood sugar (GLUCOSE) and reduces blood flow to the smaller blood vessels allowing an improved body response to threats. In the USA it is also known as epinephrine.

See also: ALKALOID

advanced gas-cooled reactor A type of NUCLEAR REACTOR used for generating heat. It is a development of the MAGNOX REACTOR and uses enriched uranium oxide contained within a stainless-steel cladding as the fuel.

adventitious roots The collective term used to describe the fibrous root system of plants.

adverse health effect Any change in the morphology, physiology, growth or development or life span of an individual that results in some form of impairment of function and is attributable to the effect of an external influence, without which the adverse effect would not have occurred.

adze A heavy hand tool used for dressing timber. It comprises a wooden handle to which a blade is attached at right angles. The adze was one of the first tools developed by humans.

aeolotropic Means the same as ANISOTROPIC.

aerobe An organism that can respire in the presence of free oxygen. If the organism requires free oxygen to respire it is known as an OBLIGATORY aerobe (i.e. it is *obliged* to utilise oxygen to respire). If the organism prefers but does not require oxygen, and can respire either with or without it, the organism is known as a FACULTATIVE aerobe.

See also: ANAEROBE

aerobic Refers to the requirement of a living organism for free atmospheric oxygen to survive.

See also: ANAEROBE; FACULTATIVE; OBLIGATORY

aerobic colony count (ACC) A microbiological term for the number of microcolonies of bacteria (known as COLONY-FORMING UNITS (CFU)) growing on the surface of a (usually AGAR) microbiological plate under AEROBIC conditions. The ACC is usually expressed as CFU/gram of material examined. It provides an assessment of the viable bacterial loading of a sample.

The reporting of bacteriological component of a sample has been known variously in the UNITED KINGDOM as the Total Colony Count (TCC), the Total

Plate Count (TPC) and the Total Viable Count (TVC), and was also sometimes known as the Standard Plate Count (SPC). All of these terms, strictly speaking, refer to different approaches to counting the presence of an overall microbiological loading. They can however be generally considered as interchangeable for practical purposes, but in absolute terms comparison should be made under the guidance of an expert microbiologist familiar with the analytical techniques employed in reference to each particular sample.

aerodynamic diameter The diameter that a spherical particle of a density of 1 g/cm^3 would need to possess to have the same terminal velocity due to gravitational force in calm air as the particle under consideration. It is used to describe how an airborne particle is likely to behave under the influence of gravity.

aerodynamic noise That created by the passage of air (wind) over or across a surface.

aerosol A finely divided suspension of liquids or solids in particulate or droplet form in air. A key consideration of an aerosol is that the suspended particles are of such a size as to resist the effect of gravity.

See also: FUME; PARTICULATE MATTER; SMOKE

aestivation The phenomenon in which some creatures reduce body activity and enter a dormant state in order to survive the summer months.

See also: HIBERNATION

aetiology The study of the causation and generation pathway of a disease. The US spelling is etiology.

affidavit A written statement made under oath or signed in the presence of a person (such as a solicitor) who is qualified to administer a legally binding oath. An affidavit is usually used as evidence in legal proceedings in cases where the individual making the affidavit is unable to be present in person.

See also: DEPOSITION

afforestation The systematic planting of trees on land that has not been used for such purpose in the recent past.

aflatoxin A particular type of MYCOTOXIN produced by certain species of fungi such as those of the fusaria (e.g. *Aspergillus flavus*) and other fungi.

See also: ENDOTOXIN; EXOTOXIN; TOXIN

aftershock One of (sometimes a series of) smaller tremors that follow the main shock of an earthquake. Smaller tremors that occur before the main shock of an earthquake are known as foreshocks.

agar A seaweed extract, also known as agar-agar, used as a solidifying or gelling agent used in cooking. In microbiology it is used to solidify bacterial growth media prior to spreading on an AGAR PLATE. One of its most useable characteristics is that it melts at around 80–90°C but only solidifies at around 40°C. It then remains solid until it achieves melting temperature again.

agar plate A piece of equipment used in microbiological examination. AGAR is mixed with a bacterial growth medium (often a meat broth base) and spread on a petri dish or microscope slide (the 'plate'). The prepared substance being examined is spread on the plate that is then incubated, dried and stained to reveal the resultant microbial growth. From this the bacterial contamination of the specimen is calculated based on the amount of bacterial growth.

See also: AEROBIC COLONY COUNT

Agenda 21 A document derived from the UNITED NATIONS CONFERENCE ON ENVIRONMENT AND DEVELOPMENT (the 'Earth Summit') in Rio in 1992. It was an important milestone in the development of a new vision for the environment and introduced specific objectives and action plans for the promotion of a 'new global partnership for sustainable development'.

agglutination In microbiology, the phenomenon in which ANTIGENS are coalesced together by ANTIBODIES to form a clump, often of sufficient size to be seen by the naked eye.

agglutinin A substance that has the capacity to cause AGGLUTINATION.

aggregate In construction terminology, the inert component, usually consisting of sand, gravel or stone, of CONCRETE. Alternative materials may be used for the creation of specialist concrete, such as in the production of more lightweight material. Aggregates help to provide bulk and act as ballast in the final product, and aid the capacity of the mixture to pour when in its fluid state.

aggregates levy A tax levied on companies that extract AGGREGATE from the ground. It is intended to penalise the use of new aggregates and thereby reduce the demand for these products in favour of recycled aggregate.

agrochemical Any chemical used to promote agricultural production. These chemicals include fertilisers, herbicides, pesticides, insecticides and fungicides.

AIDS The acronym for Acquired Immune Deficiency Syndrome, a condition created by the human immunodeficiency virus (HIV) in which the body's immune system becomes suppressed and susceptible to many serious and life-threatening infections, particularly pneumonia and rare cancers such as skin tumours of Kaposi's sarcoma. The HIV is spread principally through sexual contact (either homosexual or heterosexual) or by transplantation of the virus from an already infected individual during procedures such as blood transfusion, organ transplantation and the use of contaminated shared needles in INTRAVENOUS (IV) drug use. AIDS was first recognised in 1981 in the USA. The largest reservoir of infection is in sub-Saharan Africa where millions of people are affected, having acquired the infection predominantly through heterosexual route.

aiguillette In culinary practice, is thin strip of cooked meat, fish or poultry.

aillade In culinary practice, any of a variety of strongly flavoured garlic sauces used as an accompaniment principally for salads.

air brick A brick moulded or shaped to contain hollow passages as integral parts of its structure, these passing from one side of the brick to the other to provide or aid ventilation when the brick is in place.

air changes per hour A means of assessing or determining the ventilation capacity or requirements of a room or other air space. The number of air changes per hour is calculated as the volume of air passing through the ventilation system per hour divided by the volume of the air space. The most accurate determination of ventilation requirement is to calculate this for each space specifically based on the number of occupants, the air demand of machinery or equipment and the activities being undertaken. Most calculations are estimates based on broad guidelines, although the ultimate recommendations will depend greatly on the volume of the space in question. For example a storeroom would generally require as little as 1 air change per hour and habitable rooms up to around 6 air changes per hour. The ventilation needs for workrooms and those used for industrial processes should be assessed on merit.

air conditioning A system of controlling or maintaining the atmosphere within buildings, usually employing a single integrated piece of plant or equipment. True air conditioning manages parameters of temperature, humidity and particulate material (dust), although the term is often commonly applied to systems that control temperature and humidity alone.

See also: HUMIDITY; RELATIVE HUMIDITY

air drying The method of allowing the natural evaporation of moisture from an artefact or substance to the atmosphere. The term is applied variously to laundry, kitchen washing-up, the natural seasoning of timber and as a means of preserving certain foods such as fruits (e.g. sun-dried tomatoes) or meats (e.g. Parma ham).

air lock A chamber capable of being sealed or otherwise isolated from (usually) both of two larger air spaces between which it is located. The object of the air lock is to ensure that transmission of air between the two larger air spaces is restricted or minimised. Air locks are used to isolate contaminated air spaces from non-contaminated (such as isolating asbestos-contaminated air spaces from other areas), to facilitate decompression (as for underwater diving operations) and to reduce heat loss between internal and external spaces such as at the entranceways to buildings.

air test Used to test drainage systems, predominantly soil or waste pipes. The length of pipework is first sealed and air pressure within the sealed system is raised by means of a pump. Pressure is indicated by a MANOMETER connected to the system. A satisfactory test is one in which a constant pressure is maintained within defined limits for a specified period. The pressure used, the defined limits and the length of time are usually prescribed either by the manufacturers of the equipment used or by published guidance.

See also: COLOUR TEST; SMOKE TEST; WATER TEST

air washer A piece of equipment installed in an air-conditioning system, an air emission system or similar that employs water sprays or showers to remove suspended contaminants. The water in such systems is usually recycled after having been suitable treated or filtered.

airway permeability A measure of the capability of an air passageway of a body's respiratory system to allow the passage of air.

alabaster The pure and densely crystalline form of GYPSUM. It is usually white or translucent and can be easily carved or polished, hence its frequent use in the production of statues and containers. The term is also used to describe types of CALCITE.

alanine aminotransferase An enzyme that is thought to be activated when toxic damage is caused within an organism. Detection of elevated levels of the enzyme might therefore indicate that such damage has been caused, even in the absence of discernible clinical symptoms.

See also: ASPARTATE AMINOTRANSFERASE

ALARA An acronym for 'as low as reasonably achievable'. The ALARA principle, as it is known, ensures that risks from or exposures to a given substance or energy (such as radiation) shall be kept to a minimum. This shall be consistent with any existing standards but will also take into account any other relevant matters (such as economic or social factors) that may impinge on the decision of 'reasonableness'.

See also: ALARP, BEST PRACTICABLE MEANS

ALARP An acronym for 'As Low As Reasonably Practicable'. It is used in risk categorisation and analysis to describe the reduction of risk to the minimum within the confines of practicality.

See also: ALARA

albumen The correct term for the white (actually clear or translucent) protein component of shelled eggs.

albumin A water or SALINE-soluble PROTEIN.

alcohol Not a single compound, but one of a group of HYDROCARBONS in which a hydroxyl (i.e. a hydrogen/oxygen molecule, e.g. OH) molecule replaces a hydrocarbon (i.e. a hydrogen/carbon) bonding. Alcohols have the general formula of $CnH_2n_{+1}OH$. The simplest alcohol is METHANOL (CH_3OH), also known as 'wood alcohol'. In alcoholic drinks the alcohol content is provided by ETHANOL (C_2H_5OH).

aleotropic *see* ANISOTROPIC

algae (singular alga) Members of a group of living organisms that, like true plants, possess CHLOROPHYLL but which do not have recognisable leaves, stems or roots. The group includes DIATOMS, DINOFLAGELLATES and seaweeds.

algal blooms *see* BLUE-GREEN ALGAE

algorithm A systematic procedure that is applied to a problem and which uses a number of arithmetic or logical steps to produce a solution.

See also: HEURISTIC

alkali A BASE that is soluble in water.

See also: ACID

alkaloid One of a large group of naturally occurring (i.e. organic) nitrogenous base compounds that combine with acids to form crystalline salts. Alkaloids exhibit widely differing chemical structures. Most alkaloids are produced by plants. The most commonly known alkaloids are morphine, STRYCHNINE, CAFFEINE, QUININE and NICOTINE. ADRENALINE is one of the few animal-derived alkaloids. Many of the plant alkaloids have been used in both human and veterinary medicine for their various properties including as ANAESTHETICS, ANALGESICS, muscle relaxants, psychedelic agents, vasoconstrictors (compounds that restrict the flow of blood vessels), stimulants and tranquillisers.

alkene Another word for PARAFFIN.

alkylating agent One that has been separated from an alkyl group, such as an ADDUCT, which had formed a COVALENT BOND to proteins or DEOXYRIBONUCLEIC ACID (DNA). Alkylating agents can be carcinogenic and/or mutagenic. Some can also suppress certain activities of the immune system.

Allen key An L-shaped rod of hexagonal cross-section used, when inserted into a correspondingly sized and shaped hole in the head of a bolt, screw or similar, to rotate and thereby drive home or extract the fastening.

allergen Any substance that produces an allergic reaction when it comes into contact with a sensitive individual.

See also: ALLERGY

allergic sensitiser An irritant or ALLERGEN that, after an initial exposure, creates a sensitivity that is manifest as an allergic reaction on subsequent exposure to the same substance.

allergy A particular sensitivity on the part of an individual demonstrated on contact with a particular substance, the latter being known as an ALLERGEN. The allergic reaction is produced when allergens stimulate the production of

ANTIBODIES that in turn release substances which produce an immune response; the response is the allergy. An allergy can be produced as a result of contact with almost anything, but the most common allergies include hay fever, eczema and food allergies.

See also: ANAPHYLAXIS

allopathic In relation to a treatment for disease, refers to one that is intended to induce a different (usually an opposite) reaction in the body from that produced by the disease. For example, a medicine that lowered body temperature and was administered in response to a fever would be allopathic.

See also: HOMEOPATHIC

allotrope A variant of the same element but with a different physical form, e.g. diamond and graphite are different allotropes of carbon.

alloy A material with metallic properties made by combining two or more metals together or by combining a metal with one or more non-metallic elements.

alluvial Means of or pertaining to ALLUVIUM. An alluvial plain is a relatively flat or level area of fertile land created by the deposition of sand, silt and mud following the consecutive flooding by rivers and its subsequent abatement.

alluvium (plural alluviums or alluvia) The soil deposited as fine grains of sand, mud or silt by flowing water, especially that deposited by rivers during flood conditions on the areas of land so flooded.

alpha particle A radioactive particle equivalent to the nucleus of a helium atom, i.e. comprising two protons and two neutrons, and emitted by a RADIONUCLIDE.

See also: BETA PARTICLE; GAMMA RADIATION; IONISING RADIATION

alternating current (AC or ac) An electric current that regularly alters its direction of flow. The change from one direction to another and back again is known as a cycle. Alternating current is usually supplied at 50 or 60 cycles per second.

See also: DIRECT CURRENT

alternative treatment/medicine The collective term used to describe a wide range of therapeutic (or claimed therapeutic) techniques or remedies that are not in general use within the established practice of the Western medical world. There are numerous different therapies that come within the all-embracing term; some (such as acupuncture and yoga) are widely accepted as efficacious, whilst others are less so. Many of the therapies pre-date modern treatments and protagonists claim that the appellation 'alternative' is misleading.

alternator A device for generating alternating current.

See also: DYNAMO

alveolar region of the lung The area at the termination of the branching airways of the lung comprising the ALVEOLI (singular alveolus) where the gas exchange between the atmosphere and the blood occurs.

alveolitis *see* PNEUMONIA

alveolus (plural alveoli) A minute air sac at the end of the air passages of the lung. In humans the alveolus is around 0.3 mm in diameter. The expression also refers to the sockets in the jawbone into which the teeth fit.

See also: ALVEOLAR REGION OF THE LUNG

Alzheimer's disease A degenerative, non-infectious and progressive disorder of the cerebral cortex of the human brain characterised by premature senile dementia becoming manifest in middle or old age. The cause is unknown and various theories have been proposed, including the notion that it is linked to aluminium deposition in the brain cells, which is in turn linked to high intake of dietary aluminium as a result of using aluminium kettles and cooking utensils – this theory remains unsubstantiated.

amalgam Generally, any combination or blend of two or more substances; more specifically it is applied to an alloy of mercury with another metal, especially with silver when used as dental amalgam.

ambient Means relating to or of the immediate surroundings. The term 'ambient temperature' is often used as a synonym for 'room temperature'.

Ames test An ASSAY conducted ***in vitro*** to detect mutations in a gene. It is named after the person leading the team that developed it and uses strains of *Salmonella typhimurium* in conducting the assay.

amine An organic base that is formed when one or more of the hydrogen atoms of ammonia (NH_3) are replaced by an organic group. Amines are further classified as primary, secondary or tertiary dependent on whether one, two or three respectively of the hydrogen atoms are replaced.

See also: AMINO ACID

amino acid A compound formed by the joining of carboxyl groups and a basic AMINE. Amino acids are the basic building blocks of PROTEINS. They are also used in many processes of METABOLISM and contribute to the synthesis of a variety of secondary compounds including COENZYMES and HORMONES. There are several hundred known amino acids, but only twenty commonly occur in proteins.

ammeter An instrument used to measure electric current; the unit used is the AMPERE.

ammonia Ammonia (NH_3) is a colourless gas that is lighter than air and is highly reactive. It has a distinct pungent odour and can be detected by humans at concentrations below 5 ppm in the atmosphere. It is flammable at concentrations of around 16–25 per cent by volume in air. The eye and respiratory tract are particularly sensitive to ammonia, and irritation can occur at levels between 400–700 ppm. Elevated levels (i.e. above 1,700 ppm) will result in repeated coughing, and prolonged exposure of around 30 minutes at such levels may be fatal.

amnesiac shellfish poisoning (ASP) An illness caused by the consumption of shellfish, primarily bivalve molluscs, which have ingested and accumulated in their hepatopancreas toxin produced by algal blooms (see BLUE-GREEN ALGAE). The toxin responsible is known as Domoic Acid. The only recorded outbreak occurred in Canada in 1987. Symptoms at that time were reported as including gastroenteritis, headache and short-term memory loss.

See also: DIARRHETIC SHELLFISH POISONING; NEUROLOGICAL SHELLFISH POISONING; PARALYTIC SHELLFISH POISONING

amniocentesis A medical diagnostic procedure for detecting genetic abnormalities in the foetus. The procedure involves the removal via a hollow needle of AMNIOTIC FLUID from the uterus of the pregnant female and the subsequent examination for chromosomal or genetic abnormality of extracted cells concentrated by CENTRIFUGE.

amnion The membrane that surrounds the developing foetus of mammals, birds and reptiles, and encompasses the AMNIOTIC FLUID.

amniotic fluid That enclosed or encapsulated by the AMNION.

See also: AMNIOCENTESIS

Amoco Cadiz An oil tanker wrecked off the Brittany coast of France in 1967 spilling 223,000 tonnes of crude oil. It was one of the worst environmentally damaging oil spills of the 20th century.

See also: TORREY CANYON

amorphous Without definite or fixed shape or structure. In relation to a mineral it refers to one without a crystalline structure.

amosite One of a group of fibrous silicates known generically as ASBESTOS. It belongs to the amphibole group that, when magnified, can be seen to have short, needle-like fibres exhibiting a columnar shape. Amosite is known more commonly as 'brown' asbestos, the name deriving from its colour when seen under the microscope.

Asbestos from the amphibole group is usually incorporated into materials that benefit from their fire, heat or acid resistance capabilities such as fire cements and in some insulation materials.

ampere A unit measure of electric current. It is often abbreviated to 'amp'. One ampere is that which, if flowing in two parallel conductors, both of which are infinitely thin and long and 1 metre apart in a vacuum, will produce a force between the two conductors of 2×10^{-7} newtons per metre length.

amplitude A measure of the breadth, limits or extent of a particular characteristic. The term is specifically used in acoustics to describe a particular sound pressure level in terms of its loudness.

amplitude modulation One of the principal means of encoding a radio signal to transmit audio or visual images, differentiation being achieved by variations in the amplitude, the frequency remaining the same.

See also: FREQUENCY MODULATION

amyloid A substance comprising protein and polysaccharides. It can be deposited in some body tissues and is deposited in the brain in the form of amyloid plaques, an indicator of Creutzfeldt-Jacob disease (see CJD).

anabolic steroid One of a group of synthetic hormones that promote protein storage and tissue growth.

anaemia A decrease in the number of red blood cells (erythrocytes) or the concentration of HAEMOGLOBIN in the blood, both of which reduce the oxygen-carrying capacity causing tiredness, lack of energy and pallor.

See also: APLASTIC ANAEMIA; PERNICIOUS ANAEMIA

anaerobe A term applied usually to a micro-organism that does not require the presence of free oxygen to respire. If the organism can only respire in the absence of oxygen it is known as an OBLIGATORY anaerobe (i.e. it is *obliged* to live in the absence of oxygen). If the organism prefers but does not require an oxygen-free atmosphere and can respire either with or without it, the organism is known as a FACULTATIVE anaerobe.

See also: AEROBIC

anaerobic digestion A process whereby organic waste material is broken down in a closed vessel in the absence of oxygen. The process involves the production of CARBON DIOXIDE and METHANE (the latter can be used as a fuel) as well as solids and liquors that can be recycled in fertilisers and compost.

anaesthetic A substance that induces loss of sensation in living tissue. General anaesthetics are substances used in medical procedures to produce a reversible state of unconsciousness in a patient, usually during surgery. Local anaesthetics induce a localised desensitisation. Usage has blurred the differential meaning between anaesthetic and ANALGESIC. Original usage suggests that an anaesthetic not only induced loss of sensation but also acted as a muscle relaxant and blocked motor capacity (i.e. prevented movement). Whilst this latter remains true of general anaesthetics the term is nowadays commonly used interchangeably with the term analgesic.

analgesic A substance that induces loss of the pain sensation in an area or tissue of the body; analgesics are therefore called painkillers. Analgesics do not induce unconsciousness or affect other bodily functions such as motor capacity or co-ordination of the muscles.

See also: ANAESTHETIC; DISTALGESIC

analog *see* ANALOGUE

analogous Refers to something that is like, or has similar properties to, something else but which has a different origination or means of operation – for example the paddle of a canoe is analogous to the flippers of a seal.

See also: ANALOGUE; HOMOLOGOUS

analogue A physical entity used to represent or measure something else. An analogue clock, for example, is one that uses hands or pointers to indicate the time. In the USA the word is generally rendered as analog.

See also: ANALOGOUS

analysis of variance A test of statistical significance using an adjusted MEAN of comparable sets of data.

anaphylaxis An excessive allergic sensitivity exhibited by an individual following exposure to a certain substance such as antibiotics, food or certain animals. The acute form is known as anaphylactic shock, in which the individual exhibits severe shortage of breath, much reduced blood pressure and rash. These symptoms can manifest as shortly as a few minutes following exposure; in extreme cases such shock can lead to death.

See also: ALLERGY

andiron One of a pair of decorative metal stands used to support logs, usually positioned adjacent or near to a fireplace, allowing the logs to be stored prior to burning and also to facilitate or aid their drying.

androgenic Means capable of stimulating either the production of male sexual organs or secondary male characteristics.

See also: ENDOCRINE DISRUPTOR

anechoic Means having little or no capacity to reflect sound. It is especially applied to describe surfaces that absorb all or the majority of sound waves that fall upon them. An anechoic chamber is one constructed using anechoic surfaces and is used in acoustic testing.

anemometer Generally, any instrument that measures the rate of movement of a fluid. The term is specifically applied to those instruments that are used to measure wind or air speed, sometimes also recording direction.

anergy The condition of having a reduced hypersensitivity to a specific ANTIGEN.

aneugenic Refers to that which is of or relates to ANEUPLOIDY.

aneuploidy One of the three levels of mutation of genetic material that can affect the cell; the other levels are GENE and CLASTOGENICITY. Aneuploidy is that which relates to aberrations in respect of the number of chromosomes (i.e. the total number of chromosomes in a new cell is not an exact multiple of the normal – some have either been lost or gained).

angina A generic term, usually applied in reference to the body, which literally means choking. It can therefore be applied to any choking or blockage situation in which the free flow of a substance is restricted or prevented. Most frequently the term is used as a shortening of the medical term 'angina pectoris', which is a sharp pain in the chest caused by an inadequate blood supply to the heart, mainly due to a thickening of the walls of the coronary arteries.

See also: ARTERIOSCLEROSIS; CORONARY HEART DISEASE; INFARCTION

angiosperm A seed-bearing plant in which the fertilised seeds develop into fruit. Effectively the angiosperms are any trees except the conifers.

See also: GYMNOSPERM; HARDWOOD; SOFTWOOD

angle bar *see* ANGLE IRON

angle grinder A hand-held power tool that employs a rotating circular disc for tasks involving abrasion or cutting.

angle iron A structural bar manufactured from a length of plate steel and formed so as to create (usually) a right angle, the sides in cross-section being of either equal or unequal length dependent on requirements. An angle iron is also known as an 'angle bar' or sometimes simply as an 'angle'.

angle of incidence The angle to the perpendicular that a line of radiation makes upon contact with another medium, i.e. the angle at which the radiation meets the surface.

See also: ANGLE OF REFRACTION

angle of refraction The angle to the projected perpendicular that a line of radiation makes at its point of entering a new medium, i.e. the angle at which the radiation leaves the surface.

See also: ANGLE OF INCIDENCE

ångström or ångström unit (Å) An obsolete measure of length; it has largely been superseded by the nanometre. One ångström unit is the equivalent of 10^{-10} metres (i.e. 0.1 nanometres). Its principal use was to describe the wavelengths of various forms or ELECTROMAGNETIC RADIATION.

anhydride A compound produced by the dehydration of another compound. The term is also applied to a compound that produces an acid or a base upon the addition of water.

anhydrite Another name for anhydrous calcium sulphate ($CaSO_4$), a colourless or greyish-white mineral used in the manufacture or cement and fertilisers. It is found naturally in certain sedimentary rocks or can be manufactured from gypsum by the removal of the WATER OF CRYSTALLISATION.

anhydrous Refers to that which contains no water or from which all the water has been removed. The term is especially applied to minerals from which the WATER OF CRYSTALLISATION has been removed.

animal by-product Any product that is not for human consumption derived from an animal produced for food. Such products include meat and bone meal and blood meal.

anion A negatively charged ION.

See also: CATION

anisotropic Means having different physical properties (especially strength) or responding unequally to stimuli, in different directions. The term is synonymous with the term aeolotropic.

anneal To strengthen or toughen something by heating it to such a temperature that relieves inner stresses in its structure, holding this temperature for sufficient time to ensure the effect is evenly distributed and then cooling it sufficiently slowly so as to ensure that such stresses do not return.

annual In botany, is a plant that completes its life cycle and dies within a single year.

annual ring The pattern seen on a cross-section of the stems, trunk, branches or roots of woody plants growing in temperate climates – each 'ring' indicating the extent of one year's growth. The ring is created by the differential growth of the cells between colder and warmer periods, the former leading to slower and hence more densely packed cell growth, this appearing as a distinct 'edge' to each concentric ring. Plants growing in climates in which the growth rate is more evenly distributed throughout the year are less likely to exhibit distinct annual rings.

annuity Alternatively a fixed sum of money paid at regular or specified intervals or the right to receive or the duty to pay such a sum.

annulus The area between two concentric circles or rings.

anode The positive electrode in an electrolytic cell. The negative electrode is known as the CATHODE.

anodise To provide, by electrolysis, a protective coating of an oxidised metal (such as of aluminium or magnesium) on the surface of another metal. It is written as anodize in the USA.

anodyne Refers to that which is capable of relieving pain or distress, especially in relation to the power of certain medicinal drugs.

anorexia Classically, simply a loss of appetite. It is more commonly and popularly used these days to denote a chronic condition in which the patient loses appetite and ceases eating to such a degree that weight loss may become life-threatening. A variant of anorexia is bulimia nervosa, a condition that is thought to be psychologically induced and in which the patient feels a compulsion to over-eat, followed by induced vomiting or use of laxatives to avoid weight gain. Perhaps due to the (albeit temporary) large food intake, sufferers tend more to obesity (to which they have a pathological aversion) rather than to weight loss. The term bulimia is applied simply to the presence of an insatiable desire for food and should not be confused with the resultant induced voiding of food present in bulimia nervosa.

antagonism In the chemical sense, the phenomenon whereby the effect of two chemicals together is less than the sum of the effects of the chemicals in isolation.

See also: SYNERGY

anterior Means located at or near to the front.

See also: POSTERIOR

anthesis Refers to the stage in the floral development of a plant in which pollen is shed.

anthrax An acute bacterial ZOONOSIS caused by the spore-forming bacterium *Bacillus anthracis*. The disease is endemic in many parts of the world. In humans the disease most frequently affects the skin, exhibiting as a rash or necrotic vesicles. If untreated the disease can develop into septicaemia, meningitis or ultimately death. The bacterium is transmitted as a result of contact with infected animals, hides or blood and infection can be airborne. Traditionally the skin infection variation of the disease was common in workers handling sheep fleeces when it was known as 'Wool Sorters' Disease'.

anthropogenic Means caused by man or 'man made'. GLOBAL WARMING is said to be anthropogenic because it is thought to be caused by such human activity as the burning of fossil fuels.

antibiotic A chemical originally produced by a micro-organism, which inhibits or prevents the growth of another micro-organism. Antibiotics were originally developed as human medicines, but they have subsequently developed as veterinary medicines, as prophylactics in veterinary medicine (for which use they are sometimes referred to as GROWTH PROMOTERS) and as preservatives in food. There is global concern that the improper or uncontrolled use of antibiotics is exacerbating the development and spread of ANTIBIOTIC RESISTANCE. Antibiotics are not effective against VIRUSES.

antibiotic resistance The ability of a micro-organism to withstand or reduce the effect of an ANTIBIOTIC. Antibiotic resistance may be acquired by evolution or genetic development within a species or acquired from the transference of genetic material from other micro-organisms, either of the same or compatible species. The improper or inappropriate use of antibiotics increases the chance that antibiotic resistant organisms will develop or spread.

See also: HORIZONTAL GENE TRANSMISSION; NATURAL SELECTION

antibody A protein formed by the body as a defence mechanism in response to an attack by an ANTIGEN. Antibodies are usually specific to one antigen and the detection of the former in the blood can be used in retrospect to indicate the prior presence of the antigen. Antigens are capable of neutralising toxins and of destroying bacteria.

See also: ALLERGY

anticlockwise Means travelling in a circular pathway in the opposite direction to the hands of a clock. It is synonymous with counter-clockwise, also rendered as widdershins and withershins.

anticoagulant A substance that reduces or prevents the clotting (coagulation) of the blood. Blood clots naturally following damage to blood vessels or on exposure to air and the anticoagulant counteracts this automatic body defence mechanism. Anticoagulants are administered in controlled doses as medicines to reduce the risk of heart attacks, in heavier doses as poisons (particularly in rodenticides) and are found naturally in many blood-sucking PARASITES (such as mosquitoes and vampire bats) to prevent blood clotting whilst feeding.

anticyclone A meteorological condition in which winds rotate around an area of high pressure. In the northern hemisphere the winds rotate in clockwise direction, but anticlockwise in the southern hemisphere.

See also: CYCLONE

anti-fouling agent A chemical compound painted on that portion of the external hull of a (particularly sea-going) ship that is usually under water. It is used to prevent the colonisation of the hull by aquatic life forms such as barnacles and seaweed that can attach themselves to the hull as to any immersed solid object, thereby potentially destabilising the vessel and producing added drag to the hull. By their very nature anti-fouling agents are toxic. Many anti-fouling agents contain TRIBUTYL COMPOUNDS.

antigen Any substance (usually a foreign protein) that stimulates the production of ANTIBODIES.

See also: ALLERGY; HYPERSENSITIVITY

antihistamine A drug that acts to counter the effects of HISTAMINE. It is used especially in the treatment of allergic reactions.

antiknock compound A compound added to PETROL to reduce its capacity to pre-detonate in the combustion chamber of an internal COMBUSTION engine. The expression of the antiknock properties of petrol is expressed as the OCTANE NUMBER.

See also: KNOCK

antimicrobial Generally used as a synonym for ANTIBIOTIC, although it could equally apply to any chemical agent used to kill any form of microscopic life.

antioxidant A dietary constituent (such as vitamins C and E and non-nutrients such as flavonoids) that has the capacity to remove or reduce the levels of FREE RADICALS in the body and thereby reduce the risk of certain cancers and other chronic diseases. Although widely accepted and plausible, the antioxidant hypothesis has not been proven.

antiserum A fluid that contains ANTIBODIES.

anuria The complete cessation of the production of urine. It is usually due to a dysfunction of the kidneys.

aorta The main ARTERY carrying the blood from the left side of the heart.

aphelion The point in the orbit of a satellite when it is furthest away from the sun. The point at which it is nearest to the sun is known as the PERIHELION.

aplasia A medical term for the failure (whole or in part) of a tissue or organ to develop.

aplastic anaemia Also known as hypoplastic anaemia, a serious and potentially fatal type of ANAEMIA caused by a reduced capacity or inability to regenerate blood cells from the bone marrow.

See also: PERNICIOUS ANAEMIA

apogee The point in the orbit of a satellite when it is furthest away from the earth. The term used to describe the point at which the satellite is nearest to the earth is the PERIGEE.

apoptosis The ‘programmed’ death of a cell. It is part of normal cell growth and differentiation.

apparent temperature A subjective measure of the localised environment as perceived by the individual. In essence it is a comfort index as influenced by temperature, humidity and (any) WIND CHILL FACTOR.

aquaculture The cultivation of the plants or animals of the natural aquatic environment. Some sources distinguish between fresh and saltwater cultivation, using the term aquaculture to denote the cultivation of freshwater species and mariculture to denote the cultivation of saltwater or marine species.

aqueduct A bridge, usually incorporating a number of arches, used to convey a waterway or watercourse over a chasm, void, space or similar.

See also: VIADUCT

aqueous humour The watery fluid that fills the anterior chamber of the eye.

See also: VITREOUS HUMOUR

aquifer Any sub-surface porous rock layer that contains sufficient water which can be extracted by means of a well. In areas of the world where surface water is insufficient to meet demand, developing communities frequently tap these aquifers as the most readily available water source. Unfortunately the water in many of these aquifers may have derived from rainfall that fell sometimes centuries or even millennia ago. In such circumstances demand and usage can outstrip the rate of repletion and the aquifer can dry up, possibly leading to ground shrinkage, the subsidence of buildings and ground surface damage.

arachnid A member of the classification of animals that includes spiders, scorpions and mites.

arbitration A method of resolving a dispute between two parties by referring it to an independent person or body who can then offer an opinion as to which party is right. There are many forms of arbitration; it is important that both parties in the dispute agree and recognise the rules under which their appeals are heard. For example, arbitration can be in a form in which a solution can be proposed that may reflect aspects of the claims of both parties. Alternatively arbitration may be of the form in which a decision is only possible between one viewpoint and the other, no amendment to the claim of either party is possible – this latter is known as 'pendulum arbitration' because the decision swings either one way or the other. The person arbitrating may or may not be professionally qualified to perform the function and the decision may or may not be binding on the parties involved. In some forms of arbitration there are (usually limited) rights of appeal against the final decision.

arbovirus A virus transmitted by an arthropod (i.e. of the family *Arthropoda* including insects, crustaceans, arachnids and centipedes). Examples of diseases caused by arboviruses include dengue fever and yellow fever.

arc Generally taken as a portion of the circumference of a circle but the term is also applied to describe a single curve.

arc light A means of illumination in which the light is generated by an electric current passing between two terminals.

Archimedes' principle One of the standards of physics. It is attributed to Archimedes, a Greek philosopher who is alleged to have discovered it in Sicily in the 3rd century BC. The principle is generally rendered as 'When a body is wholly or partially immersed in a fluid its apparent loss in weight is equal to the weight of the fluid that it displaces.' Thus a body that is floating is one that has displaced its own weight.

architrave In classical architecture, the lowest section of an ENTABLATURE. In modern usage it refers to the moulding surrounding a door, window or other opening in a wall.

are One-hundredth of one HECTARE (i.e.100 m^2).

arenaceous Generally refers to SEDIMENTARY ROCKS and means composed of sand or sandstone.

See also: ARGILLACEOUS; RUDACEOUS

argillaceous Generally refers to SEDIMENTARY ROCKS and means composed of fine-grained material such as clay or shale.

See also: ARENACEOUS; RUDACEOUS

arid Commonly applied to any geographical area of land that receives little rainfall. More accurately the term strictly should be applied to an area of land that receives less than 250 mm of rainfall per year.

arithmetic mean *see* MEAN

arris A sharp edge or point produced by two surfaces meeting at an angle, such as that at the corner of a block of masonry or that formed at the junction of two architectural mouldings at a corner.

arrondissement An administrative district in France, being the largest subdivision of a department. The term is also applied to defined municipal districts in France around certain of the larger cities, especially Paris.

arteriosclerosis A condition in which the interior walls of an ARTERY become thickened and the blood supply becomes restricted. It occurs naturally with the process of ageing, but it can be enhanced if an individual has predisposing hereditary factors, high blood pressure or DIABETES.

See also: ANGINA; CORONARY HEART DISEASE; STROKE

artery The major BLOOD vessels of the body that carry blood away from the heart. The blood that they contain will generally have passed through the lungs on its way to the body's tissues and will therefore contain enhanced levels of oxygen. This is contained in the HAEMOGLOBIN component of the red blood cells (ERYTHROCYTES), making arterial blood more intensely red than the oxygen-depleted venous blood. Arteries are generally deep seated in the body with VEINS running alongside them. The vessels seen just below the surface of the skin are smaller veins returning blood to the heart.

See also: CAPILLARY; CARDIOVASCULAR SYSTEM

artesian well Originally developed in the Artois region of northern France from whence the name derives. The name describes a well into which water is forced by hydrostatic pressure from an outcrop higher than the well; as a consequence the water will rise higher than the surface of the well itself.

See also: DEEP WELL, SHALLOW WELL

arthralgia A pain that affects a joint.

arthritis An inflammation or other structural change in a joint. There are two principal types of arthritis, being osteoarthritis, which is associated with mechanical failure, usually of the cartilage within the joint, and rheumatoid arthritis, which is associated with an inflammation of the linings or covering sheaths of tissues associated with a joint.

arthropod A member of the class of animals described as being bilaterally symmetrical and having an EXOSKELETON of CHITIN and jointed appendages. The group (known as *Arthropoda*) includes crustaceans, insects, MYRIAPODS and ARACHNIDS.

artificial oestrogen *see* ENDOCRINE DISRUPTOR

asbestos The generic name given to a group of fibrous silicate minerals that are resistant to fire, heat and acid. Some types are also resistant to alkalis. There are many varieties but they are usually classified into two main types according to their physical characteristics as seen under the microscope.

The first group (which includes CHRYSOTILE) exhibits a serpentine shape. The second group is known as amphibole and has a fibrous or columnar shape. This latter group includes AMOSITE and CROCIDOLITE.

The health risks associated with asbestos have been known for over 100 years but research over the last 30 or so years has clearly shown the extent of the risks with effects including ASBESTOSIS, lung cancer, MESOTHELIOMA and cancer at sites other than the lung having being demonstrated. It is not known precisely what levels of exposure to the various types of asbestos are capable of causing disease, but there would appear to be a dose-related response. A significant drawback in being able to assess a dose-related response is the LATENCY PERIOD of around 15–60 years between exposure and manifestation of disease.

asbestosis A disease of the lung caused by the inhalation of particles of ASBESTOS.

ascertainment Any process used to detect the presence of disease conditions in a population or COHORT.

asepsis A term used to describe a (usually surgical) procedure in which instruments, equipment and contact materials are sterilised before use to minimise infection, rather than rely on the use of germicides to produce sterile conditions.

See also: ASEPTIC

aseptic Means sterile. In relation to an aseptic process the term means undertaken under sterile conditions.

See also: ASEPSIS

ashlar Squared or dressed stonework (see DRESSED STONE) used in the construction of walls and similar.

See also: RUBBLE MASONRY

askarel A generic term used to describe a group of synthetic chlorinated aromatic hydrocarbons with electrical insulating properties. They are highly toxic and their manufacture has generally been discontinued.

aspartame Aspartame ($C_{14}H_{18}N_2O_5$), also known as Nutrasweet, is an artificial sweetener of foodstuff formed from aspartic acid that was developed in the 1980s. It is used either as an ingredient or used directly as an alternative to sugar. In spite of several much publicised health scares associated with the product there has been no definitive evidence produced to indicate that the product is anything but safe.

aspartate aminotransferase An enzyme that is thought to be activated when toxic damage is caused within an organism. Detection of elevated levels of the enzyme might therefore indicate that such damage has been caused, even in the absence of discernible clinical symptoms.

See also: ALANINE AMINOTRANSFERASE

asphalt A bituminous material of high molecular weight occurring both naturally and as a residue from the distillation of PETROLEUM. It is used as a waterproofing agent or additive and as a surfacing material for roads.

asphyxia The medical term for any condition in which the lungs are deprived of air or oxygen. It is used as a synonym for suffocation.

asphyxiant Any substance that is capable if inducing ASPHYXIA by means of reducing the amount of oxygen that can be absorbed by the lungs.

aspirin Also known as ACETYLSALICYLIC ACID, an ANALGESIC drug that also reduces fever and inflammation. Many other properties have been claimed for the drug, including use as an antiseptic, anti-rheumatic and general sedative. It is used, under medical supervision, in daily therapeutic doses of between 300 mg to 1 g in the prevention of coronary thrombosis. It is not an entirely safe drug in that it can be poisonous to animals, and caution should be exercised in administering it to children.

assay Any technique used to determine the characteristics of a sample. In chemistry it generally refers to the means whereby the relative proportions of different constituents are measured. In biological terms it often refers to pathogenicity (see PATHOGENIC), VIRULENCE or immunological activity.

See also: BIOASSAY

assignment The transfer of intellectual property rights from one ownership to another, with or without payment.

asthma Medical condition in which there is an inflammation of the airways of the lung accompanied by constriction of the airways such as to reduce the passage of air through them. This produces symptoms of wheeze and difficulty in breathing, the latter of which can be serious or life-threatening. There is a strong genetic component to the condition and an attack can be triggered as an allergic response to a wide range of substances including pollen, the faeces of the house DUST MITE, cat allergens, industrial chemicals and items of diet.

See also: ALLERGY; FOOD ALLERGY

astrogliosis An abnormal enlargement and increase in number of the GLIA – the cells in the brain that fight disease.

astrovirus A small, naked RNA-carrying virus associated with enteric infections in both man and animals. They may be food or water-borne and are relatively easily cultured. They were first identified and defined in 1975. They have a smooth outer margin and a five or six-pointed star-like motif at their centre, from whence comes their name. There are seven known serotypes. In type they are more similar to plant viruses than animal viruses. They are generally less prevalent than either ROTAVIRUSES or ADENOVIRUSES. They can be difficult to identify simply from their appearance as the reported 'star-like' shape can be ill-defined and some may give the sign as of a SRSV (Small Round Structured Virus) and it is therefore possible that many astroviruses go undetected or misdiagnosed.

Astroviruses are essentially an infection of childhood with 64 per cent of children being sero-positive by the age of 4 years and 87 per cent by the age of 10 years. The illness is generally mild in adults but some infections may result in absences from work of several days. The 'at risk' groups are predominantly the elderly and the immunocompromised.

asymptomatic The term applied to the absence of clinical symptoms of a disease whether or not the PATHOGENIC organism capable of producing the infection is present within the body of the person or animal so described.

See also: CARRIER

ataxia A loss of control of the normal voluntary movements of everyday life.

atherogenic Relates to the capacity or capability of a material or substance to contribute to the thickening or degeneration of the walls of the larger arteries through the deposition of CHOLESTEROL, LIPID or calcium.

athetoid movement *see* ATHETOSIS

athetosis A condition characterised by continual slow and involuntary movements of the extremities of the limbs. The condition is also known as athetoid movement.

atmospheric pressure That exerted by the mass of air acting under gravity at any particular point. Atmospheric pressure generally decreases with altitude, but the pressure also varies according to meteorological conditions and temperature. Standard atmospheric pressure at sea level equates to 1013.25 millibars, being 101.3 kilopascals, 760 mm of mercury or 14.7 pounds per square inch (psi).

atom The smallest portion of an ELEMENT that can take part in a chemical reaction with other atoms.

atomic absorption spectroscopy An analytical technique in which non-excited atoms in a vapour state absorb radiation, each type of atom displaying a characteristic absorption pattern that is then used for the purposes of differentiation.

atomic emission spectroscopy An analytical technique used for the detection and differentiation of trace metals by means of vaporising a sample and exciting this in a flame in order to measure the different spectra emitted by the different trace metals.

atomic mass Mass of an ELEMENT expressed in units that are themselves defined as being one-twelfth that of the mass of an ATOM of carbon-12. The atomic mass is also sometimes known as the atomic weight.

atomic number The number of PROTONS contained within the NUCLEUS of an ATOM. The atomic number is sometimes designated by the letter 'Z'.

atomic weight An alternative term used to describe the ATOMIC MASS.

atopic Literally means 'unusual' or atypical. The term can also be applied to an allergic reaction associated with predisposing hereditary factors.

atopy A type of inherited sensitivity manifesting in response to certain ANTIGENS producing symptoms akin to those of asthma or hay fever..

attack rate Also known as Incidence Rate, the timed sequence of the manifestation of the first clinical symptoms of a disease or food poisoning during an outbreak. The attack rate is used to identify the potential source of infection or illness.

Attorney General The most senior law officer of and principal legal counsel of the Crown in the UK (excluding Scotland), the incumbent being a member of the Government and the House of Commons. In Scotland this position is held by the LORD ADVOCATE.

attributable risk That which describes an increased adverse health effect on a group of people exposed to a risk when compared to a similar group that was not exposed. Provided there is a cause-and-effect relationship between the exposure and the adverse health effect, it is assumed that the increase in the exposed COHORT can be ascribed or 'attributed' to the risk.

au beurre Refers to a foodstuff that has been cooked in, or is served with, butter.

au four Refers to food that has been cooked in the oven.

au gratin Refers to food that is covered with breadcrumbs and/or grated cheese and browned. The term in more modern usage has come to refer to any dish that is served with a cheese topping.

aurora An electrical phenomenon displayed in the upper atmosphere (between 40–600 miles up) in which differently coloured lights are generated by the passage of charged particles from the solar wind blowing across the earth's magnetic field. The phenomenon is more frequently seen in the polar regions. The phenomenon in the northern hemisphere is known as the aurora borealis, while in the southern hemisphere it is known as the aurora australis.

autoclave A piece of equipment used to sterilise articles placed within it by means of superheated steam. Its operation and effectiveness is based on the premise that the boiling point of a liquid (in this case usually water) can be raised significantly beyond normal atmospheric boiling point by sealing the container within which it is being boiled and raising the pressure. Correct control of the pressure can regulate very precisely the internal temperature of the autoclave.

Autoclaves are highly effective means of sterilising (particularly reusable) equipment and are widely used in hospitals and laboratories. They are not necessarily effective in denaturing chemical molecules and are not, for example, effective means of deactivating the PRIONS of either BOVINE SPONGIFORM ENCEPHALOPATHY (BSE) or CJD.

See also: STERILISATION

autoimmune disease One whose AETIOLOGY is as a body's own defective immune response to proteins or ANTIGENS from within itself.

autosome Any CHROMOSOME other than a SEX CHROMOSOME.

avoirdupois system The system of weights used in many English-speaking countries prior to the introduction of the metric scale. The system is based on the pound, which itself is comprised of 16 ounces. In the UK 112 pounds (lbs.) (in the USA 100 pounds) makes 1 hundredweight (cwt.) and 20 hundredweights equals 1 ton.

See also: SI

Aw *see* WATER ACTIVITY

axiom *see* POSTULATE

azeotrope A mixture of two or more liquids that together have a different boiling point than the liquids individually. This property of azeotrope is sometimes used in differential distillation if two liquids being distilled have similar boiling points – the addition of another substance to create an azeotrope with one of the liquids can modify the boiling point of one liquid to allow separation of the other.

azimuth In surveying, the horizontal angle measured clockwise from a standard reference point such as either due or magnetic north.

B

bacillus (plural bacilli) Rod-shaped bacteria.

Bacillus cerus A Gram positive (see GRAM'S STAIN) motile rod. It produces heat-resistant spores and one or more toxins including a heat-stable enterotoxin and a heat-resistant emetic toxin. Both of these toxins are pre-formed in food; the enterotoxin can also be formed in the gut. The bacterium is both aerobic and a facultative anaerobe. The spores survive normal cooking and can subsequently germinate and multiply in food. Sources of the organism in food include rice, other cereal products, spices, dried foods, and milk and dairy products. Environmental sources include soil, dust and sediments. The organism is spread by contaminated cooked foods, especially cooked rice and pasta dishes, which typically have been subject to inadequate temperature control during cooling and storage. Clinical symptoms include acute nausea, vomiting and stomach cramps. Diarrhoea may occur later in some cases.

back pressure The creation of undesired pressure at a point in a flow system potentially such as to both raise the pressure in the system leading to the point and to lower the pressure along the system after it. Back pressure can be caused by a blockage in the system or (in systems for moving liquids) by an air pocket (also known as an air 'bubble' or air 'lock'). Back pressure is undesirable and potentially dangerous. The effects range from creating stress on pipework and pumps, stopping the flow in the system or even, in extreme cases, causing reverse flow.

back siphonage The reverse flow of liquid through a system caused by a change in pressure. In systems such as those used for the distribution of POTABLE water, back siphonage is potentially dangerous as contaminated water could be drawn into supply pipes with possible adverse health consequences.

background noise The AMBIENT noise level as measured without the noise contribution of a new noise source that is to be introduced (or it is proposed to introduce) into the area.

backwashing A technique used to clean water filters (such as the larger sand filters used in swimming pools). The technique involves a reversal of the flow through the filter to dislodge and remove trapped debris and bacteria that are then discharged as contaminated water to waste. Backwashing needs to be carried out carefully as it is possible also to dislodge particles of the filter medium. Such particles will be lost from the filter and repeated losses will eventually deplete the medium sufficiently as to have an adverse effect on the filter's performance.

bacteraemia A condition of the body in which bacteria are present in the blood.

bactericidal Means having the capacity to kill bacteria.

bactericide A substance that has the capacity to kill bacteria.

bacteriophage One of a range of bacterial viruses. They are used in the laboratory for the selective differentiation between types of bacteria. Bacteriophages can be used in genetic engineering as vectors for cloning segments of DEOXYRIBONUCLEIC ACID (DNA).

See also: PHAGE TYPE

bacterium (plural bacteria) Microscopic single-celled organisms that do not possess nuclei. They are generally characterised by their shape with ACTINOMYCETES being filamentous, bacilli (see BACILLUS) being rod-shaped, cocci (see COCCUS) being spherical and SPIRELLI being spiral. Other classifications are based on the various organisms' ability to accept staining, growth patterns, respiratory requirements or biochemical features. All bacteria multiply by simple division.

The majority of bacteria are beneficial to humans; indeed the processes of decay of organic material, as we know it, would virtually cease without them. Comparatively few bacteria are PATHOGENIC to humans or animals although these are the ones on which much public health attention is focused. ANTIBIOTICS have been developed to combat pathogenic bacteria, but ANTIBIOTIC RESISTANCE appears to be a growing and increasingly worrying phenomenon.

See also: AEROBE; ANAEROBE; GRAM'S STAIN

badger A nocturnal hibernating species of animal of the weasel family. There are various species around the world; the one found in the UK is *Meles meles*. They live in a system of underground burrows known as a set, with some groups being recorded as having occupied the same sets for generations, reputedly, in some instances, for hundreds of years. In the UK both badgers and their sets/outliers have statutory protection. It is thought that badgers play a considerable role in the transmission of bovine TUBERCULOSIS (*Mycobacterium bovis*). They are omnivorous but their environment will largely dictate their diet. Normally earthworms are the staple food, although carrion and waste may be eaten if circumstances allow/demand.

baffle A plate installed in a pipe or expansion chamber and is used to divert or regulate the flow or pathway of liquids, gases or sound waves.

bagasse The dry fibre remains left over from the process of extracting sugar from sugar-cane. It can be used in the manufacture of hardboard.

baked enamel Another term for VITREOUS ENAMEL.

Bakelite One of the first synthetic resins to be used widely as a construction material in the manufacturing industry. It is named after its inventor L.H. Bakeland (1863–1944) and is made by condensing either cresol or phenol in the presence of formaldehyde.

Balai product A foodstuff that is regulated by the miscellaneous products Directive by the EU. The term 'Balai' is derived from the French for broom – in effect the Balai Directive 'sweeps up' all items not specifically covered directly by other animal and public health measures. The list of Balai products includes milk, milk products, pet food containing low-risk materials, rendered fats, lard, fish oil, eggs, egg products, gelatins, honey, frogs' legs and snails intended for human consumption. The main requirements of the Directive include registration of plants, veterinary checks and salmonella testing.

balanced equation *see* STOICHIOMETRIC

balcony A suspended platform projecting outwards from a wall.

balk (or baulk) In construction, variously a tie-beam holding opposing walls of a building together or any large piece of squared timber. In agriculture the term refers to a ridge of unploughed land.

ballast Material installed at or near the bottom of a ship, crane or other machine and used to weigh it down and to stabilise movement. Ballast can be either permanent (as for a crane) or temporary (such as is sometimes used for shipping to stabilise an unladen vessel on its return journey). The practice of using water as ballast on intercontinental shipping has been blamed for the transportation of the eggs and larvae of alien marine species to new habitats where, following its expulsion, it can have potentially dramatic consequences for local marine flora and fauna.

ballcock An assembly used to regulate automatically the flow of water into a cistern or similar. A float attached to a lever rises or falls with the fluctuating level in the cistern, alternatively respectively opening or closing a valve controlling the flow of water in the system in order to maintain the amount of water in the cistern to a pre-determined level. The ballcock is sometimes known as a ball valve.

ball-peen hammer A hammer with one end of the head formed into a hemisphere. It is used predominantly for working sheet metal or stone.

ball valve *see* BALLCOCK

balsa (wood) A very soft and low-density wood derived from a tropical American tree. It is used predominantly as a modelling medium or as an insulator. In spite of its soft characteristics it is classified as a HARDWOOD. Balsa wood is sometimes known as 'cork wood'.

baluster Another name for a BANISTER or alternatively the name applied to the vertical posts supporting the banister.

balustrade The complete assembly of a BANISTER and supporting members (the BALUSTER) running alongside a stairway.

bamboo The name of a wide group of plants of different nature and habitat that are all nonetheless classified as grasses. The largest species of bamboo are the largest grasses in the world and some are amongst the fastest growing plants in the world. Bamboo is used as a construction material in those parts of the world where it is indigenous and is used for furniture and ornamentation throughout the world.

band (or bandpass) filter A device that attenuates a broad range of frequencies (such as in acoustics) by limiting that which is transmitted to a smaller range, or attenuating some frequency bands to represent a required frequency response as perceived or received by a specified receptor.

band saw A mechanical saw, usually static, in which the blade is formed as a continuous and rotating loop or band. The wood or other material to be cut is pushed (usually manually) against guides towards the exposed portion of the blade.

bandwidth The range between the upper and lower limits of a frequency band. It is usually expressed in terms of cycles per second (HERTZ).

banister The handrail running alongside the length of a staircase. It is sometimes called a BALUSTER.

bar chart A graphical representation of the numbers or amounts of articles or items in a number of specified categories with the quantity in each being represented by the length of a line or bar.

See also: HISTOGRAM

bar code A device comprising parallel lines for which each individual code is made up of lines of different thickness and spacing such that each can be recognised by a bar code scanner to provide information to a computer. Bar codes are used predominantly to identify goods for the purposes of stocktaking or sale.

barding The culinary term for covering the breast of a bird with slices of fat, bacon or similar prior to roasting in order to prevent the breast drying out.

barge board A length of timber, plastic or similar positioned and fastened to protect the otherwise exposed timbers along the sloping edge of a GABLE between the roof line and the top of the wall. Barge boards are sometimes known as verge boards.

barometer A scientific instrument used to measure atmospheric pressure.

barrel A standard measure of a volume of oil in which one barrel is equivalent to 42 US gallons.

barrel bolt A device for securing doors, windows and similar in which a cylindrical TENON fits into a MORTISE to lock two articles together.

basal metabolic rate The lowest level of metabolic activity that a living creature can maintain to sustain life, i.e. when the animal is at complete rest or asleep.

See also: METABOLISM

basalt A type of IGNEOUS rock, usually of dark colouration and with a fine grain.

base A substance that has the capacity to neutralise an ACID.

See also: ALKALI

base coat The first layer of material (such as of paint or plaster) placed or installed onto a surface to form a working base and a key for subsequent coats or layers of decorative finish.

baseboard Another name for SKIRTING.

Basel Convention on Transboundary Shipments of Waste An agreement made in 1989 that introduced global controls on the transportation of waste material between countries.

bas-relief A form of sculpture in which the figures or patterns do not project far from the surface from which they have been cut. Bas-relief was historically popular for use in FRIEZES. The term is also rendered as basso-relievo or basso-rilievo.

basting The culinary term to describe the ladling of hot fat or liquid over meat or poultry at regular intervals during cooking to prevent it from drying out.

BAT The acronym for Best Available Techniques. It is a method used principally in systems of pollution control to ensure that the most effective means of control are applied or introduced by the operator of a business but balancing the cost to the operator against the benefits.

batch A quantity or group of articles or components with a common factor. The term usually applies to manufactured items that were produced at the same time or those that were produced from the same mix of raw ingredients, the essential distinguishing feature being that the batch should be HOMOGENOUS.

batholith A geological formation in which a large extrusion of IGNEOUS rock has penetrated into surrounding strata, essentially forming a stable and homogenous mass that can extend for thousands of square kilometres and to great depth in the earth's crust. Many batholiths are composed of granite. Their stability and homogeneity has resulted in their identification in some areas as possible sites for the storage of nuclear waste.

BATNEEC The acronym for Best Available Techniques Not Entailing Excessive Cost. It is a term used to describe the ways in which the release of potentially polluting proscribed substances to any environmental medium (i.e. air, water or land) are prevented or, if this is not possible, rendered harmless. BATNEEC recognises that there may be a limit to the cost that businesses can reasonably be expected to bear and that this cannot be limitless. The way in which controls are achieved in practice is usually by adhering to emission limits determined for the individual or specific type of process under consideration.

See also: BEST PRACTICAL ENVIRONMENTAL OPTION

batten Variously:

1 A strip of wood fastened between parallel structural members such as those to which tiles or slates are fastened in roof construction;
2 Thin strips of wood, to which laths are fastened prior to accepting plaster;
3 The fastenings used to secure hatches on board a ship;
4 A line of electric lights or the length of timber to which these are fastened.

battery A series of ELECTRIC CELLS connected in sequence to increase the voltage or connected in parallel to increase the current.

baulk *see* BALK

bearing Variously a support (especially one that carries a dead weight), a component of a machine that carries the friction loading or the direction of one point in relation to another as determined by an observer.

Beaufort scale A scale popularly described as being used to describe wind speed; it actually describes the effects of the wind and therefore is more accurately a scale of wind force. The scale has been amended by the World Meteorological Organization since it was originally introduced and now ranges from 0 (dead calm), through gentle breeze (3), strong breeze (6), near gale (7), strong gale (9) and violent storm (11) to 12 (hurricane). The scale was originally devised by the English Rear Admiral, Sir Francis Beaufort (1774–1857), from whom it takes its name.

Becquerel A unit of activity per second of a mass of a RADIONUCLIDE. One Becquerel is equivalent to one disintegration per second. The greater the mass of a given radionuclide the greater the activity and the greater the number of Becquerels.

beeswax The waxy secretion produced by bees and used to construct the cells in the honeycombs of their hives. It is used in the production of furniture polish and stains for woodwork.

bel A unit measure of sound pressure

See also: DECIBEL

belvedere *see* GAZEBO

benchmark Originally a line, mark or similar cut or scratched into a rock surface or on a building by a surveyor at a point that had been determined as being at a particular height or altitude. The mark was used as a point of reference from which other bearings could be taken. Nowadays the term is used to describe anything that is permanent or semi-permanent and can be used as a point of reference or against which something else can be compared.

benchmarking Originally the process of performing a task in order to establish its performance characteristics and determine its limitations. The term is nowadays also used to describe the process of identifying that which is BEST PRACTICE.

See also: BENCHMARK

bending That force applied to an object or structure that tends to distort it out of true. The bending moment is usually designated by the symbol '*M*'.

See also: COMPRESSION; SHEAR FORCE; TORSION

benign (tumour) An abnormal growth of or within a tissue that is not life-threatening. A benign tumour is one that grows locally but will not spread via the blood or lymphatic systems to form secondary growths elsewhere in the body.

See also: MALIGNANT (TUMOUR)

Bentham, Jeremy b. 1748; **d.** 1832
A London-born philosopher. He originally trained for the law but developed as a prolific author and critic of society at the time. The death of his father in 1792 gave him independence of means and a freedom to pursue his chosen path. He is perhaps best remembered for the doctrine of Utilitarianism, usually rendered as providing 'the greatest happiness for the greatest number'. He was a pioneer in recommending social change including prison reform, change to the poor law system, universal suffrage and the decriminalisation of homosexuality. Following his death on 6 June 1832 his skeleton was made into his effigy, dressed in his own clothes, provided with a wax model head and mounted in a cabinet. The cabinet and contents are still in existence and currently stand at the end of the South Cloister of University College, London.

See also: CHADWICK, SIR EDWIN

benthic Means of or pertaining to the bottom of the sea.

See also: LITTORAL

benzene The simplest of the aromatic hydrocarbons, comprising a ring of carbon atoms with a hydrogen atom attached to each (chemical formula C_6H_6). Benzene is a known CARCINOGEN and is used as a constituent in solvents and paint removers. The word benzene is often improperly used to describe BENZINE.

benzene ring The basic ring structure of the ORGANIC chemical compound BENZENE, comprising six carbon atoms each with a single and double bond to another carbon atom and a single bond to a hydrogen atom.

benzine A highly flammable mixture of hydrocarbons used as a motor fuel and in solvents.

See also: BENZENE

bespoke The term used to describe goods that are specially made to order (i.e. they are 'bespoken'). The term can be applied to virtually any goods and, although the term is sometimes taken to imply a superior finish or quality to goods that are mass manufactured, modern-day production processes may deliver goods of equal durability or suitability for purpose. Bespoke should therefore be considered as applying essentially to individuality of design or construction, although some bespoke goods may also be of superior quality.

Bessemer converter A pear-shaped furnace with a REFRACTORY lining used to convert molten pig iron into steel (known as Bessemer steel). The process was invented in 1856 by H. Bessemer and is known as the Bessemer Process.

Best Available Techniques *see* BAT

Best Available Techniques Not Entailing Excessive Cost *see* BATNEEC

Best Practical Environmental Option (BPEO) A technique used to ensure that substances released into the environment have the least impact or deliver the best benefit when this is viewed from the perspective of the environment as a whole over both long and short terms.

Best Practicable Means Those activities that fall within the control of operators whereby exposure or doses to the public in respect of substances or energy (such as radiation) are kept 'as low as reasonably achievable' within the terms of the ALARA principle. It should be noted that the control of substances or energy is considered in this context to be capable of control by operators but the exposure (usually as a result of discharge) of members of the public is not.

Best Practice The term applied to a system or method of undertaking work or performing an identified task that will deliver the optimum performance having regard to both input and output, and the current state of knowledge. In the UK local authorities have been instrumental in developing and introducing Best Practice to improve service delivery in many areas of their work.

See also: BENCHMARKING

best value A legislative requirement placed on local authorities in the UK to deliver the most effective, economic and efficient service possible in those areas prescribed. The evaluation of the value criteria of the service includes both cost and quality.

beta-blocker A substance that has the capacity to block the beta effects of the sympathetic nervous system, thereby having the effect of decreasing the heart rate and reducing blood pressure. The term applies equally to drugs prescribed to combat the effects of such medical conditions as HYPERTENSION as well as to any other substances that can have a similar effect. Beta-blockers are not therefore necessarily benign: they can cause HYPOTENSION and cardiac failure.

beta particle A radioactive particle equivalent to an electron. It has a mass 1/1836 that of a PROTON and carries a negative charge. Some beta particles carry a positive charge; if so, they are known as positrons.

See also: ALPHA PARTICLE; GAMMA RADIATION; RADIATION

bevel The junction of two surfaces that meet each other at an angle other than a right angle. A bevel is also sometimes known as a cant.

See also: CHAMFER

BHP *see* BRAKE HORSEPOWER

bias Any component that will introduce a systematic error in an epidemiological study. The term does not imply any deliberate attempt to produce inaccurate or misleading results – it does however relate to situations in which a study design leads to the use of data or inputs that are inaccurate but taken as true. The old maxim 'rubbish in – rubbish out' conveys succinctly the effect of bias on epidemiological studies. A standard evaluation of an epidemiological study will include an examination of the potential for bias. There are many types of bias, and examples are measurement bias (in which inaccurate measurements are made), reporting bias (in which inaccuracies in what is observed are recorded) and observation bias (in which what is seen is incorrectly interpreted).

See also: CONFOUNDING FACTOR

bidet An item of sanitary ware comprising a low basin supplied with water upon which the user sits to wash their genital and anal areas.

Bill The term applied to a piece of PRIMARY LEGISLATION as an item for debate before it is passed by a legislature.

See also: ACT; REGULATION

billion Now generally taken as being one thousand millions (i.e. 1×10^9). In Europe it was formerly quantified as one million millions (i.e. 1×10^{12}) but this has fallen into disuse.

bill of quantities A numbered list of items describing the individual components of work and materials that together comprise a building contract. It is usually produced (at least in the UK) by a QUANTITY SURVEYOR and is used by a contractor to produce a submitted price for a contract or TENDER.

bimetal strip A strip comprising two dissimilar metals fused together along their length. The metals are selected to have a differential expansion and contraction under changes in temperature so that when such occurs the strip bends. This property is used when a bimetal strip is incorporated into an electric circuit as a switch to make or break the circuit in response to temperature change, such as is required for a THERMOSTAT.

binding The culinary term to describe the addition of melted fat, liquid or beaten egg to a mixture to hold it together.

bioaccumulate The term used to describe the capacity whereby chemicals (such as, for example, DIOXIN) are taken up from the environment, are concentrated by accumulation and persist within organic material.

bioassay Any test of the pathogenicity (see PATHOGENIC) or VIRULENCE of a disease performed on a live animal to gauge or assess its effect.

See also: ASSAY

bioavailability Refers to the degree to which a substance or chemical is free to associate with, or affect, a living organism. It is often used to describe the proportion (if any) of a substance that is remaining with the potential for activity after it has reached the systemic circulation following administration.

biodegradable Refers to the capacity of a substance to be broken down by the action of biological agents such as plants, fungi, bacteria, worms and other microorganisms. The reduction of the proportion of biodegradable material in municipal solid waste that goes to LANDFILL is a key area in reducing the overall dependency on landfill as a means of disposal.

biodiversity The totality of life, including genetic capacity, within a defined area.

biofuel Any renewable HYDROCARBON energy source derived either directly from or by the processing of living plant or vegetative material. The increased production and usage of biofuel is likely to be a key component of SUSTAINABLE DEVELOPMENT.

biohazard classification A means of describing the levels of physical containment likely to be required in a laboratory when dealing with pathogenic microorganisms. The scheme in use in the UK is described below:

Table Biohazard classification scheme, UK

Group	*Level of risk*
Group 1	An organism unlikely to cause human disease
Group 2	An organism that may cause human disease and which may be a hazard to laboratory workers but is unlikely to spread to the community. Laboratory exposure rarely produces infection and effective prophylaxis or treatment is usually available
Group 3	An organism that may cause severe human disease and present a risk of spread in the community but there is usually effective prophylaxis or treatment available
Group 4	An organism that causes severe human disease and is a serious hazard to laboratory workers. It may present a high risk of spread in the community and there is usually no effective prophylaxis or treatment

biokinetics The science of understanding and describing the processes whereby the body deals with foreign substances, such as RADIONUCLIDES, after they have entered the body.

biological half-life (t) The time by any particular substance introduced into living tissue to be reduced by a half.

biomarker Any living organism or portion thereof that can be used as an indicator of the impact on the environment of physical, chemical or biological activity.

biomass The quantitative measure of the totality of living material in a defined geographical area.

biome A distinct environmental region of relative uniformity characterised principally by its flora. Examples of biomes are tropical rain forest, temperate forest and DESERT.

biomonitoring An evaluative technique in which the exposure to a given substance is assessed retrospectively by analysing living tissue for the presence of the substance or its METABOLITES.

biopsy The removal of a portion of tissue from the living body and subjecting it to examination for the purposes of diagnosis.

bioremediation The removal or treatment of (usually environmental) pollution or contamination through the use of biological agents – these are usually plants or micro-organisms.

biosphere Differentially defined as the region of the earth within which there is life, the areas of the earth within which life is possible (i.e. irrespective of its actual presence) or the totality of life and its immediate environs.

biotechnology Strictly speaking, any technology that relates to living things, be these animals, plants or micro-organisms. The term is however most frequently applied to technology used to alter living things in ways designed to benefit mankind.

See also: GENETIC MODIFICATION

bioterrorism The use, or threat of use, of infectious or contagious agents with the aim of creating terror or to threaten, blackmail or incapacitate an enemy. The term has been used principally to describe the intentional release – usually by air blast detonation or contamination of (particularly) water, food supplies or postal mail – of modified or attenuated bacteria or viruses.

biotype A type of bacteria described as a result of its exhibiting variable and specific defined biological characteristics (such as metabolic or physiological properties) that can be used to differentiate it from other bacteria.

bitumen Any of a variety of black tarry substances usually derived from ASPHALT, naphtha or the distillation of PETROLEUM. It is used in construction as a waterproofing agent. It is incorporated into bituminous felt, used for example as a lining under the slates of a pitched roof, or in bituminous paints, used to seal or make waterproof junctions between materials as necessary.

black bolt One used in steel construction and that has been roughly cast and is unlikely therefore to fit exactly into the holes in the articles it has been manufactured to hold together. As a consequence of the potential 'play' (i.e. freedom of movement) between the shank of the bolt and the hole, black bolts are not intended to carry as much DESIGN LOAD as would be expected of a similar BRIGHT BOLT.

black lead Another name for graphite or plumbago, such as that used for making pencils – it does not contain lead. It was also used historically to polish cast ironwork such as that of domestic stoves.

black light Light of the ultraviolet wave length. It is invisible to the naked eye but has the capacity to cause fluorescent surfaces to emit visible light. The term is sometimes also applied to describe invisible light in the infrared wavelength.

black smoke Defined by the UK's Quality Urban Air Review Group in *Urban Air Quality in the United Kingdom* (1993) as non-reflective particulate matter associated with the smoke stain measurement method.

blackhead A blockage of a pore or hair follicle of the skin caused by an excess production of SEBUM. The name derives from the adhesion of dirt and grime to the sebum creating a dark-coloured spot. Bacteria can infect blackheads and this may lead to an inflammation. Widespread inflammation of this type is known as ACNE. A blackhead is sometimes known as a comedo.

blade passing frequency The frequency with which the individual blades of a wind turbine generator pass a fixed point in their rotational cycle. For a three-bladed rotor the blade passing frequency is three times the rotational frequency.

blanching The culinary term to describe the immersion of a raw or uncooked food item in boiling water in order variously to whiten it, preserve its colour, soften it, aid removal of skin or reduce its intensity of flavour or odour.

blast furnace A cylindrical furnace with a REFRACTORY lining used to SMELT iron ore for the production of iron. The name derives from the process whereby air is 'blasted' through the furnace to raise the temperature.

bleed (or bleeder) valve *see* BLEEDING

bleeding That stage in a process of slaughtering animals for food in which a major blood vessel (often the carotid artery) is severed to drain the carcass of blood (see PITHING; STUNNING). It is also the process of extracting air from a circulatory system (such as those using water for cooling or heating) by draining off the air through a device (known as a bleed valve) incorporated into the system for the purpose. The term can be applied to the process whereby water ponds on the surface of wet cement or concrete due to the settlement of the solid portion. The term is also used to describe the exudation of resin, sap or gum that sometimes occurs from the cut surface of timber.

blind nailing The same as SECRET NAILING.

blind rivet A RIVET for which access is necessary from one side only. The rivet is hollow with a pre-formed head at one end and terminating in an unformed tube at the other. This latter end is inserted into a hole whereupon a force is exerted on a central shaft protruding through the rivet, drawing the shaft back through the hole in the rivet and deforming its originally unformed end in the process to compress the whole together. The blind rivet is also known as a pop rivet because the central shaft shears or 'pops' once the rivet is in place.

blister A collection of fluid between the Malpigian layer and superficial layers of the SKIN appearing as a raised portion on the surface. Blisters are formed usually in response to irritation or abrasion and are a body defence mechanism to relieve congestion in deeper-seated organs and to cushion the nerves against stimulation.

blood A red fluid circulating in the body in the blood system. It comprises predominantly a suspension of cells and other components in clear plasma. The composition of blood by volume is variable, although plasma is the most abundant constituent at around 50–60 per cent. Red blood corpuscles (also known as erythrocytes) constitute around 30 per cent of the blood – they contain haemoglobin and are responsible for transporting oxygen around the body to the sites where it is needed. White blood cells (various types, although predominantly leukocytes and lymphocytes) are responsible for combating infection and destroying invading micro-organisms. Blood platelets (also known as thrombocytes) are responsible for forming clots at the site of injury to prevent further blood loss. Blood has many functions in the body including as a transportation system for oxygen, nutrients, waste and hormones; regulating temperature; maintaining the correct water balance within the body; and in protecting the body from pathogenic invasion by acting as a carrier for defence mechanisms.

blood culture A laboratory procedure undertaken to detect the presence of viable bacteria in the blood.

blown can One in which microbiological activity from within has produced sufficient gas to pressurise the inside of the can and cause distortion or, more extremely, bursting. A 'typical' blown can, in the early stages of activity, will exhibit the can ends as convex rather than as flat or concave. A blown can is indicative of inadequate processing in that microbes which should have been destroyed prior to sealing still remain alive and active within the contents. Some sources include in the definition of blown cans those that are distorted due to gas produced by chemical means.

See also: HYDROGEN SWELL

blowout An uncontrolled or catastrophic eruption to atmosphere of liquid or gaseous fossil fuel from a well, borehole or similar.

blue baby syndrome *see* METHAEMOGLOBINAEMIA

blue-green algae A term used to describe certain types of potentially toxic vegetative growths of algal appearance. The term is a misnomer in that the organisms are really a type of bacteria. They exhibit some of the same properties of true algae however and are scientifically more correctly termed cyanobacteria. They have a worldwide range from the arctic to the high temperature areas around hot springs where few other life forms can survive.

Many blue-green algae are capable of fixing nitrogen and are therefore able to out-compete other algae, becoming the dominant species, especially in warm, calm conditions. They generally proliferate in waters that have high nutrient levels (EUTROPHIC) and are less abundant in waters with a low nutrient level (OLIGOTROPHIC).

At certain times abnormal growth of these organisms can lead to the production of what is known as a 'bloom', essentially the development of a dense colony. These blooms have the potential to produce toxins that, if ingested (especially) by filter-feeding shellfish, can lead to toxin accumulation and subsequent poisoning if the contaminated shellfish are eaten. Although not all blue-green algae produce toxin and not all blooms prove to be toxic it is wise to assume that all blooms are potentially toxic.

See also: AMNESIAC SHELLFISH POISONING; DIARRHETIC SHELLFISH POISONING; PARALYTIC SHELLFISH POISONING

blueprint Nowadays generally taken as a master plan (theoretical or actual) for the production of something or for the delivery of a service. The name derives from the photographic print or drawing of white on blue ferro-prussiate paper once commonly used for engineering or architectural drawings.

boiler A vessel used to raise the temperature of a liquid (usually water) by transferring heat from a fuel source (such as fossil fuels or electricity). Some boilers are designed to raise steam under pressure to generate power; other boilers, such as those used for supplying hot water for washing or heating purposes, do not actually raise the temperature of the water to boiling point.

bolt A straight bar used to fasten a door, gate or similar. It is also a device inserted into holes in two or more articles and comprising a shank with a flat head at one end and a screw thread at the other onto which a nut is tightened to join the articles together by compression.

bond The pattern or physical order in which bricks are incorporated into a wall or similar in order to ensure that they provide the appropriate structural strength. The choice of bond can also be a matter of aesthetic preference.

See also: FLEMISH BOND; STRETCHER BOND

booklice *see* PSOCIDS

boom Variously a floating barrier used to contain spillages of floating material (such as oil) in water, a barrier comprising floating logs, the structural member of a crane from which the load is suspended, a barrier stretched across a harbour or a supporting horizontal member for a sail.

borax A naturally occurring ore of boron. It is also known as sodium metaborate ($Na_2B_4O_7.10H_2O$).

bored pile A PILE that is formed by pouring concrete into a hole that has been bored into the ground specifically for the purpose. The hole is usually provided with reinforcing before the concrete is introduced.

borehole A hole drilled into the earth's crust for the purpose of exploration or to gain access to material such as oil or water that lies below the surface. Water obtained from a borehole is usually relatively free from microbial and organic contaminants as it is likely to have passed through much porous material, which acts as a filter before it reaches the site below ground from whence it is extracted.

See also: CRYPTOSPORIDIUM; DEEP WELL

borrowed light A window in an internal wall, such as that above a door, used to provide some natural lighting to another area, usually one such as a corridor, which has no external wall through which direct natural lighting could be provided.

bottle bank A facility in which individuals can deposit glass (predominantly bottles) for collection and subsequent recycling.

bottled gas *see* LPG

botulism A severe food-borne disease caused by one of, or combination of, the seven (types A–G) NEUROTOXINS produced by the spore-forming bacterium *Clostridium botulinum*. The organism is found in soil, marine sediments and the intestinal tracts of animals and fish, although there are no known human sources of the bacterium. The organism is spread by under-cooking or by under-processed food. Symptoms can initially manifest as a short period of diarrhoea and vomiting followed ultimately by loss of co-ordination, paralysis and respiratory failure.

See also: INFANT BOTULISM

boundary layer That at the junction between two materials, for example that between a warm and a cold current in the oceans at the point where the two meet.

bovine A member of the class of animals known as the *Bovidae* (cattle).

Bovine Somatotropin (BST) *see* SOMATOTROPIN

Bovine Spongiform Encephalopathy (BSE) A form of TRANSMISSIBLE SPONGIFORM ENCEPHALOPATHY that affects predominantly cattle. It was first recognised in 1985 in the UK and is characterised by loss of co-ordination, dementia and increased salivation followed by death – the symptoms giving rise to the popular name of 'mad cow disease'. In 1996 it was announced that the disease appeared to be capable of transmission to humans through the ingestion of contaminated tissue causing a new variation of Creutzfeldt-Jacob disease 'vCJD' (see CJD). The total number of human deaths associated with vCJD exceeded a hundred by the year 2001.

See also: PRION

bow saw A saw used for cutting wood and comprising a wooden frame within which a metal blade is held by the application of tension against a pivotal member, the whole having the appearance of a wooden bow or lyre. A coping saw is a bow saw with a fine blade designed for cutting intricate shapes.

BPEO *see* BEST PRACTICAL ENVIRONMENTAL OPTION

BPM *see* BEST PRACTICABLE MEANS

brace A structural member of either temporary or permanent nature that stiffens, binds or supports a framework or other structure.

bracken Also known as brake, any one of a group of large ferns with coarse stems and leaves, especially (in the UK) the species known as *Pteridium aquilinum*. The presence of a potent carcinogen known as ptaquiloside in bracken has given rise to health concerns. These are either from direct contact or from drinking the milk from animals grazing bracken, although neither fear has been proven as an actual risk. Simple contact would not be sufficient to introduce ptaquiloside to a susceptible individual and, even in experiments, grazing animals cannot be induced to eat bracken in quantities that would be likely to pose a risk.

brackish A descriptor applied to an aquatic environment with a concentration of salinity between that of salt and freshwater.

Bradford-Hill criteria Considerations that are used to evaluate the relative strength of associations detected in scientific studies. The criteria were developed by Sir Austin Bradford-Hill and are categorised in the following way:

Strength	the stronger an association the more likely it is that it is causal.
Consistency	consistent results across a number of studies suggest a true association.
Specificity	effects that are limited within a range of specific outcomes suggest a true association.
Temporality	establishing a relationship between effect and time of exposure suggests a true association.
Biological gradient	if there is a dose response this suggests a true association.
Plausibility	is there evidence to suggest that an effect could plausibly be caused by an input?
Coherence	interpretation should broadly agree with known facts.
Experiment	do results from experiments help to confirm the results of the study?
Analogy	are similar results obtained from similar inputs?

bradycardia An abnormal slowness in the heart and pulse rate.

See also: TACHYCARDIA

brain-stem The connective trunk between the main brain and the spinal cord. Brainstem death is a state in which there has been irreversible loss of brain function and complete cessation of bodily functions. In this state, an individual can only be kept alive by the aid of an artificial ventilator. The state at which clinical death has occurred has been subject to much controversy, but detailed criteria have been developed to aid clinicians in determining when the use of artificial aids can be terminated to allow the patient to die.

brake *see* BRACKEN

brake horsepower A means of measuring the effective power output of an engine in terms of describing the power necessary to brake it. For a vehicle brake horsepower is usually measured at the wheels – it therefore takes into account reduction in power due to transmission losses. Some measurements for brake horsepower are taken from the flywheel and would therefore not allow for transmission loss. Brake horsepower is measured in the same units as HORSEPOWER.

brass A metallic alloy comprising copper and zinc.

brazing The process of joining two pieces of metal together using a molten metal alloy generally of copper and zinc, sometimes with the addition of silver or occasionally using a different alloy. The melting point of the alloy is below that of the two metal components to be joined. Brazing is carried out at a temperature above that of soldering (see SOLDER) but below that of WELDING.

bread A generic name for a confection made from moistened flour, with or without additional ingredients and which is then baked. The most common variation is the addition of LEAVEN to induce FERMENTATION prior to baking in order to produce a lighter or less dense product.

breakpoint chlorination That stage in the process of the addition of chlorine to water for the purposes of DISINFECTION at which there is sufficient chlorine to combine with any organic pollutants and also to provide a combined chlorine base that will permit the presence of free chlorine in the water. The presence of free chlorine provides a safety margin to deal with those additional organic pollutants that might be introduced subsequently to the water. The technique is applied widely in the treatment of water in (especially larger) swimming pools.

breastsummer A long horizontal structural beam (nowadays generally a steel girder) supporting the main portion of the front of a building (such as the superstructure above a shop window). The name derives originally from a long wooden beam (the 'summer') used to support the front of a building (the 'breast'). Such a beam is also known as a bressumer or a brestsummer.

bressumer see BREASTSUMMER

brestsummer *see* BREASTSUMMER

brief A commission or specification given by a client to a professional in order for them to undertake work on behalf of the client.

bright bolt One used in steel construction and that has been engineered to fit exactly into the holes of the articles it is intended to join together.

See also: BLACK BOLT

bring recycling *see* DROP-OFF RECYCLING

Britain An alternative name for the UNITED KINGDOM.

See also: GREAT BRITAIN

brittleness A measure of the lack of a particular material to withstand stresses of shear or torsion although brittle materials often have good strength in compression.

broad-spectrum The term used to describe a wide range within a selected group. Broad-spectrum antibiotic drugs, for example, are those that are effective against a wide range of bacteria rather than being specific to one or two types.

broadcast Originally a method of sowing crops by manually scattering the seed widely across an area of land as opposed to sowing it in a line or DRILL. The term is nowadays used to describe widespread dissemination, especially of news or information via radio or television.

broiler A young fowl up to around 3 months old raised for the purposes of human consumption. Although most broilers are reared intensively the term does not exclusively apply to birds raised in this way.

Brompton cocktail or Brompton mixture A variable prescription concoction comprising mainly morphine, cocaine and alcoholic spirit (such as whisky or rum) used as a painkiller, especially in the treatment of terminal illnesses.

bronchial That which relates or pertains to the air passages that carry air between the trachea and the lungs.

bronchiole One of the airways of the lung, being the smallest division of the BRONCHUS. The bronchioles terminate in the ALVEOLI.

bronchitis An inflammation affecting the mucous membranes of the BRONCHUS.

bronchoconstrictor Any substance that acts in the body to cause constriction or narrowing of the airways of the lung.

bronchodilator Any substance that acts in the body to increase the diameter of the airways of the lung.

bronchospasm A narrowing of the bronchi brought about by muscular contraction.

bronchus One of the two branches of the TRACHEA, one going to each of the two lungs. The bronchus itself then subdivides into the BRONCHIOLES. The bronchus is also known as the bronchial tube.

bronze A metallic alloy comprising copper and tin, sometimes with the addition of zinc.

brown field site An area upon which buildings already stand, or have been demolished, which is intended or destined for redevelopment for domestic, commercial or industrial purposes. Redevelopment of such sites is attractive from the SUSTAINABLE DEVELOPMENT viewpoint although the cost of doing so can exceed that of development of a GREEN FIELD SITE by virtue of the expenses sometimes necessary to return the site to a condition fit to be built upon.

See also: URBAN RENEWAL

Brownian motion The random movement of particles held in suspension as caused by the constant collision of the molecules of the medium within which they are suspended.

brownstone A type of sandstone of brown or reddish-brown colouration used extensively in the eastern USA during the 19th century, particularly for the building FAÇADE. The term has come to be applied to describe any building constructed of or faced with such material.

brucellosis A potentially zoonotic (see ZOONOSIS) disease caused by one of a number of types of bacteria of the *Brucella* group. In humans the disease produces a vast range of symptoms, including fever, muscular pain and malaise, some of which can last for many months. In cattle the disease manifests as a potential cause of abortion and is caused by *B. abortus*, while in sheep and goats the disease is caused by either *B. suis* or *B. melitensis*. There is growing evidence, through the detection of *Brucella* spp antibodies in washed-up carcasses, to suggest that several species of marine mammals (such as seals, dolphins, porpoises, etc.) may harbour the infection. There may consequently be a risk to animals grazing on shoreline areas in proximity to such washed-up carcasses, but the precise risk is difficult to establish. It is considered prudent to avoid unnecessary (or unprotected) contact with such carcasses.

brùlé In culinary practice, literally means 'burnt'. It refers especially to toppings or coatings that are well browned by the application of heat.

Bruntland Commission The popular name given to the World Commission on Environment and Development chaired by Gro Harlem Bruntland. It was established in 1984 by the United Nations General Assembly and commissioned to consider the connections between the environment and development. The Commission's final report *Our Common Future* was presented to the General Assembly in 1987.

bryophyte Any of a range of primitive plants without vascular systems (i.e. they do not contain ducts for conveying sap). Such plants include the mosses and the liverworts.

BSE *see* BOVINE SPONGIFORM ENCEPHALOPATHY

buccal Means associated with the mouth (i.e. buccal cavity = mouth).

buffer solution A solution that minimises the change in hydrogen ion concentration (i.e. the PH) when an acid or base is added to it.

bugle-head(ed) screw A threaded screw in which the flat head tapers gently into the shank so that the screw has the appearance of a bugle in profile. Bugle-head screws are used to penetrate the soft surface of plasterboard and similar so as to leave a flush finish when driven home but without ripping or tearing the surrounding surface.

bulb The swollen underground bud of a plant comprising consecutive layers of scales. Its function is to act as a storage mechanism from which new growth will be generated at the start of the next growing season.

See also: CORM; TUBER

bulimia *see* ANOREXIA

buoy An anchored floating object located in a watercourse to indicate or mark a navigable waterway or mooring.

buoyance A measure of the loss in weight exhibited by a body that is wholly or partially immersed in a fluid. A body that displaces its own weight of fluid will float.

See also: ARCHIMEDES' PRINCIPLE

burette A piece of laboratory equipment comprising a vertical calibrated glass tube fitted with a tap at its lower end through which liquid can be discharged in a measured fashion. A burette is often used to measure out liquid used in TITRATION.

burlap A coarse fabric traditionally woven from hemp, jute or similar material.

bushel An old measure of volume, particularly once used as a measure of grain. One bushel is equivalent to 8 imperial gallons (0.035 m^3) or 64 US pints (0.036 m^3).

butt jointed A technique of joining two items of construction materials (frequently wood or timber) with their flat edges closely together without overlapping or creating special mouldings or shaping.

See also: CARVEL (BUILT); TONGUE AND GROOVE JOINT

buttress A supporting or reinforcing structure incorporated into a building for the purposes of resisting the (usually lateral) thrust of another part of the structure such as the walls or roof.

See also: FLYING BUTTRESS

C

Cabinet The senior committee of national or local GOVERNMENT with overall responsibility for policy. In reference to the UK Government the Cabinet comprises the Prime Minister and those MINISTERS OF STATE appointed by the Prime Minister.

cadaver A dead body.

cadmium plating Electroplating with metallic cadmium. Steel bolts for fastening aluminium plates are plated with cadmium to protect against electrochemical corrosion that would occur if unplated steel bolts were to be used.

caffeine In pure form is a white crystalline substance obtained from the coffee plant. It is a stimulant of both the heart and the CEREBRUM, and is a DIURETIC. It has been used in both human and veterinary medicine.

See also: ALKALOID

caisson A waterproof chamber open at the top and used to protect workers and facilitate working below the level of the water table or below water level.

calcareous Means containing either calcium or calcium carbonate ($CaCO_3$).

calcite The CRYSTALLINE form of calcium carbonate ($CaCO_3$).

calibre The measure of the internal diameter of a pipe or the bore of a gun barrel. The term derives from its original use as a measure of the diameter of a cannon ball or bullet.

calking The US spelling of CAULKING.

calorie A unit measure of energy. One calorie is that which will raise the temperature of one gram of water by one degree centigrade. In nutrition the term calorie is often used incorrectly to indicate the energy content of a unit mass of food – as in 'the food contains X calories per gram'. What is usually actually being quoted is the kilocalorie content; 1 kilocalorie = 1,000 calories.

See also: JOULE

calorific value The amount of heat released by the complete combustion of a unit mass of substance. It is a measure of the potential energy content of the substance and is quoted in calories or kilocalories.

See also: CALORIE

calorimeter A scientific instrument used to measure differences in heat energy during the course of a chemical reaction or process such as combustion. The differences are usually expressed in CALORIES per unit mass of substance.

cames *see* STAINED GLASS

campylobacter A potentially pathogenic curved Gram negative (see GRAM'S STAIN) bacterium. It does not form spores.

cancer A MALIGNANT growth within body tissue caused by abnormal cell division. A primary cancer might release cells to spread through the blood or lymphatic system setting up secondary cancerous growths in areas where they lodge. A cancer is also known as a neoplasm.

See also: TUMOUR

canvas An unbleached cloth with a close weave. It is made from a variety of materials including hemp, cotton and flax. It is used for clothing, curtain material and, when stretched on a frame, as a surface for artistic painting.

capacitor A device in an electric circuit that is capable of accumulating electric current.

capillary The smallest (at around 0.025 mm diameter) blood vessel in the body, forming the link between the ARTERIES and the VEINS. The body tissues are suffused with capillaries and their size and thin-walled structure allows the ready transfer of fluids and dissolved gases between the tissues and the blood.

See also: CARDIOVASCULAR SYSTEM

capillary attraction The physical force that produces the manifestation, in the absence of other forces, of a liquid rising up the internal surface of a capillary tube above the level of the liquid outside the tube when it is dipped into the surface of the liquid.

See also: SURFACE TENSION

capillary tube A glass (or similar) tube with a very small internal diameter. It is used in scientific experimentation and instrumentation.

capital In financial terms, the money, buildings and/or property used in the running of a business or other undertaking. In architecture a capital is the topmost (usually decorated) part of a column.

captive bolt pistol An instrument used to stun an animal as the first stage in one of a range of methods of slaughter. The STUNNING operation using this technique requires the pistol to be placed against the animal's head whereupon discharge of the weapon results in a metal shaft (the 'bolt' – propelled by an explosive cartridge or by compressed air) penetrating the skull, entering the brain and causing loss of consciousness. The bolt is an integral part of the pistol and is retained attached upon discharge (i.e. it is 'captive'), ready for retraction and reuse.

See also: BLEEDING; PITHING

carat A means of describing the purity of an object made out of gold – pure gold is described as 24 carat, 22 carat contains 2 parts per 24 (i.e. one twelfth) of other metal as an alloy, 18 carat contains 6 parts other metal as an alloy and so on. A carat is also a measure of the weight of gemstones – 1 carat is equivalent to 0.2 grams.

carbohydrate One of the major classes of naturally occurring organic compounds consisting of carbon, hydrogen and oxygen, and with the general formula $C_x(H_2O)_y$. They are produced during PHOTOSYNTHESIS and they allow the transport and storage of energy. There are many types of carbohydrate ranging from simple SUGARS, such as MONOSACCHARIDES and DISACCHARIDES, to POLYSACCHARIDES, such as STARCH and CELLULOSE.

See also: FAT; PROTEIN; LIPID; VITAMIN

carbon black A material produced from the incomplete combustion of petroleum or natural gas. It is used as a mineral pigment.

carbon dioxide Carbon dioxide (CO_2) is a colourless, non-combustible and odourless gas produced by the chemical combination of carbon with oxygen, usually as a result of the process of either combustion or respiration. It is a normal although minor constituent of air, but its concentration is increasing by around one quarter of a per cent annually. It is a key GREENHOUSE GAS, trapping considerable solar energy within the atmosphere. Although not the strongest greenhouse gas it is the most important by a considerable margin due to its overall volume. It has been estimated that the earth could on average be around 30°C cooler without the current atmospheric combined content of carbon dioxide and METHANE. The warming of the atmosphere is also releasing to atmosphere that carbon dioxide trapped in the permafrost layers in areas such as Siberia.

carbon monoxide A colourless, odourless and non-irritant gas. It is generated to atmosphere usually as a product of incomplete combustion. Motor vehicles, domestic heating, gas cooking and cigarettes are major sources. Carbon monoxide is highly toxic. Its molecules combine with HAEMOGLOBIN in the blood around 300 times more readily than does oxygen to create carboxyhaemoglobin, a stable molecule that is effectively removed from oxygen carrying capacity. This reduces the overall capacity of the blood to bond with oxygen atoms causing adverse health effect, particularly for people suffering from ischaemic heart disease. The effects are worse with prolonged exposure, particularly if this is at concentrations of more than 10 per cent, and death can result at higher concentrations.

carbon sink The term applied to a natural phenomenon whereby atmospheric carbon (usually in the form of CARBON DIOXIDE) is captured and locked into a storage system. The term has entered into popular vocabulary in the debate over GLOBAL WARMING and is seen by some countries, notably the USA, as a means of reducing the amount of carbon dioxide (a GREENHOUSE GAS) in the atmosphere in preference or additional to emission reduction. The most common method of creating a new carbon sink is the planting of forests to absorb carbon dioxide as part of the process of PHOTOSYNTHESIS and lock the carbon within the vegetation. The world's oceans are major natural carbon sinks for atmospheric carbon dioxide.

carbon tax The name applied to a range of taxation possibilities levied on FOSSIL FUELS proportionate to the amount of CARBON DIOXIDE emissions generated by their use. The philosophy of such a taxation initiative is both to raise the price of those fuels that contribute most towards GLOBAL WARMING and to use the money in pursuing initiatives to reduce this effect.

carbonation The process whereby calcium carbonate is slowly produced in the reaction between atmospheric CARBON DIOXIDE and compounds containing calcium.

carborundum Another name for silicon carbide (SiC). It is very hard and is used as an abrasive. It is not the same material as CORUNDUM.

carboxyhaemoglobin HAEMOGLOBIN in which the sites of the molecule that normally bind with oxygen are instead bound to carbon monoxide. This latter bond is sufficiently strong as to prevent the body from breaking it, resulting in the loss of the affected haemoglobin molecule as a carrier of oxygen. If sufficient haemoglobin molecules are affected asphyxia and death can result.

carburettor A device used to provide a suitable fuel and air mixture to an internal COMBUSTION engine.

carburising *see* CASE HARDENING

carcinogen Any substance that is capable of inducing the production of CANCER – a substance that has this capacity is known as carcinogenic. Such substances include chemical and physical agents, viruses, hormones and some internal bodily processes. In many instances the carcinogen is capable of causing cancer directly within the body; in other circumstances the body's own metabolism converts procarcinogens into carcinogens. Many carcinogens are predominantly associated with particular types or sites of cancer generation.

An important distinction when considering carcinogens is their capacity to alter directly the body's DNA. Those carcinogens that have this capacity are known as GENOTOXIC, while the remainder are non-genotoxic. When considering this aspect it is also necessary to evaluate the capacity of chemical METABOLITE to be carcinogenic.

Establishing carcinogenicity (the capacity to cause cancer) is difficult, predominantly because there is often a considerable LATENCY PERIOD (sometimes of decades) between exposure and the manifestation of recognisable clinical effects. In addition carcinogenicity can be dose-related or multifactorial and some can only become carcinogenic if acting in concert with other chemicals. These are all issues that complicate the assessment process. Most assessments use evidence derived from occupational exposure studies. Such evidence might be the best (perhaps only) evidence available but equating low dose/long exposure to high dose/short exposure is not easy or in some instances possible.

Experimentation on humans is clearly impossible but evidence for assessment from animal experimentation is sometimes available. There is however sometimes little correlation between human and animal susceptibility to a given substance or dose – some carcinogens are species-specific, some are even sex-specific. The inherent uncertainties in assessment capabilities have resulted in various classification systems for carcinogens, most of which describe carcinogenicity by degree of variability of certainty in terms of being 'confirmed', 'probable', 'possible', 'not provable' or 'no evidence for. . .'.

Categorisation varies with the increasing availability of evidence over time – it is advisable to refer to the latest definitions and categorisations for any particular reference system to obtain the most accurate information.

See also: GENOTOXIC; MUTAGENIC; TERATOGENIC

carcinogenicity The capacity or property of a substance to cause cancer.

See also: CARCINOGEN

carcinogenicity bioassay A test undertaken using laboratory animals to assess whether or not a particular substance is carcinogenic.

See also: CARCINOGEN

carcinoma A MALIGNANT tumour arising either from EPITHELIAL cells or from solid viscera such as liver, kidneys or pancreas, and sometimes the endocrine glands.

cardiovascular system The totality of all the blood vessels of the body and the heart. Cardiovascular disease is that which affects this system, usually to the effect of causing thickening of the walls of the blood vessels or their partial or total blockage.

See also: ARTERY; CAPILLARY; VEINS

carditis An inflammation of the heart.

carnivorous Means deriving sustenance from a diet consisting exclusively (or predominantly) from meat.

See also: HERBIVOROUS; OMNIVOROUS

carob The edible pod, tasting of chocolate, of an evergreen Mediterranean tree.

carotene A widely distributed natural precursor to vitamin A. It is usually found as beta-carotene (β-carotene), a yellow pigment that provides the colour of carrots and tomatoes. It has ANTIOXIDANT properties. Carotene is one of two pigments found in the skin of humans, the other pigment being MELANIN.

carriage bolt A special type of fastening bolt, only one end of which is equipped with a thread and nut. The other end is rounded and slightly domed but the shank immediately under the dome is square. As the bolt is tightened the domed end is held in position by the square shank so that the finish at the surface at that end is neater than would be presented by a traditional bolt head.

carrier A person or animal harbouring a specific pathogenic organism or INFECTION with the capability of passing that organism onwards to cause subsequent cases of the disease. The term is most usually applied to those who are ASYMPTOMATIC, either because they are themselves developing the infection during its INCUBATION PERIOD or because they have IMMUNITY against the organism. The term carrier is not usually applied to persons in whom clinical symptoms of the disease are manifest: such people are more usually described as infectious.

carse A term of Scottish derivation used to describe a flat, fertile ALLUVIAL area adjacent to a river.

Carson, Rachel Most famous as being the author of the book *SILENT SPRING*.

carte blanche An expression from the French describing the complete discretion or authority to take whatever action the person upon whom such discretion is conferred deems to be appropriate.

carte du jour An expression from the French used to describe the menu in a restaurant listing all the dishes available on a particular day.

cartel An association of businesses or business interests formed either formally or informally (and sometimes illegally) with the intention of manipulating the free market and to avoid competing with each other.

carvel (built) A boat-building term referring to the method of constructing (usually) a wooden hull in which timber planking is BUTT JOINTED smoothly without overlapping.

See also: CLINKER (BUILT)

case An individual in a COHORT or population who has been evaluated or diagnosed as having a particular disease or condition.

case comparitent study *see* CASE CONTROL STUDY

case control study Is an epidemiological study in which two matched groups of individuals are compared to ascertain whether or not statistically significant differences between the two groups can be detected to explain certain phenomena, the latter being usually determined as part of an hypothesis. The group of 'cases' usually comprises those with a disease or those receiving a treatment; the 'controls' are a group selected to be as closely comparable to the cases but without the disease or treatment. The case control study is also sometimes referred to as a case comparison study or a case referent study.

See also: COHORT STUDY

case definition A set of specified symptoms or characteristics that are produced in an epidemiological investigation against which an assessment can be made to decide whether *for the purposes of the investigation* an individual will be considered as having been affected by an event or not. In practice producing a case definition is a crucial stage in the investigation of, for example, outbreaks of disease, as this will define who is considered to be ill and who is not, and thereby dictate the accuracy or otherwise of statistical analyses.

case hardening A process of imparting extra hardness to the surface of low-carbon steel by heating the object and exposing the surface to a mixture of carbon and nitrogen in a process known as carburising.

case law The interpretation of a piece of legislation by one of the higher courts (i.e. above the level of the Magistrates' Court). The interpretation is deemed to set a precedent for subsequent cases unless it is deemed to have specific or limited application at the time the decision was announced.

case mix The range or list of the different types of disease that are managed within an individual health care facility or GP's practice.

case referent study *see* CASE CONTROL STUDY

casein A protein bound with phosphoric acid and made by precipitating it from milk by the action of RENNIN. It is used as the basic ingredient in cheese manufacture and as an ingredient in certain plastics and adhesives. In the USA casein is known as paracasein.

casement window An opening window, the opening portion of which is mounted in a frame and hinged either at one side or the top, usually to open outwards from the building in which it is located. Those casement windows that are hinged at the sides are known as 'side hung', while those hinged at the top are known as 'top hung'.

cast iron That produced by cooling in moulds the product of the process used to make iron. It is one of two traditional processes of iron manufacture, the other being WROUGHT IRON. Cast iron is less pure that wrought iron, generally having a carbon content of between 1.8–4.5 per cent. Cast iron is hard but brittle.

caster sugar Very finely ground household SUGAR.

castor bean plant The North American name for the castor-oil plant.

See also: CASTOR OIL

castor oil A thick clear or yellowish oil derived from the castor-oil plant (*Ricinus communis*), a native plant of Indian origin. In the USA the plant is known as the castor bean plant. The seeds of the plant are poisonous. The oil is used as a light lubricant and as a purgative.

catacomb Originally, in ancient Rome, an underground burial area consisting of a number of tunnels containing vaults, ledges or niches in which the bodies were laid. Nowadays the name is used for any system of underground tunnels or caves.

catalyst A substance that speeds up the rate of a particular chemical reaction but that does not itself alter the ultimate yield of the reaction or itself be consumed in the reaction.

catalytic converter A device that is incorporated into the exhaust system of an internal COMBUSTION engine to reduce the overall level of emissions. The devices convert CARBON DIOXIDE and nitric oxide into carbon dioxide and nitrogen, and they also convert organic compounds into carbon dioxide and water. Performance varies dependent on the type of catalytic converter and the emissions being dealt with but, for example, an efficient catalytic converter should be capable of removing around 90 per cent of VOLATILE ORGANIC COMPOUNDS (VOCs) (i.e. includes BENZENE, PCBs and PACs) from exhaust gas emissions.

cataract An opacity of the lens of the eye that results in sight impairment or blindness. It can arise through a number of causes, including age, trauma, metabolic disorders or hereditary and congenital factors.

catchment area In terms of water supply, the totality of the surface area of ground from which rainwater is collected to supply rivers, lakes and reservoirs. In terms of commerce or service industries the term refers to the area from which customers or clients can be expected to be drawn.

caterpillar The larval stage of butterflies and moths (*Lepidoptera*). The body is supplied with many legs on both the thorax and abdomen.

catheter A hollow tube inserted into the body to facilitate the passage of material either inwards or outwards, or to facilitate the administration of medicines.

cathode The negative electrode in an electrolytic cell. The positive electrode is known as the ANODE.

cathode ray tube A sealed tube within which there is a vacuum. At one end of the tube an electron gun emits a stream of electrons onto a fluorescent screen. Transverse electrostatic and magnetic fields control the direction of the electron stream such that the screen fluoresces to produce images.

cation A positively charged ION.

See also: ANION

catwalk A narrow elevated platform, especially one constructed to allow clothing to be displayed by fashion models parading along it.

caulking The waterproofing or sealing of a joint by pushing or inserting a waterproof agent into the gaps. The term was originally applied to the technique of sealing the hulls and timbers of ships and boats with OAKUM steeped in pitch, but the term is now applied to the operation using a variety of materials. The US spelling is calking.

causality The phenomenon whereby a circumstance occurs as a result of an action or activity, i.e. one is caused by the other. In epidemiology the question of causality is important since statistical associations do not, of themselves, prove a link between two events or occurrences: they simply show the potential for a link. For much investigative epidemiology causality often cannot be directly proven; instead one must take into account plausibility (i.e. the likelihood of one occurrence having caused the other).

See also: BRADFORD-HILL CRITERIA

caustic The adjective used to describe the property of any substance to cause rapid burning or corrosion when brought into contact with another substance, especially one that will cause such effect if brought into contact with living tissue.

caustic potash Another name for potassium hydroxide (KOH).

caustic soda Another name for sodium hydroxide (NaOH).

caveat A warning, usually appearing as a written statement attached to an artefact or piece of work, in which the owner or person responsible for the artefact or the work clarifies the degree of responsibility (if any) that they accept for that which arises subsequent to its use. The caveat is usually written as a disclaimer.

cavity wall A structure in which two solid parallel constructions (of brick, block work, stonework or similar) are built with a gap between them, this latter being known as the cavity. The cavity may be filled with insulating material or left as a void through which air is allowed to circulate. The cavity acts to prevent moisture from rainwater passing across the wall to affect the inside skin. The two skins are held together by numerous WALL TIES.

cell The smallest unit of biological life. The cell usually comprises a semi-permeable outer membrane within which a mass of protoplasm is contained. Although most cells and the cells of most organisms contain a nucleus, this is not necessarily so.

See also: ERYTHROCYTE; PROKARYOTE

cell culture A method of growing living cells away from the living organism in a nutrient media.

See also: *in vitro*; *in vivo*

cellulose A POLYSACCHARIDE carbohydrate found in the cell walls of higher plants and some fungi (see FUNGUS). It is perhaps the most abundant carbohydrate.

cement Essentially any material that sets firm and has the purpose of joining things together. In the construction sense the term is usually an abbreviation of 'Portland Cement' (also known as 'Ordinary Portland Cement' or 'OPC'). This latter is a grey material principally comprising calcium silicates and calcium aluminates produced by burning chalk or limestone with clay or shale and grinding the result into a powder.

See also: CONCRETE; HYDRAULIC CEMENT

cement mortar A MORTAR comprising a mixture of sand, cement and water.

central composting A form of recycling in which BIODEGRADABLE municipal solid waste (predominantly garden waste and organic kitchen waste) is transported to a central facility to be composted.

centre of gravity The point within any body below which the body would be in balance in any plane if it were possible to locate the body on a point. The centre of gravity is also known as the centroid or the centre of mass.

centre punch A tool with a hard point, impelled by a hammer blow or similar and used to break or indent a surface in preparation to receive a drill or screw. The centre punch allows for precise positioning of the intended hole and, on many surfaces, reduces the risk of scratching or slipping of the screw or drill.

centrifugal force A fictitious force that can be considered as that acting outwards from a body which is in rotation or travelling along a curved pathway.

See also: CENTRIPETAL FORCE

centrifuge A mechanical device that rotates at high speed with the aim of separating solid particles from suspension using the gravitational forces thereby induced. A centrifuge can also be used to generate linear motion in an object by raising its speed and then expelling it tangentially.

centripetal force A true force that acts towards the centre of a body in rotation or towards the centre of curvature on a body travelling along a curved path.

See also: CENTRIFUGAL FORCE

centroid *see* CENTRE OF GRAVITY

ceramic tile A generic term to describe a tile composed of baked clay. The term includes tiles used for walls, floors and roofs.

ceramic ware The term used in the building industry to describe collectively those goods that are (or traditionally were) made from glazed baked clay. Such goods include lavatory basins, wash basins and sinks.

cerebellum The portion at the rear of the brain of higher animals that lies underneath the lobes of the CEREBRUM. It is responsible for controlling balance, maintaining muscle tone and for the co-ordination of voluntary movement. The cerebellum is sometimes known as the 'small brain'.

cerebral cortex The outermost part of the brain.

cerebrum That tissue which forms the greater part of the brain of higher animals. It is divided into two halves or hemispheres.

See also: CEREBELLUM

cervical Means relating to, or of, the neck.

cesspit *see* CESSPOOL

cesspool In general usage, a temporary storage facility for SEWAGE. In some legal definitions the term has been taken as including a SEPTIC TANK. For the purposes of this definition a cesspool has no treatment capacity and no outlet, the container requiring regular emptying to remain in use. A cesspool is also known colloquially in some areas as a cesspit.

cestode A parasitic flatworm belonging to the classification *Cestoda*, which includes tapeworms.

See also: HELMINTHS

cetacean A marine mammal of the order *Cetacea*, which includes whales, dolphins and porpoises.

CFC *see* CHLOROFLUOROCARBONS

CFU *see* COLONY-FORMING UNIT

Chadwick, Sir Edwin b. 1800; **d.** 1890 A Manchester-born public health pioneer of the Victorian era. He moved to London at the age of 10 years. He became secretary to Jeremy BENTHAM and was appointed Assistant Commissioner to the Royal Commission, set up in 1832 to enquire into the workings of the antiquated Poor Law statutes. In 1842 he published his famous report 'Survey into the Sanitary Condition of the Labouring Classes in Great Britain'. In spite of much opposition at the time this report resulted in the passing in 1848 of two major pieces of public health legislation, the Public Health Act and the Nuisances Removal and Diseases Prevention Act. He was awarded the KCB in 1889.

chain An obsolete measure of length, equivalent to 66 ft. (approximately 20.12 m).

chalet Originally the term that was used to describe the small wooden huts used by herdsmen as summer lodgings in Switzerland when cattle were traditionally moved to higher ground to take advantage of summer pasture. The term has now become debased in use and is taken to refer to any (generally small) cottage or bungalow built with a resemblance to a Swiss chalet and especially to self-contained units of accommodation at holiday camps.

challenge In a microbiological or immunological sense, any demand made upon the body's defence mechanisms by microbial or chemical agents.

chamber study The generic name used to describe the experimental study of the effects of controlled exposure to various gases or aerosols on a group of volunteers within a sealed or controlled air space.

chamfer A flat surface cut at an angle of around 45° at the corner of a timber beam.

See also: BEVEL

chancellor In the UK, the honorary head of a university (the chief executive or chief administrator is known as the vice-chancellor). In several European countries the chancellor is the Head of State. In the USA a chancellor is variously the chief administrator or president of a university, although sometimes the name is also applied to the presiding judge of a court of chancery or equity.

Chancellor of the Exchequer The cabinet minister in the UK Government who carries the responsibility for financial matters.

chancery One of the Divisions in the United Kingdom of the HIGH COURT.

Chartered Institute of Environmental Health A professional and educational body for England, Wales and Northern Ireland founded in 1883 for the promotion of public and environmental health. Full and qualified members are known as ENVIRONMENTAL HEALTH OFFICERS. The equivalent body in Scotland is known as the ROYAL ENVIRONMENTAL HEALTH INSTITUTE OF SCOTLAND (REHIS).

CHD *see* CORONARY HEART DISEASE

chemiluminescence That light which is produced by chemical reaction. Its generation does not result in the production of heat, a characteristic recognised in its alternative name of 'cold light'.

chemotherapy The medical treatment of a disease through the use of chemicals. The term is most frequently used in application to the treatment of cancer (cancer chemotherapy) but can equally be applied to the treatment of bacterial infections through the use of antibiotics (antibiotic chemotherapy).

Chernobyl A town in the Ukraine, formerly part of the USSR. It is notorious for being the site of an accident that occurred on 26 April 1986 within a nuclear reactor. The resultant RADIATION leakage was distributed worldwide and monitoring of contaminated land and grazing livestock continues in some areas of the UK over 15 years after the event. The accident itself killed thirty workers on-site, resulted in the hospitalisation of hundreds and exposed in excess of 5 million people to FALLOUT.

CHI An acronym for the COMMISSION FOR HEALTH IMPROVEMENT (UK).

chi-squared test A statistical technique used in EPIDEMIOLOGY to indicate the probability of an association between an incident or activity and an outcome by testing the difference between proportions.

See also: SIGNIFICANCE LEVEL

chicken wire A material consisting of light galvanised steel wire, usually formed to create a hexagonal pattern, and used as a fencing material for enclosing animals and (particularly) poultry.

chimney The totality of the conduit and ancillary structures (such as supporting structure – the chimney breast – and the exit point – the chimney pot) for the passage of exhaust gases from a boiler, domestic fire or similar. It is often used as a synonym for FLUE.

china clay Another term for KAOLIN, a pure white form of aluminium silicate. The name refers to its use in the production of china porcelain and not to any geographical location where it can be mined. In the UK, Cornwall has traditionally been associated with china clay mining.

China syndrome A theoretical (possibly apocryphal) outcome should a nuclear reactor go into MELTDOWN. The theory holds that the resultant molten mass could burn its way entirely through the earth to emerge on the other side ('to China' – hence the name).

See also: THREE MILE ISLAND

chine The spine and associated parts (such as the ribs) of an animal used for food. A chine saw is one used for cutting or splitting the backbone.

chining The culinary term to describe the process of severing the ribs from the backbone of an animal being prepared for food.

See also: CHINE

chipboard A constructional material based on timber comprising chips of wood held together by an adhesive binder, the whole being assembled under the application of heat and pressure. It is not generally a strong structural material in shear but it has greater strength under COMPRESSION and is used widely as, for example, a relatively cheap flooring material and for furniture such as cupboards. Many forms of chipboard are not proof against dampness and lose their strength catastrophically when wet. For this reason chipboard is used predominantly for applications indoors.

chisel A linear-shaped tool with a cutting blade. The cutting process is achieved by applying sustained pressure or impact blows.

chitin A flexible but tough POLYSACCHARIDE, primarily containing nitrogen. It is an essential component of the EXOSKELETON of insects and crustaceans. It can also be found in the cell walls of some fungi (see FUNGUS).

See also: CELLULOSE

chlamidia Intracellular bacteria, increasingly recognised as human pathogens with sexual transmission becoming increasingly a cause for concern. *Chlamidia pneumoniae* causes pneumonia and respiratory problems, *Chlamidia psittaci* causes PSITTACOSIS and *Chlamidia trachomatis* is sexually transmitted, causing a range of infections including infant pneumonia and non-specific urethritis (NSU).

chloracne An occupational skin disorder. It is a distinctly different reaction from OIL ACNE, being a response to specific types of polychlorinated aromatic HYDROCARBONS rather than to PETROLEUM or its products. The symptoms of chloracne include the formation of blackheads or yellowish cysts on the temples, body or limbs, or behind the ears. In more serious cases inflammation may result. Once established the condition may become chronic, even in the absence of the initiating substance. Chloracne is a response to systemic absorption rather than a skin surface reaction. It can be difficult to treat and medical advice should be sought.

chlorination The process of disinfection by exposing contaminating microorganisms to an aqueous solution containing chlorine. Chlorination is used widely in the chemical treatment of swimming pools and in disinfection procedures in the food industry. The aqueous solution is usually generated by introducing liquid or solid compounds containing chlorine to water. Chlorine gas is dangerous to handle and the generation of chlorinating solutions using elemental gas is not normally practised.

See also: DISINFECTION

chlorofluorocarbons (CFCs) A group of chemicals that contain chlorine, fluorine and carbon. They are highly stable and of low toxicity, and have historically been widely used as propellant gases in aerosol cans and in refrigeration and air conditioning systems. They have been found to react with OZONE in the atmosphere and have been cited as being largely responsible for the depletion in the OZONE LAYER. Their use worldwide is being phased out.

chloroform ($CHCl_3$) (also known as methane trichloride, trichloroform and trichloromethane) is a colourless, volatile liquid with a characteristic sweet odour. It is non-flammable. It is carcinogenic and is sometimes found as a contaminant of water, arising principally as a by-product of disinfection or as a result of reactions with residual disinfectant in the distribution system. It is a strong suppressant of the central nervous system (CNS) and may cause liver and kidney damage. It was used historically as an anaesthetic but has now largely been superseded in this use.

chlorophyll A green pigment found in plants and cyanobacteria (see BLUE-GREEN ALGAE). There are two main types, being *chlorophyll a* ($C_{55}H_{72}MgN_4O_5$) and *chlorophyll b* ($C_{55}H_{70}MgN_4O_6$). Chlorophyll is the chemical that enables the process of PHOTOSYNTHESIS to proceed.

cholera An acute diarrhoeal disease caused by the comma-shaped bacterium *Vibrio cholerae*. It is primarily associated with water contaminated by sewage, and its ability to survive in aquatic environments has contributed to its ability to cause serious and widespread epidemics in circumstances, such as flooding, in which normal water supplies are disrupted and people have to rely on alternative, polluted natural sources. Prevention of the disease during an outbreak in London in 1854 is a notable and celebrated early success for epidemiology when JOHN SNOW famously determined that water from a pump in Broad Street was the source of the infection. He is reported as removing the pump handle to render the pump inoperable, thereby preventing the consumption of sewage-contaminated water and ending the outbreak.

cholesterol A waxy substance produced by the liver and adrenal glands. It is present naturally in the blood, bile, nervous tissues, skin and in the brain. It is an essential chemical for the production of the sex hormones and in the body's repair mechanism. Excess cholesterol in the blood (i.e. above 6 m.mol per litre) is thought to lead to deposition on the internal walls of the blood vessels, especially in the arteries, restricting the blood flow and potentially leading to CORONARY HEART DISEASE. Contrary to popular belief the amount of cholesterol *per se* consumed does not appear directly to affect blood cholesterol levels because most body cholesterol is manufactured in the body from saturated fatty acids. It is the consumption of foods high in these (e.g. animal fats) that seems to be the key determinant of blood cholesterol levels.

See also: STEROL; THROMBOSIS

chorea A jerky muscular spasm that can affect large portions of the body, such as in the condition described as St Vitus's dance, although there are various differentiated types.

CHP *see* COMBINED HEAT AND POWER

chromatography A method used to analyse the respective chemical components in a mixture. There are two main variations. Gas chromatography uses a volatised sample absorbed into a liquid to produce differentiated peaks (corresponding to individual chemicals) on a chromatogram – the heights of which indicate the concentration of the different components. Liquid chromatography (also called high-pressure normal-phase liquid chromatography) separates liquids on the basis of the time they differentially take to pass through a column packed with coated tiny spheres or beads.

chromium plating The name given to the process whereby a final layer of chromium is deposited by electroplating on an article of steel composition for the purpose of providing a decorative and protective finish. The process usually involves deposition of a first coat of copper, followed by one of nickel and finally one of chromium.

chromosome An individual structure within a cell comprised of a linear set of GENES. Chromosomes are arranged as a double helix structure. Not all life forms have the same numbers of chromosomes; humans have 23 pairs (22 pairs of AUTOSOMES and 1 pair of SEX CHROMOSOMES).

chromosome aberration A generic term used to describe the damage that can be caused to the DNA by EXOGENOUS chemicals or by physical agents.

chromosome probe A length of DEOXYRIBONUCLEIC ACID (DNA), tagged for ease of identification (usually with a fluorescent marker), which hybridises with its complementary region of a chromosome.

chronic Means long standing or of long duration. It often implies dull and persistent pain. When applied to disease it relates to recurring or persistent symptoms, often to those diseases that are of gradual onset. The term does not imply anything about the severity of the disease. The term chronic is sometimes taken to mean the opposite of ACUTE but this would be inaccurate: chronic and acute in this instance reflect temporal differences in disease or symptoms; they do not reflect presence or absence.

Chronic Obstructive Pulmonary Disease (COPD) A disease condition in which the air passageways of the lung become increasingly unable to perform their function. The condition can be caused by a range of diseases, including asthma, chronic bronchitis and pulmonary emphysema.

chrysalis *see* PUPA

chrysotile One of a group of fibrous silicates known generically as ASBESTOS. Under the microscope it appears as long, white, pliable fibres that exhibit a serpentine shape. Its colour under the microscope has given it the common name of 'white' asbestos but this can be misleading as the material within which it can be incorporated need not necessarily appear as white to the naked eye. Historically it has been used primarily for fire protection (e.g. in fire curtains and fire blankets).

chymosin *see* RENNIN

CIE The acronym for the Commission Internationale de l'Éclairage, the international co-ordination body for issues relating to research into light. The Commission is based in Vienna.

CIEH *see* CHARTERED INSTITUTE FOR ENVIRONMENTAL HEALTH

cilium (plural cilia) Minute hair-like projections on the outer membrane of certain cells. Their function, generally, is to induce either movement of the cell through a medium or the movement of a medium over a cell.

circadian rhythm The body's fluctuating physiological response to variations in the DIURNAL cycle.

circuit breaker A device in an electric circuit that automatically operates as a switch to cut off the flow of electricity under certain preconditions. They are generally used as safety devices and can be operated, for example, by increased temperature (e.g. if current flow exceeds design criteria) or by creating an electromagnet.

circular saw A cutting device in which a circular blade equipped with serrated teeth is used as the cutting face.

cistern A small tank or reservoir used to store water.

civic amenity site A designated facility provided by local authorities in the UK for the reception and subsequent onward transmission of household waste. Although such sites were once predominantly simply waste transfer sites used to hold waste until it could be collected for disposal, many such sites now offer facilities for waste separation and recycling.

CJD The acronym for Creutzfeldt-Jacob disease, a TRANSMISSIBLE SPONGIFORM ENCEPHALOPATHY disease in which the degeneration of the tissue of the brain leads to dementia. The disease classically affects people of 40 to 65 years of age and death usually follows within 12 months of the first manifestation of symptoms. A new variant of CJD (vCJD – reported in the *Lancet* on 6 April 1996), which generally affects younger people (late teens and early twenties), has been identified. The cause is attributed to eating infected meat infected with BOVINE SPONGIFORM ENCEPHALOPATHY (BSE).

Classical CJD can arise in three ways. Sporadic CJD is the most common form and in the UK occurs in around one in every million people. Inherited CJD is very rare and is passed through families. Acquired CJD is transmitted through the donation of contaminated tissue derived from a patient suffering from CJD or, at least theoretically, by transmission of contaminated material during medical or dental procedures.

See also: TRANSMISSIBLE SPONGIFORM ENCEPHALOPATHY

cladding A material that has been added as a cover to either an internal or external surface of a building to provide a decorative and/or protective finish.

clapboard The US expression for what is known in the UK as WEATHERBOARD.

clarifying The culinary term to describe the process of making clear or purifying an ingredient used in cooking. It is used mainly to describe the process of melting fat (such as butter) or separating fat from other ingredients so that it can be used for frying or making pastry.

clastogen A substance or agent that can produce a break or other structural damage in a chromosome. Chemicals, viruses and physical agents can all be clastogens.

See also: CLASTOGENICITY

clastogenicity One of the three levels of mutation of genetic material that can affect the cell; the other levels are GENE and ANEUPLOIDY respectively. Clastogenicity is that which relates to aberrations in respect of the structure of the chromosomes. Clastogenicity can be important in the development of some types of TUMOUR.

claw hammer A hammer in which one face of the head is elongated into a curved and forked shape for the purpose of extracting nails.

clay puddle *see* PUDDLE

clean in place (CIP) A technique applied to (usually) continuous-flow food-processing plant by the application of successive wash and rinse cycles without the need to dismantle the plant.

See also: DEEP CLEAN

clean-to-dirty flow A system used predominantly (but not exclusively) in food manufacturing in which employees are only allowed to travel through the system backwards along the flow of work, i.e. from one operation to the next. The object is to ensure that a contamination at a point in the flow cannot be transmitted forwards. Employees are only allowed to travel forwards along a flow of work if they undertake any precautions that have been deemed to be necessary to safeguard hygiene (e.g. hand washing and boot sterilisation). The system is often used in combination with HAZARD ANALYSIS CRITICAL CONTROL POINT (HACCP) systems to maintain the integrity of CRITICAL CONTROL POINTS.

cleaning The removal by physical or chemical means of deposited matter from surfaces, equipment or materials.

See also: DISINFECTION; STERILISATION

clearance The distance between two objects that might otherwise touch.

clinical governance An organised framework within which continuing improvement in clinical care is sought.

clinical indicator A measurement of the quality of any particular aspect of clinical care.

clinical waste That waste arising from medical, nursing, dental, veterinary or pharmaceutical facilities that, for the public protection and in light of its potential contamination with infected or infectious material, needs special handling and disposal. The classification currently used in the UK to identify the varying levels and degrees of risk associated with clinical waste defines the waste as falling within one of five groups. Group A comprises all human tissue, including blood, and waste materials presenting an infection risk to staff who handle them. Group B comprises mainly contaminated disposable sharps. Group C includes microbiological cultures and potentially infected waste from pathology departments. Group D comprises chemical waste and Group E includes those items used to dispose of urine, faeces and other bodily secretions or excretions not falling into Group A.

See also: SHARPS

clinker (built) A boat-building term referring to the method of constructing (usually) a wooden hull in which the external timber planking is overlapped downwards to form a watertight seal.

See also: CARVEL (BUILT)

clockwise Means rotating in the same direction as the hands of a clock.

clone A cell, group of cells or complete organism that is genetically identical and is derived from a single common ancestor. The term is also used to describe genetically identical individuals produced asexually from a single ancestor by the processes of GENETIC MODIFICATION.

Clostridium difficile A bacterium that is carried by between 1–3 per cent of normal healthy adults although this figure may rise to between 10–20 per cent of patients in hospital as a result of infection acquired in such institutions. The most common symptom of infection with *Clostridium difficile* is diarrhoea. Patients using antibiotics are generally considered to be most at risk. Human illness associated with *Clostridium difficile* is mostly restricted to cases in hospital and it is not considered that people in the normal population are at risk. *Clostridium difficile* is not considered to be a significant cause of food poisoning.

Clostridium perfringens The name now applied to the organism previously known as *Clostridium welchii*. It is an anaerobic spore-forming bacterium that causes food poisoning. The spores survive normal cooking. Ingestion of large numbers of vegetative cells results in enterotoxin production in the small intestine. Usually more than 10^6 micro-organisms (i.e. 1 million) are required to cause illness (the infective dose). Sources of the bacterium include the faeces of animals and man, soil, sewage, dust and feeds of animal origin. The organism is spread most frequently through contaminated bulk (anaerobic) cooked meat and poultry dishes that have been left at ambient temperature at cooling and storage. Clinical symptoms produced by infection are diarrhoea and acute abdominal pain; vomiting is uncommon.

cloud A visible aggregation of water droplets and/or ice crystals suspended in the atmosphere. Clouds are primarily classified by their shape as discerned from the ground. Cirrus clouds are those that swirl, cumulus clouds appear as a pile or heap and stratus clouds appear in layers. The term nimbus describes clouds that appear ready to release rain (the term is often combined with others, e.g. cumulonimbus).

In defining clouds by height the prefix 'cirr' refers to high altitude and the prefix 'alto' refers to middle altitude. High-level clouds are those that form above around 6,000 m (20,000 ft.) and, because of the low temperatures at this altitude, are composed mainly of ice crystals. Mid-level clouds are those between around 2,000–6,000 m (6,500–20,000 ft.) and are generally composed of water droplets, although ice crystals will form if the temperature is low enough. Low-level clouds are those below around 2,000 m (6,500 ft.) and again are generally composed of water droplets, although ice crystals will form in these clouds if the temperature is low enough.

clout nail A nail with a large head. Clout nails are used for fastening sheet metal and similar materials.

cluster A term applied to a grouping of cases of disease or illness within a given time and space that is statistically greater than would be expected in the same population whilst allowing for chance fluctuations.

coal A material comprising mainly amorphous carbon together with various organic and inorganic constituents. It is formed mostly from fossilised plants, although some coal derives from fossilised algae. Coal is used primarily as a fuel, either directly or as a production source for gas (especially methane – CH_4) and coke. Coal continues to develop over time, generally increasing in carbon content. The primary classification of coal relates to its carbon content; the higher this is the higher the calorific value and consequently the more economically valuable the coal. Peat is a precursor to coal, generally having less than 55 per cent carbon, lignite (or brown coal) has around 70 per cent carbon, bituminous coal has around 85 per cent carbon and anthracite has up to around 95 per cent carbon.

coal gasification The process of heating coal to drive off volatile gases such as METHANE (CH_4), which can then be used as a fuel source.

See also: GASIFICATION

coastal erosion The abrasion and subsequent EROSION of the landscape by the action of the water and sediment impelled by the waves and the tides of the sea or ocean. The erosion can be gradual or catastrophic, the latter occurring as land slips or landslides, particularly during extreme weather conditions such as storms. Coastal protection schemes to prevent or reduce erosion are very expensive and require a clear understanding of the geology and tidal patterns over many miles of coastline and can require actions over extensive areas. Limited local schemes have proved to be of variable success. The REGOLITH is especially vulnerable to coastal erosion.

coaxial cable A type of cable used for communication purposes. A central conducting cable is insulated from an outer, surrounding cable, the whole being encased in an insulating material.

cob A collation of materials used for the construction of walling in which predominantly unburnt clay is combined for structural strength with straw and (possibly) stones, the whole being covered in a protective coating to protect against damage by rainfall.

coccus (plural cocci) A spherical-shaped BACTERIUM.

Cochrane Collaboration An international organisation that conducts and maintains a systematic review of published literature on the effectiveness of health care interventions. The database of systematic reviews is known as the Cochrane Database and each review weighs the methodology used in conducting the research to aid in assessing the relative validity of the findings.

cockle The term used in the leather industry to describe the small hard nodules formed on the skin of sheep that have been infected with KEDS. Cockle has also been reported as being caused by lice. If they are numerous the nodules can make the skin commercially worthless.

coddling The culinary term to describe a particular method of producing soft-boiled eggs in which the egg is placed into boiling water, removed from the heat and allowed to stand for around 8–10 minutes to achieve the required degree of cooking.

co-disposal landfill That practised at a LANDFILL site in which municipal solid waste is deposited along with other waste, such as certain categories of industrial waste, with the aim of effecting some form of treatment. It is likely that this practice will be banned in the European Community as a result of legislation.

coefficient of friction *see* FRICTION

coefficient of haze A measure of how absorbent of visible light are the particles suspended in a given atmosphere.

coenzyme A relatively small organic molecule that assists an ENZYME in acting as a CATALYST in the processes of METABOLISM. Unlike an enzyme the coenzyme is not a true (or in itself a complete) catalyst as it may be changed chemically during the reaction but can subsequently be regenerated.

cogeneration *see* COMBINED HEAT AND POWER

cognitive An adjective that refers or pertains to the voluntary mental processes such as recognition, memory, reasoning and judgement.

cohobate The act of the redistillation of a DISTILLATE, usually following its reintroduction to that which has been left over from the first process.

cohort A defined or identifiable group of people. The term originated from the ancient Roman descriptor for a division of troops equating to one-tenth of a legion.

See also: COHORT STUDY

cohort study An epidemiological study of a defined group of persons (the 'study group' or 'cohort') used to compare the frequency of an event in those members of the group 'exposed' to a certain phenomenon with those 'unexposed'. Cohort studies are frequently undertaken in relation to outbreaks of food poisoning in trying to identify potentially causal foodstuffs by describing the 'attack rate' in persons who ate a specified food (the 'exposed' group) with those who did not eat the food (the 'unexposed' group). A cohort study is also known as a prospective study, a follow-up study and a LONGITUDINAL STUDY.

See also: CASE CONTROL STUDY

coke The porous, solid residue that remains when all the volatile material, such as gas and tar, has been removed from bituminous coal or other carbonaceous material. It comprises mainly carbon in quantities of up to around 90 per cent but also contains a number of impurities dependent on source. It has a high calorific value and is used in metallurgical industries for making steel. The removal of the volatile components also makes it useful as a smokeless fuel.

cold chisel A chisel with an edge designed as being sufficiently hard to cut cold metal.

colic A vague term applied to symptoms of abdominal pain. Strictly speaking it refers to a disturbance in the colon or large intestine but, through common usage, it has become almost entirely reserved for digestive disturbances and is applied indiscriminately to conditions that vary widely in their causes. The terms 'gripe' and 'stomach ache' are popularly synonymous with the term 'colic'.

collagen The most abundant PROTEIN in animals and humans. It is found in many parts of the body as a structural component in, for example, skin or hide, tendons, blood vessels and bone. When such tissues are boiled the collagen produces GELATIN.

collar beam A horizontal beam joining two opposing rafters in a pitched roof at points around halfway along the length of the rafters.

colloid A substance comprising very fine particles generally taken as of a size under 100 nanometres in diameter (i.e. usually between 5–5,000 ÅNGSTRÖM units).

See also: COLLOIDAL

colloidal Refers to a solid that is very finely divided and suspended in an EMULSION. A colloidal suspension may appear to be a solution but the distinction can be tested, as suspended particles are unable to pass through a membrane whereas those in solution can.

See also: COLLOID

colonisation In relation to micro-organisms refers to the establishment of a group of the organisms within a specific environment without necessarily giving rise to clinical manifestation of their presence. The term is often used to describe the presence of COMMENSAL organisms, such as certain types of bacteria in the intestinal tract of the host.

colony-forming unit (CFU) An identifiable individual growth unit created by the multiplication of bacteria on a microbiological plate.

See also: AEROBIC COLONY COUNT

colophony *see* ROSIN

colorimeter A device for measuring the intensity of a colour by comparing it to a standard colour slide. The device is also known as a tintometer.

colostrum Also known as 'first milk', the first fluid secreted by the breasts of a mammal following PARTURITION. It is secreted for up to 3–4 days after giving birth and is particularly rich in protein, vitamins and ANTIBODIES to give the new offspring rapid nourishment and biological protection in its start in life.

colour test Used to test drainage systems using dye to colour water that is then passed through the system. At its simplest coloured leakage indicates defective pipework. Using the coloured water in measured amounts can also provide an indication of the flow rate (and thereby the fall) between two points in the system. Additionally pooling of water within the pipework can be detected if clear water flushed through the system results in gradual rather that immediate loss of colour, the reason being that pooling will retain coloured water in the system to be gradually diluted until clear.

See also: AIR TEST; SMOKE TEST; WATER TEST

coma A state of deep unconsciousness in which a sufferer is incapable of exhibiting a voluntary response to external stimulation. The condition is not necessarily irreversible although medical understanding and treatment of the condition is often not sufficiently adequate to enable the outcome in many cases to be determined with precision.

combined drainage system A wastewater disposal system in which both foul water and surface water are carried together to the point of treatment or disposal. Although the presence of surface water both dilutes the foul water and aids flow through the system it increases the volume of water for treatment and adds to the overall cost of disposal.

See also: SEPARATE DRAINAGE SYSTEM

combined heat and power (CHP) An efficient technology in which useable electricity and heat are generated together in the same process. A CHP plant is usually one in which the heat generated by a plant primarily used for electricity generation is recovered and used for a variety of purposes such as community heating, industrial processing or space heating. CHP is also known as 'total energy' or 'cogeneration'.

combustion The chemical reaction that causes a substance to bind with oxygen with the associated generation of heat.

COMEAP The acronym for the Committee on the Medical Effects of Air Pollutants.

comedo *see* BLACKHEAD

commensal The term applied to describe those micro-organisms that live either in the body (e.g. in the intestinal or respiratory tract) or on the skin of an individual or animal but which do no harm as a result of their presence. A commensal relationship is a type of SYMBIOTIC relationship in which one organism benefits and the other is not adversely affected.

commercial waste A classification of waste used in the UK to describe waste arising from trades or businesses but specifically excluding that which arises as a result of domestic or industrial activity.

Commission for Health Improvement (CHI) A UK body charged with the improvement in the quality of patient care through quality assurance, monitoring and review of current practice.

Commissioner A member of the European Commission with responsibility for a defined policy portfolio. The post is a European equivalent of the UK Cabinet Minister.

Committee on the Medical Aspects of Air Pollution A non-departmental public body set up in the UK in 1992 to advise the Government, through the Chief Medical Officer, on a wide range of issues relating to the effects of air pollution on health.

common area A shared space within, adjacent to or surrounding individually occupied space in which all parties share the common area and have equal rights over it or usage of it. Common areas include corridors, passageways, lobbies and communal gardens.

common carriage The term used in relation to the water industry for the practice whereby raw or treated water is introduced by a new entrant to the marketplace into an existing water system operated by a statutory water company. This has the aim of abstracting that water at some point further down the distribution chain for use by the new entrant's customers.

communicable disease *see* INFECTION

co-morbidity The presence of two or more health disorders in the same person at the same time.

competent authority An agency or organisation determined by the Government as having the necessary expertise and capability (and thereby authorised according to statute) to undertake statutory functions on behalf of the State. The concept of competent authority is one that is particularly enshrined within the European Union where such are usually determined as being answerable, in respect of the deemed area of competence, to the European Union on behalf of the individual Nation State.

competitive exclusion A phenomenon in which one species of bacterium is able to gain sufficient ascendancy in a given environment (possibly limited to a discrete area or individual foodstuff) sufficient to be able to out-compete other bacteria that might also be present. The success of the dominant bacteria can lead to a decline or even exclusion of other bacterial species.

compost That which is produced by the process of COMPOSTING. It is used as a soil improver or as a media for growing plants.

composting The controlled decomposition of BIODEGRADABLE material, primarily under AEROBIC conditions, by the action of micro-organisms, fungi and/or worms in a process that produces heat. This acts to sanitise the final product, which is known as compost. Vermicomposting is a type of composting that actively uses the action of earthworms to assist the process.

compression A force applied to a structure or object in the manner so as to squash or crush it.

See also: BENDING; SHEAR FORCE; TENSION; TORSION

compression ration Refers to the ratio between the volume of a cylinder in an internal COMBUSTION engine when the piston is at the bottom of the cylinder compared to the volume of the cylinder when the piston is at the top. The higher the compression ratio the more likely the engine is to KNOCK and the greater the need to use a fuel with a higher octane rating (see OCTANE NUMBER).

concomitant Refers to those things that exist or occur together and have a degree of association with each other.

concordat An agreement, pact or treaty between two parties.

concrete A construction material produced by mixing CEMENT, AGGREGATE and sand with water and allowing this to set into a hard stone-like substance. Various grades, strengths and properties of concrete can be created by altering the proportions of the constituents or by incorporating special additives.

concrete cancer A popular expression used to describe the deterioration of a (usually) reinforced concrete structure caused by the penetration of moisture and subsequent and consequential breakdown of the internal strengthening members by rusting. The process of rusting ultimately causes the reinforcing to expand and results in the typical appearance of fracture and shedding of the surface covering of the concrete. This exposes the internal members, which leads to the breakdown of the integrity and loss of the structural strength of the affected area.

condensate The liquid formed by the process of CONDENSATION.

condensation Liquid returned from gaseous, via vapour, to liquid form (also known as condensate) or the process whereby this is achieved. It is a natural phenomenon, although many commercial and industrial processes (such as the separation of petroleum and the DISTILLATION of spirits) employ the technique. Clouds, mist and fog are natural examples of this phenomenon whilst the vapour remains suspended in the air when the liquid in question is water.

Condensation can be a problem in buildings as the water can cause damage and lead to events (such as providing conditions suitable for DUST MITES) that can adversely affect health. The air naturally contains water that has evaporated. The amount of evaporated water the air can hold is finite. The air can however hold different amounts of water at different temperatures. As a general rule, the higher the air temperature, the more evaporated water can be held.

The amount of water the air holds at any given time compared to the maximum it could hold at that temperature is a measure of the HUMIDITY. When air cools, for example by contact with a cool surface, the humidity increases, possibly beyond the limit (or ABSOLUTE HUMIDITY) for that temperature – this is known as the dewpoint. At this stage the water vapour reforms (or condenses) and condensation results.

Most condensation in buildings occurs on the surface of either the inside structure or on items contained therein. It is also possible for condensation to occur within the fabric of the structure, for example within the internal structure of brickwork. This is known as 'interstitial condensation'.

condominium Sometimes shortened to 'condo', a US expression used to describe individual apartments in a block in which each apartment is wholly owned separately (i.e. not leased) but in which the common areas are jointly owned.

conduit Variously:

1 A protective casing through which cables or electric wires pass;
2 A pipe or channel used to convey fluids;
3 A facility through which something can be conveyed.

cones *see* RETINA

confidence intervals Also sometimes known as confidence limits, the upper and lower limits around a statistically determined value within which it is determined the true value might lie. It is usual for confidence limits to be determined at the 95 per cent level – in other words 95 times out of 100 the true value would be expected to lie within the stated limits. The narrower the range between the confidence intervals, the more sure the alignment of the statistics with the true position. Confidence intervals are associated with sample size – the more individuals that are sampled, the narrower the confidence intervals become.

confidence limits The same as CONFIDENCE INTERVALS.

confounding factor A variable in an epidemiological study that changes the outcome of a cause-and-effect relationship between what is being examined and the principle factor of concern. This distortion is important since the cause-and-effect relationship can either be enhanced or masked as a result of the effect of the 'confounder', thereby potentially profoundly affecting the outcome and viability of the study.

See also: BIAS

congener Any one of a number of chemical compounds with different molecular structures but that are grouped together because they exhibit similar biological properties or effects.

congenital abnormality A distinct variation from normal physical configuration exhibited by a foetus as it develops during pregnancy. Many congenital abnormalities can be detected in the womb but some may only be apparent at, or even after, birth. There are several reasons as to why congenital abnormalities occur; the most common are thought to arise as a result of genetic factors or as a result of exposure of the mother to toxic substances. The rôle of exposure to environmental chemicals in the development of congenital abnormalities is poorly understood in many instances but this is an area in which research is continuing.

See also: DEVELOPMENTAL EFFECTS

congestive heart failure An inability of the heart adequately to pump the blood through the circulatory system due to back-pressure caused by the blood in the lungs and the liver.

conifer A tree whose sexual reproduction is achieved through seeds produced in cones. The wood from all CONIFERS is described as SOFTWOOD.

coniferous Describes any tree that bears cones.

See also: CONIFER

conjoint analysis A method of assessing preference or value within a defined population by asking a sample of the population to rate different potential scenarios according to a variety of criteria.

conjunctiva The transparent membrane covering the SCLERA, also known as the 'white' of the eye.

See also: CORNEA

constitutional law In the UNITED KINGDOM comprises the whole totality of law as being that made by PARLIAMENT (statute law), together with that having derived from the higher courts (CASE LAW) and the unwritten rights of citizenry generally referred to as 'common' law. This latter generally comprises a set of constitutional conventions that, although not having statutory authority, nevertheless is considered as binding. There is no written constitution for the United Kingdom.

construction and demolition waste A significant component of the proportion of waste that currently goes to landfill. The waste comprises predominantly brick, concrete, hardcore and topsoil that many landfill site operators find valuable for stabilising sites, especially for further development. However, the waste can also contain large quantities of timber, metal and plastics, some of which may require special handling or disposal.

contact A person or animal that has potentially been in proximity to an infectious agent or CARRIER for sufficient time to be at risk themselves of developing the infection or be capable of carrying it for onward transmission.

contagious An adjective that refers or pertains to an infection that is capable of being easily transmitted from person to person by direct contact, via FOMITES or by the inhalation of infected aerosols.

contaminant A substance, the presence of which is not desired, which is contained within a product or environment. It may be present, for example, as a result of environmental contamination, manufacturing or process faults, or as a result of malicious or fraudulent activity.

contingent valuation An assessment of how prepared people are to pay for a non-marketed commodity (such as the National Health Service) based on the responses about preferences for a number of different scenarios. The evaluation of preferences may also include the use of DISCUSSION GROUPS. Contingent valuation is also sometimes known as 'Stated Preference'.

contrail A trail of ice crystals visible at high altitude and arising from the rapid (i.e. within a few seconds) freezing of water vapour discharged to atmosphere by the exhausts of the jet engines of aircraft. Contrails are also known more fully as condensation trails.

control In the epidemiological sense, a person who as closely as is reasonably possible resembles a CASE but who did not have the exposure to the causal factor(s) under consideration.

See also: ATTRIBUTABLE RISK

controlled waste Any waste for which an operator requires a waste management licence for transfer, treatment or disposal. Certain highly specialised waste such as that which is radioactive or explosive is dealt with under special legislation. In the UK most waste is considered to be controlled waste, although notable exceptions include waste from mines, quarries and farms.

cook chill A system of food production in which the food is cooked, cooled rapidly and then maintained in chilled storage until required, whereupon it is reheated to a pre-specified temperature.

See also: COOK FREEZE

cook freeze A system of food production in which the food is cooked, cooled rapidly and then maintained in frozen storage until required, whereupon it is reheated to a pre-specified temperature. The system requires high standards of hygiene and precise control over the process since the products have a relatively long shelf life and any failure in the process could leave viable (particularly spore-forming) bacteria in the food; these would not necessarily be deactivated on subsequent reheating.

See also: COOK CHILL

concave The term that describes a curved surface of a solid object (such as a mirror or lens) that bends inwards. A curved surface that bends outwards is deemed to be CONVEX.

consortium An association of individuals or companies formed to further or pursue a specific purpose.

constant Something that is considered to be unchangeable or immutable. In physics the speed of light is considered to be a constant.

contact adhesive An adhesive that, when applied to two surfaces, will form a bond between the surfaces at, or closely following upon, their being brought together.

contaminant A substance that is contained within or incorporated into another substance in which the presence of the former is considered to render the latter impure or unclean. The contaminant need not necessarily possess the characteristic of actually causing harm.

See also: POLLUTANT

contiguous The term used to describe two things that are physically adjacent to each other, which share a common boundary or which are touching. For example land that is contiguous to a river is that which bounds or runs alongside the river.

continental seating An arrangement of seats in an auditorium in which access to the rows is from the ends only, and there are no aisles.

contingency planning The process of considering, preparing and devising alternative potential courses of action to those considered as optimal so that an alternative can be implemented should circumstances or conditions change sufficiently to prevent the preferred option being progressed.

contingency sum A provisional amount of money allowed for in a contract to cover an activity or item, the cost of which cannot accurately be determined until the contract is underway. Contingency sums would be incorporated, for example, in relation to works to areas that can only be exposed and examined in detail once other work has started.

continuing professional development (CPD) A system developed in varying ways by many professional organisations to ensure that their members maintain professional competence throughout their careers by undertaking continuing training and improvement. Many organisations allocate points to various training or professional activities and specify what level of training is deemed appropriate (i.e. how many points should be accrued) per specified time period. For many organisations the maintenance of the specified minimum continuing professional development requirement is a condition of membership of the professional body.

convection The process whereby material in a fluid (such as air or water) moves due to its differential density relative to other parts (generally) of the same fluid. The difference in density is caused by a difference in temperature between one part of the fluid and another. Warmer parts are less dense than cooler ones and tend to rise, being pushed upwards by cooler dense areas. The movement of portions of the fluid results in the movement of heat energy, hence convection is often described as the process of the movement of heat within a fluid by particle movement.

Convention on Biological Diversity A major agreement signed by over 150 countries in 1992 at the UNITED NATIONS CONFERENCE ON ENVIRONMENT AND DEVELOPMENT. The signatory countries agreed each to identify and monitor the genetic resources of their countries and to prepare national plans to protect BIODIVERSITY.

convex The term that describes a curved surface of a solid object (such as a mirror or lens) that bends outwards. A curved surface that bends inwards is deemed to be CONCAVE.

COPD The acronym for CHRONIC OBSTRUCTIVE PULMONARY DISEASE.

coping The topmost course of brick, stone, concrete or similar on the top of a wall positioned as a decoration or to protect the wall beneath from the effects of the weather. The coping may finish flush with, or proud from, the wall's vertical surface.

coping saw A BOW SAW with a narrow blade capable of cutting tight curves and intricate designs.

corbel That specially constructed portion in a course of brickwork or stonework in the construction of a wall that sticks outwards from the vertical surface. In a 'corbelled arch' successive courses project ever further outwards until one side of the arch meets the other being constructed from the opposing wall. A succession of corbels is known as 'corbelling' and can be used to support the weight of a superstructure such as overhanging windows.

COREPER The acronym for Comité des Représentants Permanents (translated as Committee of Permanent Representatives), a standing committee of the EU with a role in decision-making.

Coriolis force The force hypothesised to account for the deflection observed from the earth of a body moving along its otherwise linear course. Bodies moving across the northern hemisphere deviate to the right and those in the southern hemisphere deviate to the left. It is thought that the deviation (also known as the 'Coriolis effect') is due to the rotational movement of the earth. The Coriolis effect is suggested as being the primary force responsible for inducing secondary rotational direction in the ocean currents and the major air streams of the earth.

cork The light porous material that makes up the bark of the cork-oak tree. It has good heat insulation properties and is used in compression to seal bottles, especially of wine (see CORKED). A 'cork' is therefore the term used to describe a seal for a bottle made from cork, although the term is also nowadays used for sealing 'bungs' made of more modern materials such as plastics.

corked The expression used to describe wine that has deteriorated such as to be unfit for consumption due to its having been exposed to air whilst in the bottle. The expression derives from the exposure usually being due to the cork shrinking *in situ* thereby allowing air to enter the bottle. 'Corked' is sometimes used incorrectly to describe poured wine with portions of the cork floating in it – such having broken away during opening. Such debris does not render the wine unfit (i.e. truly 'corked') and can safely be removed prior to drinking.

cork wood *see* BALSA WOOD

corm A swollen and rounded portion of the underground stem of certain plants (such as crocuses) from which new growth is generated. Although a corm has a superficially similar appearance to a BULB it is actually a completely different structure.

See also: TUBER

cornea The convex, colourless and transparent anterior surface of the eye.

See also: SCLERA

cornice In classical architecture, that portion of an ENTABLATURE which projects outwards. The term is also applied to a moulding at the apex of an external wall or at the junction of an internal wall and the ceiling.

coronary heart disease (CHD) An acute condition caused by the complete or partial blockage of an ARTERY of the heart. It is one of the major causes of death in Western societies. There appear to be hereditary factors that may predispose a person to the disease, but smoking, sedentary lifestyle and high blood CHOLESTEROL levels also appear to increase the risk. The disease is also known as ischaemic heart disease or coronary artery disease and onset is often referred to as a 'heart attack' or simply as a 'coronary'.

See also: ANGINA; INFARCTION

corrosion The process whereby a material, especially a metal, is degraded by chemical activity, such as the oxidation of the iron component in steel in the presence of water by an electrolytic reaction.

corrugated Means having a cross-section of an otherwise flat material formed in alternative peaks and troughs so as to present such a profile as a series of waves. Corrugated cardboard or paper usually has either one or both sides of the corrugations covered with an additional flat layer, either to add strength or to enable it to act as a protective packaging material. Corrugated steel (also known as corrugated 'iron' or 'tin') is used as a construction material to clad external walls or roofs, the corrugations adding strength and aiding in the displacement of rainwater from the surface.

corundum Another name for aluminium oxide (Al_2O_3). It is very hard and is used as an abrasive. It is not the same material as CARBORUNDUM.

cosmic rays High-energy IONISING RADIATION originating in outer space.

cost–benefit analysis A method of evaluating the worth prospectively of a potential course of action, or retrospectively of a given activity, by considering what advantages or improvements accrue to the action for a given expenditure of money, time, effort and/or resources. Cost–benefit analysis is undertaken to inform the decision-making process as to whether, or to what degree, a proposed course of action should be pursued. In practice it can be very difficult to identify all costs and benefits, and many analyses confine themselves to comparing financial expenditure with tangible or demonstrable benefits.

See also: COST EFFECTIVENESS ANALYSIS

cost effectiveness analysis A method of evaluating the outcome of activity by considering how effective an action has been in achieving its objectives. It differs principally from COST–BENEFIT ANALYSIS by expressing benefits in terms of effectiveness only and by introducing an implied element of efficiency into the considerations. Cost effectiveness analysis does not question whether the stated objective is worth pursuing; it can only indicate how effective a particular course of action might be in pursuing an objective.

coulomb A unit measure of electric charge. One coulomb is that which is produced by one AMPERE flowing for one second.

Council of the European Union The main decision-making body of the EUROPEAN UNION, it is usually referred to simply as the 'Council' or the 'European Council'. It is convened regularly and comprises political representation of Member States at Ministerial level. The Council shares legislation-making power with the EUROPEAN PARLIAMENT, co-ordinates the economic policies of the Union, agrees international agreements, shares budgetary authority with the Parliament and co-ordinates activities in relation to police and judicial matters at the European level in respect of criminal matters.

counterclockwise A synonym for ANTICLOCKWISE, i.e. in the other rotational direction to that of the direction of the hands of a clock.

countersink The activity of making or creating a depression in the surface of an object in order to cause the head of a screw, rivet or similar to be recessed level with the surface when it is fully driven home.

counterweight A weight applied to the other side of a simple machine such as a lever or pulley in order to balance the weight (usually the load) on the other side. Counterweights are used, for example, in the frames of SASH WINDOWS to achieve a balance with the weight of the window sash as it acts on sash cords connected to a pulley arrangement. This balance facilitates the safe movement up and down of the sash and ensures that the window remains in whichever open position it is put.

County Court In the UNITED KINGDOM, the lowest of the COURTS hearing civil cases. These courts are also known as Courts of First Instance.

Court of Appeal (Criminal Division) *see* HIGH COURT

Court of First Instance An alternative name in the UNITED KINGDOM judicial system for the COUNTY COURT.

courts (UK) The principal bodies for conducting trials and for arbitration in the UNITED KINGDOM. The UK comprises four countries with three jurisdictions being England and Wales, Scotland and Northern Ireland. Each has its own court system.

The lowest criminal courts in England and Wales are the MAGISTRATES' COURTS – these deal primarily with minor offences and, in some instances, refer more serious matter to the higher courts for decision. The Crown Court hears these more serious matters in front of a judge and a jury. County Courts (also known as Courts of First Instance) hear civil cases. More serious civil matters are referred to the High Court. This latter also acts as a court of appeal from the County Court. The High Court is divided into three Divisions, being Queen's Bench Division, Family Division and Chancery Division. The most senior court in the United Kingdom is the House of Lords and acts as the supreme court of appeal. In addition to the court system there are also specialist tribunals that generally are appointed to hear appeals on behalf of various public bodies and Government departments.

In Scotland the lowest courts are the Sheriff Courts and the District Courts. The supreme criminal court is the High Court of Justiciary, the supreme civil court is the Court of Session and appeal is to the House of Lords. The principal law officer is the Lord Advocate.

See also: CONSTITUTIONAL LAW

covalent bond A join in the make-up of a molecule involving sub-atomic particles. A covalent bond consists of two electrons that are shared between two adjacent atoms.

cowl A protective covering or hood located over an opening, such as of a ventilation shaft or chimney, in order to aid draught and to protect the opening against the ingress of rainwater.

CPD An acronym for CONTINUING PROFESSIONAL DEVELOPMENT.

cracking The process in the refining of PETROLEUM whereby larger liquid hydrocarbon molecules are transformed into smaller molecules. During the process some more unstable molecules combine to create molecules of tar or carbon (the latter is known as 'coke'). There are two main types of cracking. Thermal cracking requires the use of heat and pressure to complete the process, while catalytic cracking, as the name indicates, uses a CATALYST.

crampon A device equipped with spikes that is fitted to the soles of boots or shoes to facilitate grip when climbing rocks or walking over ice or compacted snow.

crank A double right-angle bend in a rod or shaft to produce an angular 'S' shape such as to provide leverage to facilitate the rotational movement of the main member. The crankshaft in an internal COMBUSTION engine is the main shaft of the engine that, by the operation of the 'cranks', converts the reciprocal motion of the pistons in the cylinder into the rotational movement necessary for propulsion.

crazing The development of a pattern of small cracks on and within the surface covering of an article or structure. Crazing is caused by expansion or contraction of the surface over time or in response to temperature changes, the cracks occurring because the brittle nature of the surface covering will not allow elastic movement.

creosote The name applied to one of two different substances. Wood tar creosote is obtained from the distillation of wood (especially that of the beech tree) and is used as a constituent in some expectorant and disinfectant preparations.

Coal tar creosote is a mixture of over 160 different chemical compounds in varying proportions, and is used primarily as a wood preservative, although, in the past, it has been used for spraying fruit trees and as an active ingredient in sheep dips. Coal tar creosote is potentially harmful to humans since it contains many substances that are irritants.

crepuscular Refers to that which is of, or appears in relation to, twilight. It is applied, for example, to those animals that become active at dusk.

Creutzfeldt-Jacob Disease *see* **CJD**

critical control points Those stages in the production of goods or services at which specific HAZARDS or RISKS can be eliminated or reduced to acceptable levels. The control of these points is essential in maintaining QUALITY ASSURANCE, especially in relation to systems such as HAZARD ANALYSIS CRITICAL CONTROL POINT (HACCP).

critical mass The minimum quantity needed to produce an effect. The term is used in relation to nuclear physics to describe the minimum amount of fissionable material that is necessary to sustain a nuclear chain reaction at a constant rate.

critical path (or pathway) That series of actions within an overall activity that are sequential to each other and which are crucial to the ultimate success of the operation.

critical path analysis The detailed consideration of a series of individual activities that are together necessary to the successful completion of an activity. It is also the identification of those actions that must be conducted sequentially and cannot be rescheduled without extending the shortest possible completion time. The critical path in this analysis is that sequence of activities that will dictate the shortest possible time within which the overall activity can be completed. Those activities that are crucial for the completion of the activity, but which could be scheduled within a broader time 'window', are not considered to be 'critical' in terms of defining the critical path.

crocidolite One of a group of fibrous silicates known generically as ASBESTOS. It has short, needle-like fibres and is classified as a member of the amphibole asbestos group because it exhibits a fibrous or columnar shape when magnified. Asbestos from the amphibole group is usually incorporated into materials such as fire cements and in some insulation materials.

Crocidolite is also known as 'blue' asbestos, the name deriving from its colour when seen under the microscope. All varieties of asbestos are potentially hazardous to human health, but crocidolite is considered to be especially dangerous.

cross-pollination The transfer of pollen between the flowers of two different flowering plants.

cross-sectional study An epidemiological examination of either the whole or a sample of a defined population or COHORT at a single point in time. Its aim is of establishing the status of that group in respect of the matter under consideration.

See also: COHORT STUDY

crosscut saw One in which the teeth are set at an angle to facilitate cutting across the grain of a piece of timber.

crossover trial A study in which individuals both receive an intervention (usually a form of therapeutic or preventive treatment) and subsequently do not receive it over a period of time (or vice versa). The study is conducted by random assignment of individuals to a 'get' or 'don't get' group, administering the intervention to one group and then, at some predetermined time, switching the groups over so 'get' becomes 'don't get' and VICE VERSA. Comparison is then made between the groups to ascertain what difference the intervention makes to the health status of the individuals. Crossover trials are useful because the individuals act as their own CONTROLS and reduce BIAS due to varying susceptibility, limiting the effect of seasonal or temporal differentiation.

Crown Court The next highest court from the MAGISTRATES' COURT in the UNITED KINGDOM system of legislation in respect of criminal legislation. The Crown Court sits as a judge and jury. It hears the more serious criminal cases as referred to them from the Magistrates' Court or as a court of appeal from that level on matters of fact.

See also: COURTS

Crown dependencies The name applied to the islands known as the Channel Islands and the Isle of Man, off the coast of the UK. These territories are not part of the UK itself, but are self-governing dependencies of the Crown. Each has its own legislative assembly and administration. Although they do not have representation within the UK parliamentary system the UK Government maintains a responsibility for matters pertaining to their international relations and for defence.

cruck One of a pair of large curved timbers that together form an arch used to construct the cross-sectional frame for an early type of barn or similar building. In the USA they are known as 'crutches'.

crude oil A naturally occurring bituminous liquid containing a complex mixture of organic chemicals. It is a fossil fuel and was formed between 100–200 million years ago. It is derived from the decomposition of organic material, principally microscopic marine plants and animals. It is found below the surface of the earth in many areas of the world and is extracted by drilling to reach the deposits. Crude oil is usually found in layers of porous rocks such as limestone or sandstone, capped or sealed from the surface by an impervious layer of shale or clay to create a reservoir.

See also: PETROLEUM

cryogenic That which relates to very low temperature and the resultant effects.

cryoprotective broth A medium designed to protect the viability of microorganisms when they are stored under conditions of refrigeration or freezing.

cryosphere The term used to describe the totality of the earth's frozen environment including ice, snow and permafrost.

cryostat A device for maintaining very low temperatures.

cryptosporidium A MONOXENOUS protozoan parasite. Infection can lead to the manifestation of the disease known as cryptosporidiosis, the clinical symptoms of which include watery diarrhoea, bloating and cramping abdominal pain with vomiting and fever reported as present in around half of the cases. There is a high prevalence of infection in young children, especially in day nurseries. Cryptosporidium is spread principally via contaminated water and SEWAGE SLUDGE, and infected animal faeces are significant vehicles of transmission. The parasite can cause serious disease in immunocompromised patients, and AIDS charities have expressed concerns that if oocysts can be detected from time to time in treated water supplies there could be implications for AIDS sufferers or other immunocompromised groups. The infective dose can be as low as one oocyst. In some areas cryptosporidium is the fourth most common cause of gastroenteritis after campylobacter, salmonella and viruses.

An outbreak in the UK, in spring 1989, affected thousands in the Oxford and Swindon areas, causing the Government to set up an expert committee of enquiry chaired by Sir John Badenoch. The Committee reported in 1990, and this was followed by a second report in 1995. A subsequent outbreak in North London led to a further report, under the Chairmanship of Professor Ian Bouchier, this being published in November 1998. The spreading of animal manure or the disposal of sewage sludge on CATCHMENT AREAS presents a clear risk of contamination. The largest known outbreak of cryptosporidiosis in the world occurred in spring 1993 in Milwaukee, on the western side of Lake Michigan, USA. The outbreak is thought to have occurred as a result of local farmers spreading contaminated animal manure in the water catchment area, resulting in oocysts being carried into the source water. Researchers estimate that around 400,000 people were affected from this single event.

crystal A solid mineral in which the atoms are arranged in a regular and ordered structure with a repeating pattern.

crystalline Means having the configuration of a CRYSTAL.

crustacean A member of the classification of animals that includes crabs, lobsters, prawns and shrimps.

CS gas The acronym for a chemical the full name of which is 2-chlorobenzylidene malonitrile. CS gas is also known as 'tear gas'. It is a contact irritant of nerves in the skin, eyes and mucous membranes, and is used in riot or person control sprays.

cull The act of reducing the numbers of creatures or plants in a defined group by selective slaughter or picking.

cultivar A cultivated plant that has been produced through a selective breeding programme and therefore did not arise in nature by the processes of evolution or NATURAL SELECTION. Cultivars are produced to emphasise a particular characteristic or trait in the PHENOTYPE but the process can result in plants that are less well adapted to survive or compete in other ways with wild plants. Consequently many cultivars can only thrive (or survive) through the continuing husbandry of humans.

culture The laboratory-based growth and/or enhancement of micro-organisms from a sample in order that they can better be examined, analysed or described.

culture medium Any substance or combination of substances designed to accelerate or promote the growth of micro-organisms to facilitate their examination, analysis or characterisation.

culvert Classically a drain constructed under a road or railway. The term is also used to describe larger underground drains and sometimes to describe drainage ditches that run alongside roads or railways.

cumulative error A component within a study or operation that is essentially false or mistaken and is attributable to a number of smaller errors that together combine to create a larger error. Many cumulative errors arise as an effect of SYSTEMATIC ERRORS that exist in a wider study or activity.

cupola In architecture, a small dome on a roof.

curtain wall A non-load-bearing wall incorporated into a structure of a building for the purpose of separating one area from another. In modern buildings many curtain walls are built as STUD PARTITIONS.

curvilinear Relates to the condition of comprising or being contained by curved lines.

cusp The point at which two curves meet.

cutaneous The term used to describe that which is of, or relates to, the skin.

cut-in wind speed The air velocity at which the rotors of a wind turbine generator will begin to rotate.

cutwater The term used either as an alternative for the prow of a ship or to describe the wedge-shaped projection of a pier or jetty, designed to deflect the force of waves impacting on the structure.

cyanobacteria *see* BLUE-GREEN ALGAE

cyanosis The clinical symptom of a bluish tinge around the lips and extremities. It is associated with the reduced oxygen-carrying capacity of the blood.

See also: METHAEMOGLOBINAEMIA

cybernetics The scientific study of the processes of automatic communication and control in either animals or machines, such as the operation of the nervous system or the operation of computers, respectively.

cyclone A meteorological condition in which winds spiral inwards towards an area of low pressure. The term is also applied to a wind of strength 12 on the BEAUFORT SCALE, i.e. as a synonym for hurricane.

See also: ANTICYCLONE

cytokine Any of a large group of peptide or protein MEDIATORS.

cytokinesis The change that is undergone by the CYTOPLASM during cell division.

cytoplasm The soft albumen-like material that comprises the largest proportion of cells.

See also: EUCARYOTE

cytotoic Means destructive to cells. The term is often applied to therapeutic drugs used for the treatment of CANCER and for the suppression of the immune system to reduce the chances of organ rejection during transplant procedures.

See also: CHEMOTHERAPY

D

d-value The time (usually in minutes) at a specified temperature that it takes to reduce the number of micro-organisms or their SPORES to one-tenth their original number.

See also: Z-VALUE

dado The lower portion of a wall when it is covered differently from the rest of the wall, such as by panelling. The dado is often finished at its upper edge with a decorative moulding, which is known as the dado rail. A dado is also the plinth of a column or the base of a pedestal.

dais A low platform, especially one that is located at one end of a hall or auditorium.

DALY The acronym for Disability-Adjusted Life Year. It is a means adopted by the World Bank of trying to provide an estimate for the burden of disease. It is derived from a combination of disability and premature mortality.

See also: QALY

damper A plate housed within a duct or flue that can be moved either manually or automatically to alter the flow of air through the passageway.

damping A reduction of the peaks or overall level of energy in the pathway of a sound, vibration or similar by the introduction of an energy-absorbent material between transmitter and receiver.

damp-proof course (DPC) Any layer of impermeable material incorporated into the structure of a wall or similar for the purpose of excluding water. The term is applied most frequently to the thin layer of impermeable material positioned near the foot of a wall for the purpose of preventing RISING DAMP.

See also: DAMP-PROOF MEMBRANE; VAPOUR BARRIER

damp-proof membrane A sheet of impermeable material incorporated into a structure for the purpose of excluding water.

Darwin, Charles *see* NATURAL SELECTION

data (singular datum) Recorded values in a collection or series. Often incorrectly rendered as singular, data provide the raw information for analysis. The location where data are stored is known as a database or databank.

daub and wattle *see* WATTLE AND DAUB

daughter An adjective applied to describe the resultant offspring or product from an event or process, e.g. 'daughter nuclei' are those resulting from radioactive decay and 'daughter cells' are those deriving from cell division.

See also: MUTAGENIC

dB The symbol for DECIBEL.

de facto A Latin term which, when used in English, refers to that which is actual (i.e. that which exists). The term does not imply that something that is *de facto* is necessarily right or proper, for example the *de facto* occupier of a property is one who is actually living their, irrespective of any right so to do.

de jure Latin, meaning literally 'from the law'. It is used to describe something that is done by right or is performed in accordance with the law.

de minimis Latin, meaning 'concerning trifles'. It is used in hearings, tribunals or other legal proceedings to describe something that is so trivial as to have no significance or importance to that which is under consideration.

de novo Latin, meaning literally 'from (the) new'. It is used to describe something that is done over again or which is redone but as if for the first time.

de rigueur A French expression taken in use in English to describe that which is considered to be demanded or required according to etiquette or fashion.

dead light A window, no portion of which is capable of being opened.

dead load One whose whole weight is borne by that which is supporting it, i.e. the load is not partially supported against or by another structure or similar. The term is sometimes rendered as a dead weight.

deadlock A mechanism for securing a door or window in which the TENON of the lock can only be withdrawn from the MORTISE by means of a key. The tenon cannot be pushed back into the lock as can the type that is spring-loaded.

dead shore *see* SHORE

dead weight *see* DEAD LOAD

death rate *see* MORTALITY

decay sequence The successive disintegration of the atom of a radioactive element leading to the creation of an element with a smaller structure. Such elements may themselves be radioactive, thereby producing a sequence of disintegration events.

See also: RADIOACTIVITY

decibel (dB) A measure of the loudness or intensity of a sound. The term actually stands for one tenth of a bel, the larger and less frequently used unit. The decibel is a tenfold logarithm to the base 10 of the ratio between two amounts of power and described by the formula $10\log_{10}(p_1/p_2)$. Because decibels are measured on a logarithmic scale it is not possible directly to add decibels together. Each step of 10 dB in the scale reflects a doubling of the sound intensity. The decibel is also a ratio of the difference in power of two electric signals.

An acoustic level of 0 dB is the normal threshold of hearing for a healthy youth (note that it is possible to have a sound pressure level below 0 dB). Quiet conversation would measure around 40 dB, normal conversation at 1 foot away would measure around 60 dB, heavy vehicle traffic at around 85 dB and the threshold of pain at around 135 dB. A jet engine at close proximity could generate around 155 dB.

deciduous The adjective used to describe a tree that loses its leaves in winter.

See also: EVERGREEN

declination Variously:

1 The angle between the true north and the magnetic north;
2 The angle between the elevation of a star or other celestial body and the celestial equator;
3 A downwards bend.

deep clean The application of an intensive CLEANING regime to (usually) food processing or manufacturing facilities. The process usually involves the dismantling of all equipment and the application of rigorous physical or chemical means to remove dirt and debris. The process may incorporate elements of DISINFECTION or STERILISATION.

See also: CLEAN IN PLACE

deep well A well in which the water supply is obtained from water-bearing strata at least one level below an impermeable layer. In theory the deep well will deliver water of higher POTABLE quality as it is more likely to exclude microflora deriving from surface contamination. To maintain its integrity the well must be constructed so as to exclude water entering from layers above the impermeable stratum.

See also: ARTESIAN WELL; SHALLOW WELL

defibrilation The medical process of applying a large electric shock to the thoracic cavity of a patient with the aim of returning an irregularly beating heart to its normal rhythm.

definitive host In relation to a PARASITE, the host within which the adult parasite lives.

See also: INTERMEDIATE HOST

deforestation The large-scale clearance of an area of forest in order that the land can be transferred to other use. Globally at the turn of the millennium it was estimated that around 17 million hectares (equivalent to around 170,000 sq. km or 65,000 sq. miles) were being cleared annually, predominantly for farming and resource exploitation.

See also: FOREST DEGRADATION

deformation The alteration of the physical shape of an article or substance from that of the original.

degradation The breaking down of an article or the molecules of a substance into smaller constituent units or parts.

dehumidification The process whereby the level of HUMIDITY in an atmosphere is reduced.

dehydration The process whereby moisture or water that was incorporated into or was part of a substance or article becomes lost.

deletion In reference to cell microbiology refers to the loss of a proportion of a chromosome.

deliquescent An adjective applied to a substance that is capable of absorbing moisture from the air sufficient to become dissolved in that which it absorbs – the process is known as deliquescence.

See also: HUMECTANT

dementia A loss or reduction of mental function. There are many different types, not all of which are irreversible. Dementia is usually characterised by memory loss (especially short-term memory loss) and impairment of mental faculties such as speech and co-ordination.

demersal The term applied to describe those fish or other sea creatures that live on or near the seabed. Within the fish family they can be further classified as round fish (e.g. cod, haddock and whiting) or flat fish (e.g. plaice, sole and turbot).

See also: PELAGIC

demulsifier A substance that when added to a mixture or emulsion of oil and water promotes their rapid separation.

dendrochronology The science of dating historic climate changes through the examination of the annular growth rings of trees.

denominator That portion of a vulgar fraction which appears below the line. Used extensively in STATISTICS, the denominator is essentially the totality of that which is being assessed.

See also: NUMERATOR

dental amalgam *see* AMALGAM

denudation The lowering of the level of the surface of the land due to EROSION.

deoxyribonucleic acid (DNA) A complex organic polymer classified as a nucleic acid. It has a double helix structure. It is self-replicating and stores the genetic information necessary for cell replication. DNA is located in the CHROMOSOMES of the cell and passes on genetic information to RIBONUCLEIC ACID (RNA) in a process known as transcription. The RNA is then directly involved in the synthesis of PROTEINS. All living things, except RNA viruses, require and contain DNA for cell replication, development and function.

deposition A statement made under oath, or a document certified to record such a statement.

See also: AFFIDAVIT

depreciation The reduction in value ascribed to a physical asset due to wear, tear and the passage of time.

deprivation That status of an individual or group which fails to achieve parity with another due to circumstances beyond their control. Deprivation occurs because of differences in a range of determinants such as social status, ethnicity, age, sex, sexual orientation and mental capacity. Deprivation is a key determinant of HEALTH INEQUALITY.

depth of field In photography, the distance between the closest object in focus and the furthest away. It is used in a similar way in other applications to describe the useful range of data, equipment or systems.

depuration A process whereby potentially pathogenic bacteria in the digestive tract of filter-feeding molluscan shellfish are removed or reduced in number. This is done by flushing them from the system using cleaner water taken in during the natural metabolic processes. Depuration can be undertaken by transporting and relaying the shellfish in cleaner waters or by keeping them in specially designed tanks in which (usually) sterilised clean water is resterilised and recycled for a predetermined period.

derelict The term used to describe a condition of decay or ruination caused by the abandonment of that which has fallen into a state of dereliction.

dermis *see* SKIN

derrick A crane in which the lifting arm (the boom or jib) is steadied from the main supporting post or tower by an arrangement of guy ropes.

desalination The process of removing SALT from seawater or other SALINE water usually with the intent of providing drinking or POTABLE water. Desalination is usually achieved through DISTILLATION or REVERSE OSMOSIS, although both ELECTRODIALYSIS and freezing have also been used. Freezing reduces the capacity of the liquid to carry dissolved salts and these precipitate out as temperatures are lowered.

See also: SALINE; SALINITY

desert Commonly any large area of land that receives little or no rainfall. More precisely the term is usually applied to an area of land that receives less than 100 mm of rainfall per year.

See also: ARID; BIOME

desiccant A substance that has the property of being able to remove moisture from, or to dry, something else. The term is often used to describe a substance that has the power to adsorb moisture from the atmosphere.

desiccate To dry up or to preserve by drying.

design load The load that a structure or entity has been designed to cope with during an expected working life. Although the stated design load may have been determined, allowing a tolerance between that point and its point of destruction, the design load should never be exceeded. This latter will not only be likely to invalidate any warranty but it will also place an unacceptable risk on those using the structure or entity and is likely to contravene health and safety legislation.

destruction testing The examination of the performance of a structure or entity up to and including the point at which it is destroyed. The aim of the examination is to establish the point at which the structure or entity ceases to operate satisfactorily in order to determine a safe working tolerance in practice. The point of physical destruction itself may not be the point at which the structure or entity ceases to function satisfactorily. For example, a structural component that loses elasticity beyond a certain point may do so long before the point at which it collapses.

detergent A substance, other than SOAP, which is used as a cleaning agent. They vary in chemical composition but contain SURFACTANTS to lower the surface tension of liquids and thereby improve their ability to dislodge dirt adhering to surfaces. The term does not imply any capacity to destroy micro-organisms such as bacteria, and a detergent is therefore distinct from either a DISINFECTANT or a STERULENT.

determinant Something that affects an outcome. It is used extensively in PUBLIC HEALTH assessment where 'determinants of health' are those things that impact on health leading to a particular health state.

See also: EPIDEMIOLOGY

detritus Classically, the collected naturally abraded portion of a rock face or solid body. The term is now often used to describe any collection of waste material.

detritus tank A container in which solid material is allowed to separate out from suspension in a liquid. The term is synonymous with the term 'settlement tank'.

deuterium A non-radioactive ISOTOPE of hydrogen, possessing one proton and one neutron in its nucleus. It comprises around 0.015 per cent of natural waters and is used in some nuclear reactors as a MODERATOR. Deuterium, like hydrogen, forms water on combination with oxygen, the chemical formula being rendered as D_2O, the compound being known as 'heavy water'.

See also: TRITIUM

developmental effects Those deviations from the normal physical configuration of a living organism manifest as it grows. The causes can be due either to factors arising after birth or to exposure during pregnancy.

See also: CONGENITAL ABNORMALITY

deviation The variance from the MEAN of a set of observed values of one observed value in the series. A standard deviation is the root of the squares of the differences from the mean of a number of observations – it is a measure of the spread of observations from the mean.

devolution In the UNITED KINGDOM, the process that created the Scottish Parliament and the National Assembly for Wales after referendums in each country following the re-election of the Labour Government in 1997. Northern Ireland by this time already had its own Assembly, although constitutional and other disputes have variously disrupted its operation as intended. Both Scotland and Northern Ireland have their own legislature. The National Assembly for Wales is only empowered to pass Statutory Instruments (i.e. it has no powers in relation to PRIMARY LEGISLATION) and these powers do not cover the areas of foreign affairs, defence, taxation, overall economic policy, social security and broadcasting.

The Scottish legal system is separate from that of England and Wales. The Scottish Parliament can legislate for domestic policy, but foreign affairs, defence and national security, economic and monetary policy, employment and social security remain the purview of the Westminster Parliament.

dewpoint (or dew point) *see* CONDENSATION

dextrose Another name for GLUCOSE.

diabetes The name commonly given to a condition for a disorder characterised by a raised blood sugar (GLUCOSE) level. There are, however, different types and causes of diabetes.

See also: DIABETES INSIPIDUS; DIABETES MELLITUS

diabetes insipidus A disease caused by HORMONE deficiency or by damage to the urinary system, which results in the passage of large amounts of urine and a commensurate excessive thirst. It is a distinct and separate condition from DIABETES MELLITUS.

diabetes mellitus A disorder of the body that results in raised levels of blood sugar (GLUCOSE) and caused by a deficiency in the production or effect of INSULIN. The disease is classified as 'insulin-dependent diabetes' (or Type 1 diabetes) in which the patient is usually affected from childhood and 'non-insulin-dependent diabetes' (or Type 2 diabetes), which usually has an onset after the age of 40 years. Diabetes can lead to complications including cataract, blindness and kidney damage, and increases the risk of diseases of the heart and circulatory system.

diagonal A straight line running across a rectangle to join the two opposing angles. In building, a diagonal is usually a structural member acting as a brace or strut and running at an angle to join both vertical and horizontal main members or surfaces.

dialysis A process whereby a SEMI-PERMEABLE MEMBRANE is used to separate substances in solution. The process works because the membrane allows smaller molecules to pass through it, but prevents the passage of larger ones. Dialysis is the natural process whereby the kidneys remove waste products from the blood stream.

See also: ELECTRODIALYSIS; HAEMODIALYSIS; OSMOSIS; PERITONEAL DIALYSIS

diamond The crystalline form of carbon. It is a very hard and durable material, and is used in both jewellery and for the work facing of industrial tools such as diamond drills and diamond saws in which the cutting edge is faced with industrial diamonds.

diapause A resting phase describing the period of dormancy of (particularly) some insects usually (but not exclusively) triggered by the onset of cold weather.

diaphragm A thin membrane or partition separating one portion or side of a body or structure from another. In anatomy it is the muscular dividing structure separating the thoracic and abdominal chambers. In architecture a diaphragm is generally a rectangular plate or similar used as a component for strengthening or stiffening a structure. In some types of telephone or microphone the diaphragm is the vibrating disc-like membrane used to convert electrical waves to sound waves or vice versa. A diaphragm in contraception is a thin plastic or rubber cap that fits over the entrance to the cervix.

diaphragm pump A type of pump in which a reciprocating or moving DIAPHRAGM is used to impel the fluid that is being moved.

diarrhetic shellfish poisoning (DSP) An acute diarrhoeal illness caused by the consumption of shellfish that have accumulated toxin in their hepatopancreas. The toxin responsible is known as Okadaic Acid. The illness was first recorded in the UK in 1974. It usually lasts around three days. There is no known antidote.

See also: AMNESIAC SHELLFISH POISONING; PARALYTIC SHELLFISH POISONING

diastole The relaxation of a hollow organ. It is applied particularly in relation to the rhythmic period of relaxation and dilation of a chamber of the heart during which it fills with blood.

See also: DIASTOLIC BLOOD PRESSURE

diastolic blood pressure The minimum pressure in the circulatory system of the blood as measured against the arterial wall when the ventricle of the heart is in DIASTOLE.

See also: SYSTOLE

diatom A form of microscopic (largely) unicellular algae. Diatoms have cell walls composed of silica consisting of two interlocking symmetrical valves. They contain CHLOROPHYLL and are therefore capable of PHOTOSYNTHESIS. They can contribute to the totality of PHYTOPLANKTON in a body of water.

See also: DIATOMACEOUS EARTH

diatomaceous earth A mineral deposit derived from the silica (silicon dioxide, SiO_2) content of the cell walls of DIATOMS. It is used in the manufacture of fertilisers, detergents and some industrial chemicals; it also has insulating and fireproofing properties.

dichotomous category One in which data or items are arranged into either one of two mutually exclusive categories, e.g. either male or female.

diclinous Any flowering plant that carries only either male or female reproductive organs on its flowers.

See also: MONOCLINOUS

die A tool used to give form to a material upon which it is applied. Dies are used to form extruded wire, cut metal surfaces (e.g. making screws) or to impress a design or form onto the surface of another material.

dielectric An adjective used to describe something that does not conduct electricity. Such substances are sometimes used to provide electrical insulation.

dielectric strength Also known as the breakdown voltage, the minimum voltage needed to produce an electric arc across a liquid DIELECTRIC insulator. There are standardised laboratory methods for determining this point in samples and such insulating material is usually supplied with its strength specified. Contamination of the product can significantly reduce its insulating capacity.

diesel A fuel used to power internal COMBUSTION engines. It is derived from the distillation of PETROLEUM. It has a higher distillation range (at between 200–370°C, i.e. between 392–698°F) than does PETROL, a characteristic that is used to separate it out during the refining process. Diesel fuel is ignited in use by the heat generated from compression, rather than by an electric spark such as in the petrol engine. In the USA lighter grades of diesel are known as kerosene. Diesel fuel, much more so than petrol, is a significant source of airborne particulate matter.

dietary reference value (DRV) A term used to describe any one of a range of indicators or standards produced or used to enable an assessment of dietary intake to be made. There are many different ways of expressing dietary reference values; each might potentially assess different nutrients, levels or target groups. The absence of international standardisation in this general field means that each value should be considered in respect of how it has been derived.

dietetic food Food intended to satisfy a specified need or nutritional requirement of a particular (defined) section of the population.

differential diagnosis A way of trying to explain the known symptoms of a disease through the examination or consideration of what alternative conditions or diseases could have caused the symptoms.

diffraction Alteration in the otherwise normal path of an electromagnetic wave, such as a wave of light or acoustic wave, caused by the contact or influence of an obstacle. A diffracted light wave can either be deviated from its course (such as by passing at an angle between air and water) or split into its constituent wavelengths (such as by a prism) – this latter effect is known as interference.

diffuser A device located around or near a light source to produce DIFFUSION.

diffusion The physical dispersion, without chemical reaction, of one fluid material within another by the action of the movement of atoms or molecules of the two substances concerned. In terms of light, diffusion is the random dispersal of light waves caused or effected by a medium through which they have just passed. Diffusion is used as a technique for altering the light from a light source in order to reduce glare or to produce a more even light distribution. The device used to produce diffusion is known as a DIFFUSER.

DIN The acronym for Deutsche Industrie Normung, a German Standard.

dinoflagellate Any of a numerous order of unicellular protozoans (sometimes considered as ALGAE) possessing two FLAGELLA. They are mainly (but not exclusively) found as constituent members of the totality of marine PLANKTON.

diode A device with two terminals fitted into an electric system, its purpose being that it allows electric current to flow in one direction only.

dioxin The generic term applied to members of a group of chemicals known as chlorinated hydrocarbons (i.e. HYDROCARBON molecules with the addition of chlorine). They are formed as contaminants during certain industrial processes and as a result of the incineration of material (especially plastics) containing chlorine. Dioxins exhibit varying degrees of toxicity and carcinogenicity to humans, and the most toxic of the group is known as TCDD (2,3,7,8-tetrachlorodibenzo-*p*-dioxin). They are persistent chemicals in the environment and BIOACCUMULATE in fatty tissue.

See also: TOXIC EQUIVALENCE

diploid number The number of chromosomes in a normal cell. Sex cells carry half the normal chromosome complement (the HAPLOID NUMBER) of a normal cell.

direct current (DC or dc) An electric current that flows only in one direction. As a consequence it has a generally even flow and is not transmitted in pulses.

See also: ALTERNATING CURRENT

direct resistance heating *see* OHMIC HEATING

Directive *see* EC DIRECTIVE

disability A permanent or temporary reduction in the capacity or capability of an individual to perform or undertake the normal functions of life.

Disability-Adjusted Life Year *see* DALY

disaccharide A relatively simple SUGAR containing twelve carbon atoms. SUCROSE ($C_{12}H_{22}O_{11}$) is an example of a disaccharide.

See also: GLUCOSE; MONOSACCHARIDE

discount rate An annual percentage rate at which an ascribed value of something in the present day is expected to devalue over time due to wear, tear and obsolescence.

discussion groups Assemblages of individuals selected as being representative of a wider group or population. The groups are used to assess the likely attitudes or responses of the wider population through directed discussion and evaluation of various scenarios or options. They are used to assess attitudes across a number of disciplines but are used especially in relation to consumer and political preference.

disease Any abnormality arising in relation to a bodily function, organ or structure that is not simply the primary manifestation of physical injury.

See also: PREVENTION OF DISEASE

disinfectant A chemical agent that has the power to act to produce DISINFECTION.

See also: DETERGENT; STERULENT

disinfection The process whereby (usually chemical or heat) means are used to kill or remove micro-organisms to a level such as to ensure safety from infection in respect of the use of the article(s) being disinfected. Disinfection is not an absolute process in that it cannot guarantee that all organisms or (especially) their spores are killed or inactivated.

See also: CHLORINATION, CLEANING; STERILISATION

Distalgesic A proprietary brand name for a tablet painkiller containing paracetamol and propoxyphene.

See also: ANALGESIC

distemper A cheap paint comprising pigment suspended in a glue or SIZE. It is usually capable of dilution with water, although some distemper incorporates oil as an additive to provide some degree of being washable. In veterinary science distemper is any of a range of infectious diseases of animals, especially the disease known as canine distemper – a highly contagious viral disease of dogs, characterised by a discharge from eyes and nose, and accompanied by fever.

distillate That which is produced by the process of DISTILLATION.

distillation The process of evaporation of a liquid and its subsequent condensation. Distillation occurs both naturally and in industrial processing. The hydrological cycle is an example of natural distillation, whereas distilling spirits, the differential distillation of, for example, crude oil (a process using different evaporation temperatures to separate different constituents of the original liquid) and DESALINATION are examples of industrial application.

See also: ETHANOL

diuretic A substance that increases the production of URINE by the kidneys. Many diuretics such as alcohol and CAFFEINE are ingested as part of the diet, but they may be administered therapeutically for the treatment of diseases such as OEDEMA. In susceptible individuals diuretics may contribute to adverse health conditions such as GOUT.

diurnal A term that is applied to animals and plants meaning that they are active during the daylight hours. The term is sometimes also used to describe activities performed during daylight hours or to those that occur within the period of one day.

See also: NOCTURNAL

DNA *see* DEOXYRIBONUCLEIC ACID

DNA hybridisation The process whereby a DNA PROBE is matched and attached to a target sequence of DNA in order to detect or identify the organism. Each type of DNA probe will only lock on to its target DNA sequence so recording attachment equates to positive identification of the sequence.

DNA probe A fragment of DNA that has been labelled with a marker to aid its detection and used in molecular biology to lock on to a specified sequence of DNA. It can be used for the differential identification of certain types of bacteria within a single species.

See also: DNA HYBRIDISATION

dog iron A short steel bar, the ends of which are bent at right angles and sharpened to a point, like a large angular STAPLE. The pointed ends are driven into two pieces of adjacent timber in order to fasten the two together.

dog tapeworm *see* ECHINOCOCCUS GRANULOSUS

dolly A (usually shaped) wooden anvil used to cushion an impact or blow (such as from a pile driver) or one such as to act as a former to shape sheet material by hammering. In cinematography a dolly is a moveable platform upon which a cine camera is mounted.

dolomite The common name given to the naturally occurring sedimentary rock predominantly comprising mineral calcium and magnesium carbonate (chemical symbol Ca(or Mg)CO_3). Petroleum is sometimes found in a reservoir comprising dolomite deposits.

domestic waste A classification of waste used in the UK to describe that which arises solely as a result of normal domestic activity. It is distinct from both commercial and industrial waste, and some authorities consider it to be distinct from garden waste. Domestic waste is sometimes known as household waste.

doorpost The vertical side member of a doorframe.

dopamine A precursor of noradrenaline, which is used as a NEUROTRANSMITTER, especially in the brain.

Doppler effect A real or an apparent change in the frequency of a wave (e.g. of sound or light) manifest when the wave source and an observer move relative to one another. Convergence of source and observer produces an increase in frequency, and divergence creates a decrease in frequency. For example, for a sound, during convergence the observer would perceive a sound of a higher frequency and during divergence the observer would perceive a sound of lower frequency. If the source moves the effect is real, and if the observer moves the effect is apparent. The Doppler effect is also known as the Doppler shift. The name derives from the Austrian physicist Christian Johann Doppler who discovered the phenomenon in 1842.

dormer window A vertical window incorporated or installed into a sloping roof.

dose A quantitative measure of the amount of a designated substance or exposure to which an individual or group has been subjected.

dose effect curve A graphical representation of the outcome as measured by the differential effect on the individual of varying amounts of a foreign substance.

See also: DOSE RESPONSE CURVE

dose response curve A graphical representation of the percentage of a given population that responds to varying amounts of a specified foreign substance.

See also: DOSE EFFECT CURVE

dose response effect A phenomenon in which the exposure to, or application of, increasing amounts of a substance leads to a correspondingly increasing strength or severity of outcome. The principle is used in the evaluation of THRESHOLDS for determining safety, recognising that for some substances there is no 'safe' minimum level.

double blind trial An assessment study in which a treatment under test is administered to one group but not to the CONTROL group. The study is known as 'double blind' because neither the patient nor the person administering the treatment knows who is receiving it. This is achieved by providing a treatment and a PLACEBO that are indistinguishable from each other. The researchers allocate which group should receive treatment and which the placebo, and this information is then kept confidential to prevent psychological factors affecting the outcome. It is important that those administering the treatment/placebo are also kept in ignorance to prevent them inadvertently (by language, body language or similar) indicating to the patient which group is which.

double-glazing A form of constructing the glazed portion of windows and doors in which two pieces of glass, separated by a void or sealed space, are used to improve thermal and/or acoustic insulation. As a general rule the inter-glazing space used in thermal double-glazing is less than that used in acoustic double-glazing. This is because the thermal resistance to transmittance is greatest between surfaces so the size of the gap is not paramount for thermal considerations, whereas a larger gap improves sound insulation.

Secondary double-glazing is the addition of another glazed panel over an existing single glazed unit. Since such secondary panels do not create sealed installations, although improving thermal insulation, they are not as efficient as purpose-built units and they do not generally eliminate problems of condensation. Secondary double-glazing, if correctly fitted, can however be very efficient in improving acoustic insulation. In especially cold situations, or where greater thermal efficiency is needed, it is possible to install triple-glazed units.

See also: U-VALUE

dovetail joint A method of joining the ends of two pieces of wood at 90° to each other. The joint is formed by a number of interlocking fan shapes, cut-outs and projections that fit closely together when the two pieces are assembled. Dovetail joints were once commonly used in joinery and cabinet making although nowadays the availability of alternative fasteners and sealants (together with increased cost associated with such joints) has reduced their use.

dowel A headless pin of wood or metal inserted into a pre-formed hole in each of two pieces of a construction in order to strengthen the whole, especially against the action of a SHEAR FORCE.

dower A widow's share of her husband's estate.

dower house A smaller house built as part of a DOWER, usually being within the grounds of, or in close proximity to, the original house occupied during marriage.

down time Any period during which equipment is unavailable for use due to breakdown or maintenance, other than that which has been planned for outside normal operating periods.

downer cattle Those that have fallen and are unable to regain their feet because of the effects of disease or injury.

downstream assessment An evaluation of the events that follow on from the occurrence of a given circumstance; essentially it is an analysis of effects or outcome.

See also: UPSTREAM ASSESSMENT

DPC *see* DAMP-PROOF COURSE

drainage system Any construction, mechanism or pathway whereby fluid can be conveyed from one location to another. A sewerage system is a drainage system designed and installed specifically for the purposes of conveying SEWAGE.

draw Culinary term to describe the removal of the intestines or internal organs of an animal by pulling them out through an incision made in the skin, usually made around the pelvic area.

dredge The widening or deepening of a body of water by the physical removal of mud and silt from the bottom. In fishing it is a method of catching fish and shellfish by dragging a net along the bottom of the sea to catch the animals living there. Dredging is also the culinary term to describe the action of sprinkling with a light and even coating, such as of flour or sugar, prior to cooking or afterwards as a flavouring (such as adding sugar to pancakes).

dressed stone Rough quarried stone that has been shaped either manually or by cutting so that it is squared and can be incorporated into a building or similar as a constructional unit without further working.

dressed timber SAWN TIMBER that has been planed or sanded to a smooth finish *INTER ALIA* for decorative or aesthetic purposes. It is important to recognise that the dimensions quoted for dressed timber at the point of sale are usually those of the sawn timber *before* finishing is undertaken, the actual size consequently being slightly smaller than that quoted.

drill In agriculture, a linear depression pushed or scratched into the ground for the purpose of accepting seeds. Usually the surrounding soil is brushed over the seeds to protect them after sowing.

drilled pile An alternative name for a bored PILE, i.e. one that is located by being screwed through the ground until its final position has been achieved.

driven pile A variety of pre-formed PILE that is impelled into the ground by force, such as by a pile driver.

drop-off recycling Refers to the type of recycling that uses designated collection sites where individuals can deposit items to be recycled. Such sites include CIVIC AMENITY SITES and BOTTLE BANKS.

dry ice Solidified (i.e. frozen) CARBON DIOXIDE. It is used for CRYOGENIC storage and to generate the cloud-like mist at floor level as a stage effect in entertainment.

dry rot A condition of decay in wood or timber primarily brought about by fungal attack arising as a result of dampness. The term dry rot is used in building science to distinguish the condition from 'WET ROT', another decay process, brought about by different fungal attack in circumstances in which the wood is even more damp. Dry rot can be caused by a number of different fungal species although they all have similar requirements for growth and the techniques of prevention or treatment are the same. In the UK most instances of dry rot are caused by the fungus *Serpula lacrymans*.

dry (stone) wall Any wall in which the units of construction are positioned and retained solely on the basis of their locking together to form a cohesive whole; no MORTAR is used to create the finished product. In the USA such constructions are termed as 'drywall'.

dual flushing system A type of FLUSHING SYSTEM sometimes incorporated into a WC (water closet – also known as a lavatory) in which the user can opt to discharge fully or partially the water from the cistern. The device can potentially make a considerable saving in overall water usage.

duct A hollow passageway such as a tube, conduit or pipework through which fluids, liquids, gases or cables, etc. can be conveyed.

ductile Refers to a solid, especially a metal, and indicates a capacity to be worked or bent without breaking. It particularly refers to the ability of a metal to be drawn or pulled out into wire.

See also: MALLEABLE

dumb waiter A small mechanical lift used for the vertical transportation of goods. They are especially used in hotels and restaurants for the transportation of food between floors.

duodecimal Refers to anything that is based upon or reckoned by units of 12.

duplex The US term for a MAISONETTE. In the UK the term also refers to a property in which there are two self-contained separate living units in a building of two floors, either:

1. each being of two storeys on either side of a party wall or;
2. each occupying a separate floor of a two-storey construction.

dura mater The outermost and strongest of the MENINGES, the membranes covering the brain and the spinal cord.

dust mite A small mite living predominantly in houses and obtaining sustenance from a range of organic detritus such as the scales shed from human skin. The most common dust mite in the UK is *Dermatophagoides pteronyssinus*. Although they are directly harmless to most humans, allergic reactions such as asthma, eczema and rhinitis can be triggered in people who are sensitive, particularly to their faeces. It is unlikely that the complete eradication of mites from homes in which they have become established can be achieved, but taking appropriate remedial action can reduce their numbers. Up to 2 months may have to elapse before the effects of any remedial action are felt. Dampness in houses favours their multiplication.

duty of care A legal obligation imposed on an operator of a business in specific circumstances to ensure the operation of the business does not result in damage to the environment, wildlife or humans.

dynamic pressure The internal pressure within pipework when that which it is used to convey is flowing. The term also refers to the pressure generated when the flow of a fluid through the pipework is reversed.

dynamic range A description of the breadth of field that a scientific or measurement instrument can accommodate ranging between maximum and minimum values.

dynamics The study of the interrelationships between forces and the motions that they produce.

dynamo A device for generating DIRECT CURRENT.

See also: ALTERNATOR

dyne The centimetre/gram/second unit of force. One dyne is that necessary to produce an acceleration of one centimetre in a mass of one gram. One dyne is equivalent to 1×10^{-5} newton.

dyscrasia A medical term used to describe generally any abnormal state or condition in the body, especially one including disease or disorder.

dyspnoea A medical term used to describe difficulty in breathing.

dystrophic In medical terminology refers to that in a state of dystrophy – i.e. any of a number of different disorders characterised by the degeneration of tissues such as the muscles. In ecology the term refers to freshwaters such as lakes and ponds to describe acidic waters with low nutrient content and, consequently, small plant or animal populations.

E

E. coli *see* ESCHERICHIA COLI

ear muff A hearing protector that fits over the external part of the ear (the PINNA) and fits closely against the sides of the head.

ear plug A hearing protector that is inserted into the ear to fit closely inside the ear channel (the meatus).

earth radius (RE) The earth radius is not precise since the earth is not a perfect sphere. It is generally taken to be approximately 6,371 km (3,960 miles).

Earthwatch Institute An international non-profit organisation, originally founded in 1972 in Boston, USA. It describes itself as promoting the sustainable conservation of natural resources and cultural heritage by creating partnerships between scientists, educators and the general public. By the turn of the millennium it had field projects in over fifty countries around the world.

See also: FRIENDS OF THE EARTH (INTERNATIONAL); GREENPEACE

easement An area either of land or within a building over which a person other than the owner or occupier has a legal right of passage for a specified purpose, such as for access or maintenance.

eaves (singular eave) That portion of the roof that overhangs the wall.

Ebola and Marburg viruses The only known members of the filovirus family, the name describing their filamentous appearance. Marburg viruses were first identified during an outbreak in 1967 when laboratory workers were exposed to infected tissue from African green monkeys imported to Germany from Uganda. Ebola viruses were first isolated from humans during outbreaks in northern Zaire and southern Sudan in 1976. There have been several resurgences of these diseases in Africa. Two sub-types of Ebola virus, Ebola-Sudan and Ebola-Zaire, are recognised.

Symptoms include fever, headache, chills, myalgia and malaise. These later develop to include severe abdominal pain, vomiting and diarrhoea. Massive haemorrhagic manifestations usually occur in fatal cases. In reported outbreaks 50–90 per cent of cases are fatal. The natural reservoir for the viruses is unknown. Transmission to secondary cases occurs through close personal contact with infected blood or other bodily fluids or tissues. Aerosol spread has not been documented in humans but has been demonstrated amongst non-human primates.

EC Formerly the acronym for the European Community; it now more properly refers to the European Council.

EC Directive A legal instruction from the European Council to a Member State. It is legally binding on the Member State and requires (or 'directs') them to enact the provision(s) of the Directive by an article of their own national legislation and to do so within a specified time period.

Echinococcus granulosus The dog tapeworm, a parasitic CESTODE that lives in the intestines of dogs. The adult tapeworm is between 3–9 mm in length with a body consisting of three to four segments known as proglotids. The terminal segment is GRAVID and is voided in the faeces. In the natural life cycle the eggs pass onto pastureland where they are eaten by grazing animals. The eggs hatch and eventually form cysts in body tissues such as the liver, lung or brain – at this stage the parasitised animal is known as the INTERMEDIATE HOST. If the cysts are subsequently eaten, such as by scavenging dogs, they mature and the tapeworm emerges into the intestine of the DEFINITIVE HOST to start the cycle over again. If humans inadvertently eat the cysts they can lead to a condition known as HYDATID DISEASE.

eco-labelling A system of consumer advice that provides information about the goods in relation to their comparatively less harmful effects on the environment than other equivalent products.

economics The study of how and why a population chooses and uses finite resources to produce various commodities and to distribute them for consumption amongst the various groups in that society both now and in the future. Economics can include consideration of the use of money in this context, but this is neither exclusive nor inevitable.

ecosystem The totality of life within a given interrelated and self-sustaining system existing in a defined geographical area.

ecotourism An imprecise term used predominantly to describe the use of the natural environment as a place of recreation for activities such as wildlife safaris and bird watching that promote awareness of the environment without adversely affecting it. An often quoted example of the philosophy is in relation to African wildlife parks in the axiom that one live elephant will bring in more money to the local economy than the money derived from the ivory of the dead elephant. Although this might be a simplistic view, and it is arguable whether tourism can truly be said not to have an effect on the local environment, it is nevertheless much more environmentally friendly and in tune with SUSTAINABLE DEVELOPMENT than rampant exploitation.

ecotoxic The term used to describe a substance (or substances) that can cause a toxic reaction to an ECOSYSTEM or to the living organisms within it.

ecowarrior An imprecise appellation used to describe those environmental activists who involve themselves in direct (possibly illegal) action to prevent or disrupt developments that they see as causing damage to the environment.

ectoparasite A PARASITE that lives on the external body of the host. Examples are fleas, lice, leeches and ticks.

See also: ENDOPARASITE

ectopic Means 'out of place'. An ectopic pregnancy or ectopic gestation is one that relates to the development of the foetus outside the womb. This most frequently happens when the fertilised egg develops within one of the fallopian tubes (the tubes that carry the ovum from the ovaries to the womb).

eczema Classically an allergenic skin reaction characterised by the eruption of blisters on the surface of the skin accompanied by intense itching. The term is now commonly applied more widely to include a range of skin conditions in which rash, redness or itching is present.

effective dose The minimum amount of a substance or numbers of pathogenic organisms that are capable of producing an effect. In terms of radiation the effective dose is that calculated to have been received by a person allowing for differential absorption and effects rates of different tissues.

See also: EQUIVALENT DOSE

effectiveness An expression of the relationship between an input and the production of that which the input was intended to deliver. It is usually assessed as a relative concept and described as more or less effective than an alternative.

efficacy Means having the power to produce a result.

See also: EFFICIENCY

efficiency In mechanical or engineering terms, an expression of the ratio between energy or power input and output, usually expressed as a percentage. In systems management it is an expression of input to output, but usually assessed as a relative concept and described as more or less efficient than an alternative. Efficiency is essentially about either maximising the benefit from the expenditure of a resource or minimising the cost of any defined level of benefit.

efflorescence In building terminology a deposit, usually of crystalline or powdery appearance and usually white in colour, which forms on the surface of some plaster, brickwork, concrete or similar constructions. It is due to the action of moisture leaching salts from the substance of the construction, transporting these in solution to the surface and depositing them on evaporation of the moisture. Efflorescence does not damage the structural integrity of the construction, but it can reduce the adherence of paint, wallpaper and similar finishes, and can look unsightly. Although efflorescence can be brushed away from the surface where it is deposited, its recurrence can be indicative of a continuing problem of damp penetration or interstitial CONDENSATION. In health terms efflorescence is a reddening of the skin. In botany it is the process of the flowering of plants.

El Niño A term originally applied to a periodic warming of the surface temperature of the Pacific Ocean located off the coast of Peru. The term is now more commonly used to describe the more general periodic warming of the surface temperature of the ocean in the Pacific Basin, a phenomenon that occurs, on average, once every 3–7 years and is usually accompanied by a reduction in the strength of the trade winds. It is claimed by some scientists that GLOBAL WARMING has increased the frequency of 'super El Niños' in which global weather patterns are disrupted to catastrophic effect.

The term La Niña is applied to the converse effect, whereby the ocean surface temperature drops and the strength of the trade winds increase.

elastic limit That point beyond which an elastic substance that has been stretched is no longer able to resume its original configuration.

elasticity A property exhibited by some substances to deform their shape under a load and to regain their original configuration when the load is removed.

elastin A fibrous PROTEIN found in elastic body tissues such as the arteries, lungs and some ligaments.

elastomer A material of either natural or synthetic origin with rubber-like properties. It is used in the manufacture of products that require elasticity in use, such as vehicle tyres, hoses, elastic couplings and footwear.

electric cell A device used to create electricity from chemical activity. A simple cell comprises two electrodes embedded or immersed in an electrolyte. Chemical reaction causes current to flow between the electrodes when they are connected so as to form an electric circuit. A number of electric cells connected together is known as a BATTERY.

electric current A continuous flow of electrons or ions through a conducting material.

electrical resistance A measure in an electric circuit of the ratio between the voltage and the ELECTRIC CURRENT. The unit of measurement is the OHM.

electrodialysis The process of DIALYSIS, in which an electric potential is applied across a membrane to make it permeable to either positive or negative IONS. The process has been used in the DESALINATION of seawater.

electrolysis The process of the chemical decomposition of an ELECTROLYTE by the passage of an electric current. Because an electrolyte is a solution of ions, the passage of an electric current causes the positively charged anions to migrate to the cathode and the negatively charged cations to migrate to the anode.

electrolyte A solution or melt (usually of a naturally solid) of an ionic compound that undergoes chemical decomposition when electricity is passed through it.

See also: ELECTROLYSIS

electromagnet A device in which magnetism is induced by passing an electric current through wire coiled around an iron core.

electromagnetic field (EMF) An energy field created around any material through which electricity passes. The strength of the field is determined by the amount of electricity passing any given point at a given moment in time. There have been numerous health concerns associated with EMFs, notably in relation to mobile telephones and radio transmission masts. Opinion is divided in relation to whether or not the claimed effects are real, but most sources agree that minimising exposure is a sound precautionary measure.

electromagnetic radiation Any radioactive energy transmitted as electric and magnetic waves, and that has the capability of being transmitted through a vacuum. Examples of electromagnetic radiation include LIGHT, X-RAYS, gamma rays and RADIOFREQUENCY RADIATION.

electromagnetic spectrum The full range of all wavelengths of electromagnetic radiation including ULTRAVIOLET RADIATION, visible LIGHT, INFRARED RADIATION, heat and RADIOFREQUENCY RADIATION.

electromotive force That which creates the motion of electrons in an electric current. It is a function of the difference in potential between two points in an electric circuit. The unit of measurement is the VOLT.

electron An elementary particle with the mass 1/1836 that of a PROTON and carrying a negative electric charge. When emitted from the NUCLEUS of a RADIONUCLIDE the electron is known as a BETA PARTICLE.

electron microscope Uses electrons rather than light to examine a specimen. Magnification is achieved by focusing the electrons with magnets rather than lenses. Electrons have a much shorter wavelength than light. This enables much higher magnification, of up to around 300,000 times the original sample size, to be achieved. The image is displayed via a fluorescent screen or by photograph.

electrophoresis The movement of suspended particles within a fluid that is produced in response to the application of an electromotive force.

electroscope A simple device used to indicate the presence of an electric charge. The instrument consists of two leaves of metallic foil hanging next to each other in a transparent container, such as a glass jar. The presence of an electric charge causes the foil leaves to spread apart. The instrument can be used as a rough indicator of the prevailing level of ionising radiation by observing the rate at which it loses its electric charge in dry air.

electrostatic precipitator A device that induces an electric charge on suspended particles in a gas in order to attract and precipitate them onto charged plates for the purposes of cleaning the gas within which they were suspended. Electrostatic precipitators are used principally for removing particulate material from exhaust gases from larger industrial processes such as coal-fired power stations or from the flues of furnaces. Electrostatic precipitators are very efficient at removing even very fine particulate material but they can be expensive to run.

element A substance whose ATOMS all have the same ATOMIC NUMBER.

elevation A type of drawing used in architecture in which the details of the sides of a building or similar construction are reproduced with mathematical accuracy. This is done without attempting to produce a three-dimensional effect, such as that which would be achieved through incorporating shading or perspective into the drawing.

See also: PLAN

ELISA The acronym for Enzyme Linked Immunosorbent Assay, an analytical test that uses ANTIBODIES to detect proteins in solution. It was developed for detecting mammalian protein in animal feed as part of the BOVINE SPONGIFORM ENCEPHALOPATHY (BSE) control programme. The test is sensitive enough to detect animal protein at a level of around 1 part to 400. The test will also detect SOYA proteins and wheat GLUTEN.

eluviation The process of the removal of fine particles from the upper layers of soil by the action of water. The process of their subsequent deposition somewhere else is known as illuviation.

embolism A blockage of any of the small blood vessels of the body by material that has been transported by the bloodstream through the larger vessels until it lodges in a vessel that is too small to allow its passage. The material that causes the blockage is most frequently a fragment of a blood clot from elsewhere in the body, although portions of tumours, rafts of bacteria, protozoan parasites and air are also able to cause embolism. The effect of the blockage is incapacity or destruction of the area of tissue deprived of blood, the degree or extent depending on the extent and severity of the blockage.

See also: STROKE; THROMBOSIS

emerging pathogen A micro-organism, PATHOGENIC in nature, which is newly recognised as presenting a threat to health. The term is usually applied to new strains of existing organisms that are thought to have recently evolved or developed the capacity to cause disease.

See also: ESCHERICHIA COLI

emery A very hard mineral and a form of CORUNDUM. As a powder it is used for polishing and/or grinding. It is used to coat paper or cloth (emery paper/cloth), which can then be used for polishing or for coating individual wood or card strips (emery board) that are used for manicure.

emetic Means that which induces vomiting either as in the case of infection, chemically induced or as part of a medical treatment process. The application of emetic substances to induce vomiting in clinical treatment is now not often practised but where it is carried out it is most frequently associated with trying to rid the stomach of its contents following certain cases of poisoning. There are significant dangers associated with this type of approach and medical supervision should be considered as mandatory.

emphysema A chronic disease of the lungs in which the lung tissue becomes destroyed and the affected area is unable to participate in the transfer of oxygen to the body. In advanced form sufferers have great difficulty breathing and may require regular and/or frequent use of a respirator to deliver pure oxygen to the remaining functioning lung tissue in order to sustain life. Emphysema is closely associated with cigarette smoking.

emulsification The process whereby two immiscible liquids are mixed together to create a homogenous whole known as an 'emulsion'. In practice the emulsion is usually created by dispersing one liquid in the form of fine droplets within the other liquid.

emulsifier A substance that promotes the creation and maintenance of an EMULSION by altering the SURFACE TENSION of the constituent droplets to prevent them from aggregating and precipitating.

See also: COLLOIDAL

emulsion A COLLOIDAL suspension of a liquid in another liquid.

emulsion paint A suspension of synthetic resin and pigment in water. It is used primarily for decorative purposes indoors. Following application the evaporation of the water leaves the pigment embedded in a synthetic resin film.

en broche (or en brochette) Refers to food that has been roasted or grilled on a skewer or on a spit.

en coquille Refers to food that is served in its shell or served in a covering made to resemble a shell.

en tasse Refers to a foodstuff (usually soup) that is served in a cup.

encephalitis An inflammation of the brain; most commonly this is caused by infection with a virus.

See also: MENINGITIS

endangered species One whose worldwide population is assessed as being at, or marginally above, the minimum viable population level. If the population level falls below this level the population is assessed as being incapable of recovery in the natural state – a condition that is likely to lead to EXTINCTION.

See also: RED DATA BOOKS

endemic Means 'ever present' and is applied to an entity within a defined locality, geographical area or parameter (such as people, foodstuffs or given environment). The term is usually used to describe the persistent presence of animal or plant species, or diseases in a given locality or population.

endocannibalism The cannibalism of family or relatives.

endocrine disruptor Any external chemical that is capable of disrupting the normal activity of the body's HORMONE messengers. The term is often applied to pollutants that have disruptive effects on reproductive tissue. Those that produce feminising effects in males are termed oestrogenic (also known as 'artificial oestrogens'), while those that produce masculinising effects in females are called androgenic.

See also: PHYTOESTROGEN

endocrine gland A structure in the body of all vertebrates and some INVERTEBRATES that produces HORMONES which are then secreted directly into the bloodstream.

endogenous Refers to that which is produced or arises from within something.

See also: EXOGENOUS

endometrium The mucous membrane lining of the uterus.

endoparasite A PARASITE that lives within the body of the host. Examples are tapeworms, lungworms and the malaria parasite *Plasmodium* spp.

See also: ECTOPARASITE

endorphin Any one of a group of substances known as PEPTIDES with opium-like properties produced by the brain and the pituitary gland primarily as painkillers. They are also known as opiate peptides. It has been suggested that ACUPUNCTURE might stimulate the production of endorphins and that this might be a plausible mechanism to explain its pain-relieving action.

endoskeleton A skeleton (the support structure for the body of animals) that is formed within the body and is surrounded by tissues and skin. Endoskeletons are found in the higher animals and man.

See also: EXOSKELETON

endosperm The nutritive tissue surrounding the embryonic sac in the seed of a flowering plant. It provides the first food for the seedling as it begins to grow.

endothermic reaction A chemical reaction in which heat is absorbed.

See also: EXOTHERMIC REACTION

endotoxin A toxin produced by certain bacteria and fungi, but which is not released until the cell that produced it ruptures and dies.

See also: EXOTOXIN

energy from waste (EfW) An all-embracing term to describe technologies used to recover useable energy from waste materials. These technologies usually involve the direct burning of the waste (such as incineration) or the generation of fuels that can subsequently be burnt.

See also: GASIFICATION; PYROLYSIS

English bond A type of BOND of a brick wall in which each alternate course is made up of either HEADERS or STRETCHERS.

See also: FLEMISH BOND

enriched uranium Uranium in which the proportion by weight of the ISOTOPE uranium-235 has been increased beyond that in its naturally occurring state (i.e. above 0.7 per cent).

See also: NUCLEAR REACTOR

enrichment The practice of adding additional nutrient to a substance to promote growth. The term is commonly encountered in microbiology where growth media are frequently enriched to enhance bacterial multiplication.

entablature In classical architecture, the (usually) decorative portion above a supporting column or at the apex of a wall comprising (in descending order) the CORNICE, FRIEZE and ARCHITRAVE.

enteric Means related to the intestines or intestinal tract. An enteric disease is therefore one that affects the whole or part of the intestines.

enteropathogenic The term used to describe the capacity or potential of a micro-organism to cause disease in the intestines.

enterotoxigenic Means capable of producing a toxin that has the capacity to adversely affect the intestines.

entomology The branch of zoology dedicated to the study of insects (*Insecta*).

enveloping The term used to describe a technique of URBAN RENEWAL popular in the 1990s in the UK. The principle was based on the renovation, primarily of the exterior, of a set of houses (often a whole terrace) to a uniform standard using a single contractor to reduce unit price. The use of enveloping had significant and relatively immediate impact on the area under development, and ensured that all properties were brought up to a suitable standard of fitness, thereby promoting uniformity and equality within the neighbourhood.

Environment Agency A Government Agency for England and Wales providing a national regulatory service for issues relating to the environment including in relation to waste regulation, recreational and drinking water, and environmental pollution. The Agency was established in April 1996. The Agency does not undertake environmental protection duties exclusively; other bodies, such as local authorities, play an important role. The work of the Agency is broadly paralleled in Scotland by the Scottish Environment Protection Agency and in Northern Ireland by the ENVIRONMENT AND HERITAGE SERVICE.

Environment and Heritage Service The public body responsible in Northern Ireland for taking the lead in implementing the Government's environmental policy. It is broadly analogous to the ENVIRONMENT AGENCY of England and Wales.

environmental capital The term used to describe the totality of a geographical region or area's complement of natural assets.

environmental health A term that is not easy to define since it is used in various parts of the world to describe different activities. The principal variations in definition relate to whether the term is intended to convey a relationship to the health of the environment generally (i.e. without reference to a particular species such as *Homo sapiens*) or whether it is more precisely aligned to considering the relationship that the environment has with human health. The distinction is probably not as important as some might claim, provided the term is understood in the context within which it is used and there is a recognition that the term can mean different things to different people.

The World Health Organization Regional Office for Europe defines Environmental Health and Environmental Health Services separately as follows:

> Environmental health comprises those aspects of human health, including quality of life, that are determined by physical, chemical, biological and psychosocial factors in the environment. It also refers to the theory and practice of assessing, correcting, controlling and preventing those factors in the environment that can potentially affect adversely the health of present and future generations.

Environmental health services are those services which implement environmental health policies through monitoring and control activities. They also carry out that role by promoting the improvement of environmental parameters and by encouraging the use of environmentally friendly and healthy technologies and behaviours. They also have a leading role in developing and suggesting new policy areas.

Environmental Health Officer A professional appellation used in the UK to describe an individual who has achieved a recognised academic qualification entitling them to full membership of the CHARTERED INSTITUTE OF ENVIRONMENTAL HEALTH or the ROYAL ENVIRONMENTAL HEALTH INSTITUTE OF SCOTLAND.

Environmental Impact Assessment (EIA) An evaluation of the effect that a (proposed) development is likely to have on the area surrounding the site of the development and also sometimes including the impact it might have on the wider environment. Environmental Impact Assessments are increasingly being called for as part of the process of obtaining planning permission, especially in relation to major developments.

See also: HEALTH IMPACT ASSESSMENT, SUSTAINABLE DEVELOPMENT

environmental sanitation Concerned with the promotion of hygiene and the prevention of disease. It covers both environmental factors such as housing, waste and sewage together with sanitation practices such as personal hygiene, household cleanliness and community cleanliness. The World Health Organization has been involved in many programmes around the world to promote environmental sanitation citing that at the turn of the millennium around 66 per cent of the world's population did not have access to safe methods of disposal of excreta and 25 per cent did not have access to a safe water supply.

environmental tobacco smoke (ETS) That component of tobacco smoke that disperses to atmosphere directly from burning tobacco together with smoke expelled following inhalation by smokers. Tobacco smoke contains a complex cocktail of chemicals, over 4,000 of which have been identified. At least forty of these chemicals are known or suspected human carcinogens.

enzootic Means that a specified ZOONOSIS is persistently present in a defined animal population within a designated geographical locality or population.

enzyme A PROTEIN that acts as a biological CATALYST in that it promotes (usually specific individual) chemical processes in the living body. Enzymes have an important role in controlling the many thousand chemical reactions that take place in all living entities. Enzymes are also used in numerous industrial processes for the manufacture of food, drinks and pharmaceuticals.

See also: COENZYME

eosin A red dye.

eosinophil Any cell in the body that contains granules and is readily stained with EOSIN. Approximately 2 per cent of the human body's component of white blood cells are eosinophils.

epicentre That point on the earth's surface immediately below which an earthquake or sub-surface nuclear explosion is deemed to be focused.

epidemic An extensive and abnormal outbreak with a high INCIDENCE of a particular disease within a defined population or area within a specified time. The designation of a particular phenomenon as an epidemic can be emotive and some purists have tried to claim that such can only be officially declared when incidence reaches a particular predetermined numerical rate. This is unrealistic in practice as different diseases will be manifest at different rates and public or media perception is unlikely to be swayed significantly by what can be perceived as facile argument.

See also: EPIZOOTIC

epidemic curve An epidemiological tool in which each case of disease in an outbreak is plotted according to when symptoms began. The identification of when cases occurred can be used to determine whether the outbreak was due to a point source (i.e. most symptoms were manifest at around the same time) or whether the outbreak is spread over time. If due to a point source, knowledge of the INCUBATION PERIOD can help narrow down the possible time of infection – this is particularly useful in the investigation of point source food poisoning outbreaks.

epidemic parotitis An alternative name for MUMPS.

epidemiology Has been defined as 'the study of the distribution and DETERMINANTS of health-related states or events in specified populations and the application of this study to control PUBLIC HEALTH'. The important elements are that epidemiology uses STATISTICS to relate those things that impact on health with the population as the DENOMINATOR.

See also: PUBLIC HEALTH

epidermis The outer layer of SKIN that covers the body. It comprises four layers being the 'stratum corneum', or horny outer layer, the 'stratum lucidium', or clear layer, the 'stratum granulosum', an intermediate layer between the fourth and second layers and the 'Malpigian layer'. The last itself comprises two layers being respectively the 'stratum spinosum' or prickly cell layer and the 'stratum basale'.

epidural Refers to that which is situated over the DURA MATER.

epiglottis The flap of cartilage that forms a flap over the GLOTTIS to protect the windpipe from the ingress of food or drink when these latter are being swallowed.

epinephrine *see* ADRENALINE

epiphyte A plant that grows on another plant but which does not derive nourishment from it – the plant is therefore not a PARASITE.

See also: SAPROPHYTE

epithelial Means of or pertaining to the EPITHELIUM.

epithelium One of the simplest tissues of the body. It comprises the cellular layer that forms the epidermis, the internal lining of the alimentary tract and the internal lining of ducts and hollow organs within the body such as the air passages of the lung and the internal surface of the bladder respectively.

epizootic An extensive and abnormal outbreak of disease in an animal population within a given time frame. It is the animal equivalent of an EPIDEMIC.

equilateral Means having all sides of equal length.

See also: ISOSCELES

equinox The time of year when the sun is directly overhead at the equator. These events happen twice per year at around 21 March (the VERNAL equinox) and 23 September (the autumnal equinox).

See also: SOLSTICE

equivalent dose A measure of the amount of radioactive energy received by a body and determined by multiplying the ABSORBED DOSE by a differential factor to allow for the differentially adverse effects that such a dose could have on different tissues. The equivalent dose can also in turn be multiplied by a differential factor to take account of the different rates at which different tissues can absorb the radiation; the measure thus determined is known as the EFFECTIVE DOSE. Both the equivalent dose and the effective dose are measured in units known as the 'sievert' (symbol Sv).

erg The centimetre/gram/second unit measure of work. One erg of work is done when a force of one DYNE is moved one centimetre in the direction of the force.

See also: JOULE

ergonomics The study of human beings in the work environment leading to the development of improved ways of working and operations that are more suitable for efficiency, health and safety.

ergot A generic term for a group of fungi that attack stored crops, particularly those such as cereals or grass that are normally kept dry but which have become damp due (particularly) to incorrect storage conditions. The term is applied commonly to the parasitic fungus *Claviceps purpurea* that attacks rye and other cereal crops. The fungi produce strong neurotoxins that can be passed on unchanged to foodstuffs such as bread. Symptoms include delirium, convulsion and hallucination. In the Middle Ages bread made from rye was a dietary staple. Bread made from stored crops affected with 'ergot of rye' frequently resulted in poisonings and death. Death due to poisonings from this source was possibly a major limiting factor in the growth of populations at the time.

erosion Removal of surface layers of materials such as soil, sediment or rock from the land by the action of water, ice or wind. The process is distinct from WEATHERING as the former simply transports materials from their site of origin to another place, while the latter requires some prior action to abrade or otherwise strip such surface material from a firm attachment. Erosion is sometimes known as DENUDATION.

See also: COASTAL EROSION; REGOLITH

erythema A medical term to describe a general condition in which the skin becomes increasingly reddened due to the presence of an excess of blood. It has a number of causes including exposure to the sun, exposure to industrial processes or chemicals and exposure to IONISING RADIATION.

erythrocyte Another term for the red blood cell. They do not have nuclei and are produced by the bone marrow. They are very numerous: each cubic millilitre of BLOOD contains around 5 million erythrocytes.

See also: LEUCOCYTE

erythrogenic Refers to that which induces a rash or redness of the skin.

erythrogenic toxin A TOXIN that is produced by certain haemolytic streptococci. It is responsible for the rash of scarlet fever.

Escherichia coli A bacterium, the name usually being abbreviated to *E. coli*. *E. coli* is part of the normal microflora found in the intestinal tracts of humans and most warm-blooded animals. The presence of *E. coli* in either foodstuffs or water is usually taken as an indicator of sewage or faecal contamination and hence potential contamination with sewage or faecally transmitted pathogens.

The strains of *E. coli* that normal colonise the human intestine are harmless COMMENSAL organisms. Some strains of *E. coli* are, however, PATHOGENIC. In developing countries pathogenic *E. coli* are important causes of ENTERIC infections, particularly in children and infants; such infections are often transmitted by contaminated water supplies. In the UK in 1982 a strain known as *E. coli* O157, H7 was identified as an EMERGING PATHOGEN.

ester The organic equivalent of a SALT in inorganic chemistry, that is to say it is the product of an organic chemical reaction in which a hydrogen atom in an organic ACID is replaced by an organic unit. Esters tend to be insoluble in water but they may be soluble in organic solvents.

estimate An imprecise indication of what a builder or other tradesman is likely to charge for undertaking a particular job. It is very difficult legally to hold someone who provided an estimate to the estimated figure as it is unlikely to have been considered in depth and is unlikely to have included items for contingency. Wherever possible it is desirable that the customer receives a priced QUOTATION with any qualifying items clearly identified.

See also: BILL OF QUANTITIES; QUANTITY SURVEYOR

estrogen The US spelling of OESTROGEN.

ethanol Ethanol (C_2H_5OH) is a relatively simple member of the ALCOHOL group of chemical compounds, but in common parlance is often referred to simply as 'alcohol', being the alcoholic component of alcoholic beverages. It is produced by the FERMENTATION of natural SUGARS contained (predominantly) in grains and fruits. The resultant broth may undergo a process of DISTILLATION if the intention is to manufacture spirits.

etiology The US spelling of the UK AETIOLOGY.

ETS The acronym for ENVIRONMENTAL TOBACCO SMOKE.

EU *see* EUROPEAN UNION

eucaryote Sometimes eukaryote, any organism the cells of which have a distinct nucleus and CYTOPLASM. The term therefore applies to all life forms except bacteria and viruses.

See also: PROKARYOTE

eugenics The study and application of techniques intended to manage and control the development of populations in a predetermined direction. Originally the concept was seen as a way of improving the human race through such advances as the elimination of genetic disease. The concept proved to have a darker side when pursuit of supposed racial purity in Germany during the Second World War resulted in the genocide of the Jewish population during the Holocaust. Subsequently genetic or human experimentation has been seen potentially as having similar dimensions (albeit lesser in scale) and there is much debate about the ethics of pursuing such courses.

See also: SUSTAINABLE DEVELOPMENT; UTOPIA; ZERO POPULATION GROWTH

euphoria A heightened or exaggerated feeling of well-being, especially one in which the feeling is irrational or without foundation.

European Central Bank Responsible within the EUROPEAN UNION for framing and implementing European monetary policy. It runs the payment systems of the Union and conducts foreign exchange operations.

See also: EUROPEAN INVESTMENT BANK

European Commission The driving force of the institutional system of the EUROPEAN UNION. It has the right to initiate draft legislation, represents the Union at international level and has responsibility for ensuring that the programmes of the Union (i.e. legislation, budgets and other programmes) are implemented. The Commission consists of a President and Members, all of whom are appointed by Member States after prior approval by the EUROPEAN PARLIAMENT.

European Committee of the Regions Responsible within the EUROPEAN UNION for ensuring that the regional identities and prerogatives throughout the Union are respected and taken into account as part of the decision-making process. It has to be consulted on matters affecting regional policy, the environment and education.

European Court of Auditors The body of the EUROPEAN UNION that has responsibility for ensuring that the receipt of revenue and its expenditure is conducted in accordance with sound financial principles and that the financial management of the Union's affairs is in accordance with requirements.

European Court of Justice Responsible for ensuring that the legislation of the EUROPEAN UNION is interpreted uniformly and applied universally. It has jurisdiction in disputes between Member States and between Union institutions as well as in respect of business interests and individuals.

European Economic and Social Committee The main body of the EUROPEAN UNION responsible for ensuring that the views of the citizenry are being represented within the dealings of the Union and its various institutions. It has to be consulted on matters affecting the economic or social policy of the Union.

European Investment Bank The financial institution of the EUROPEAN UNION responsible for overseeing the proper implementation of the processes of investment projects within the Union in respect of a balanced development programme.

See also: EUROPEAN CENTRAL BANK

European Ombudsman The main appeal body of the EUROPEAN UNION to which individuals, businesses or organisations can appeal in respect of any harm caused to them arising from maladministration by any body or institution of the Union.

European Parliament The directly elected parliament of the EUROPEAN UNION elected every five years by universal suffrage. It shares power of legislation and budgetary authority with the COUNCIL OF THE EUROPEAN UNION and supervises the activities of the EUROPEAN COMMISSION.

European Union (EU) An institutional system comprising a number of Member States, all of which designate sovereignty for certain matters to a number of independent institutions that together represent the interests of the Union as a whole. The European Union was created by the twelve members of the European Community by the Treaty of Maastricht on 1 November 1993. By 2002 the Union comprised fifteen countries, being Austria, Belgium, Denmark, Finland, France, Germany, Greece, Ireland, Italy, Luxembourg, the Netherlands, Portugal, Spain, Sweden and the UNITED KINGDOM. A number of countries are applying to join the Union and it seems certain that it will continue to grow for the foreseeable future. Citizens of the European Union are citizens of all Member States under the provisions of the Treaty of the European Union.

EuroQol/EQ5D A European index for assessing health-related quality of life based on responses over five dimensions.

See also: QALY

eutrophic Refers to high levels of organic and mineral nutrients (generally) in bodies of fresh water sufficient to promote and sustain abundant plant life.

See also: EUTROPHICATION; OLIGOTROPHIC

eutrophication The process in which a body of water receives enhanced nutrients over a period of time. Eutrophication is both a natural and artificial process. Natural eutrophication tends to be slower and occurs due to the leaching of naturally occurring ground nutrients by rivers and streams, and deposition in lakes and coastal areas. Artificial eutrophication occurs when nutrients from, for example, man-made fertilisers and sewage, are washed into bodies of water.

evaporation The process whereby a liquid is changed into a vapour without the liquid necessarily reaching its boiling point.

evergreen The adjective used to describe a tree that does not lose all of its leaves in winter. Evergreen trees do shed their leaves, but they do so throughout the year, and thus they carry some covering of leaves throughout their lives and can thereby give the false impression that they never lose leaves.

See also: DECIDUOUS

evidence-based decision-making The process of deciding upon a course of action by systematically and impartially weighing the available published evidence and its relative validity to differentiate the merits between various options. Although the decision should theoretically be based solely on the available evidence, it is rare in practice completely to disregard other criteria or considerations such as personal or political belief, or preference. It is perhaps more realistic to describe evidence-based decision-making as a process in which there is a transparency or logic in the way in which the decision is determined and that the outcome is compatible with the available evidence.

evisceration The process of removing all the abdominal organs, including the intestines, from the abdominal cavity.

ex ante Means 'before the event', i.e. that which is expected or intended to happen.

See also: EX POSTE

ex gratia Means conducted as an act of thanks or favour committed without pre-existing obligation and without liability, such as in *ex gratia* payment.

ex hypothesi Means derived from or as a result of a hypothesis.

ex parte A legal term used to describe an action undertaken on behalf of or involving a single party to a legal matter, usually without notifying the other party and therefore usually in their absence.

See also: INTER PARTES

ex poste Means 'after the event', i.e. it is the result or outcome of what preceded it.

See also: EX ANTE

Exchequer A department of State charged with responsibility for REVENUE. The name derives from the practice historically of the accounts of State ceremonially being reckoned on a table covered by a chequered cloth.

exfoliate To come away from or be shed by the parent surface in the form of thin layers such as flakes or scales. The process is known as exfoliation.

existing chemical Defined as one of over 100,000 chemicals that are listed by the European Inventory of Existing Chemicals (EINECS) between January 1971 and September 1981. All other chemicals are defined as 'new chemicals'.

exittance The ratio of the amount of light reflected by a surface to the amount of light arriving at the surface.

exogenous Refers to that which is due to external causes.

exomphalos A CONGENITAL ABNORMALITY in which an umbilical hernia is manifest.

exoskeleton A skeleton (the support structure for the body of animals) formed outside the main tissues of the body. Exoskeletons are found in such animals as insects and crustaceans.

See also: ENDOSKELETON

exosphere The outermost layer of the earth's atmosphere extending outwards at around 400 km from the earth's surface beyond the THERMOSPHERE.

exothermic reaction A chemical reaction in which heat is released.

See also: ENDOTHERMIC REACTION

exotoxin A TOXIN produced by certain bacteria and fungi that is released or secreted from the living cell.

See also: ENDOTOXIN

expansion bolt A fastening device, the end of which is supplied with tapered grips designed to be inserted into a tightly fitting (either drilled or pre-formed) hole in a wall or similar. The action of tightening the bolt progressively forces the tapering grips apart, increasingly wedging the bolt into place.

expansion joint A gap designed into a structure or similar to allow for the unavoidable physical expansion or contraction of the adjoining components such as that due to changes in temperature. The gap is usually filled with an elastic material such as MASTIC to ensure continuity in the structure.

experimental study A type of epidemiological study in which the researcher or experimenter controls one or more of the VARIABLES.

See also: OBSERVATIONAL STUDY

expiration The stage in the respiration cycle in which air is expelled from the lungs. In a normal respiratory cycle expiration takes less time than the INHALATION phase; if this pattern is reversed it might suggest the individual is suffering from a pathological condition such as ASTHMA.

exponential A characteristic in which the speed of change from one condition to another continues to accelerate or to increase. The overall effect is therefore cumulative.

exposure A measure of the actual contact between an individual and a potentially harmful biological, chemical or physical agent or energy source. For many causes of ill health there is a direct relationship between the effect (outcome) and the DOSE received, and the duration of exposure.

expressed need A requirement or desire of a patient that is translated into some form of positive action, such as a visit to a doctor.

See also: FELT NEED

expression The translation of instructions into actuality.

exsanguinate To remove or to drain all of the BLOOD from the body.

extermination Defined variously dependent on source. The term in its most general sense has a similar meaning to EXTIRPATION or even EXTINCTION. More specifically usage suggests that extermination is most frequently applied to the elimination of an identified species within a defined environment and (crucially) as part of a deliberate or planned campaign.

extinction The complete elimination from all environments anywhere on earth of all the individuals in a single species. Extinction is essentially considered to be an irreversible process, although the storage of DEOXYRIBONUCLEIC ACID (DNA) and the establishment of SEED BANKS may hold out some hope of remediation and, in the longer term, the possibilities of genetic engineering could conceivably reverse this position in respect to some species. Whilst extinction is also a natural process, it is generally accepted that the effect of mankind on the environment, especially over the last few hundred years, has increased the rate of extinction by several thousand times.

See also: EXTERMINATION; EXTIRPATION

extirpation The elimination of a species of animal or plant from a discrete environment in which it was once present as a residential species. The process of extirpation is not to be confused with EXTINCTION as in the former the species still survives elsewhere. Extirpation can occur as a result of many different factors, such as environmental change, alteration in the food supply chain, disease or human intervention.

See also: EXTERMINATION

Exxon Valdez An oil tanker that ran aground in Prince William Sound, Alaska, on 24 March 1989. The resultant oil spill was calculated at nearly 39,000 tonnes of oil, which, although not the largest oil spill in the world by any means, was certainly one of the most damaging to the environment. The spill affected around 1,300 miles of coastline, around 200 miles of which was reported as heavily contaminated. The cost of the clean-up exercise was put at around $2.1 million.

F

façade The external front wall of a building.

facultative A term indicating an optional or enabling capacity. It is usually applied to the gaseous respiratory requirement of micro-organisms. A facultative AEROBE, for example, is one that can respire either with or without the presence of free oxygen.

See also: ANAEROBE; OBLIGATORY

fall The amount of downwards slope (usually expressed as unit fall in X unit distance) engineered or built in to structures, equipment or systems to ensure that liquids flow towards a pre-designated location. The amount of fall can be critical, especially in systems (such as SEWERS) in which solids are conveyed in liquids, as too little or too much slope can cause deposition of the solids and consequent blockage.

fallout The precipitation from the atmosphere (caused by the influence of meteorological or gravitational influences) of RADIONUCLIDES derived either from nuclear weapons or accidental discharges from processes such as nuclear reactors.

fallow A state or condition of arable (i.e. crop-growing) land in which it is left untended for a period, usually for at least a year. The state is an integral component of a crop rotation system in which the farmer works with the land, allowing natural recovery and replenishment of the land to take place in order to provide a sustainable future for the resource. Modern agricultural practice has reduced the reliance on leaving land fallow and the application of artificial fertilisers has been used to compensate for diminution in the reliance on natural recovery.

See also: GREEN REVOLUTION; SUSTAINABLE DEVELOPMENT

false negative result *see* TYPE II ERROR

false positive result *see* TYPE I ERROR

Family Division One of the Divisions in the United Kingdom of the HIGH COURT.

fanlight Originally a fan-shaped window, commonly situated over a doorway and usually provided with GLAZING BARS radiating outwards like the ribs of a fan from the mid-point of the lower frame member. In modern use the term is applied to any window situated over a doorway.

farm scale evaluation Part of a research programme being conducted in the UK in relation to the evaluation of GENETICALLY MODIFIED ORGANISMS. The programme relates particularly to the assessment of the effect, if any, on wildlife in or around farms of broad-spectrum herbicides used to kill weeds in areas in which genetically modified crops have been planted.

See also: SEPARATION DISTANCE

fascia (or facia) In classical architectural design, a horizontal plain band on the vertical surface of a building. The term 'fascia board' is commonly applied to the horizontal board fixed externally at the EAVES or at the junction of a pitched roof and the wall. The term is also applied to describe the fibrous coating of muscles.

See also: ENTABLATURE

fascioliasis The disease caused by infestation with the parasitic liver fluke *Fasciola hepatica*. It is a disease principally of cattle, sheep and other herbivores (where the condition is also known as 'liver rot') but man can become infected. The eggs of the fluke are voided in the faeces and infect wild water snails, which in turn pass the infection on to aquatic plants, particularly wild watercress. In humans extensive outbreaks have been reported through the consumption of contaminated wild watercress, although infection through the consumption of contaminated lettuce has also been reported.

fat A compound that contains one or more TRIGLYCERIDES. In the widest nutritional sense the term 'fat' includes all such compounds, but in practice the classification is usually sub-divided into fats and OILS. The difference is that whilst oils are liquid at room temperature, fats are solid. Fat is used as a food storage tissue in animals and a few plants.

See also: ADIPOSE TISSUE; LIPID

fatal familial insomnia A form of Creutzfeldt-Jacob disease (see CJD) that is known to occur within family groups. The disease is characterised initially as a persistent MORBID insomnia.

fathom A nautical measure of distance (usually applied to depth) equivalent to 6 ft. (1.829 m).

fatigue In the engineering sense, the condition reached by a component at its point of collapse or fracture caused by repeated stress induced during use.

fatty acids Those nutrients that consist of a chain of carbon atoms with a complement of attached hydrogen atoms and a terminal acid group. They are constituents of fats and oils, and may be either fully saturated or unsaturated (i.e. contain one or more double bonds in the carbon chain). Cardiovascular disease has been linked to the high intake of some types of saturated fatty acids.

faucet The US equivalent of the UK water tap.

fault line *see* GEOLOGICAL FAULT

febrile Means affected by fever.

feedstock Any raw material or substance that is used in a (usually industrial) process of manufacture, such as metal ore in a furnace or petroleum in an oil refinery.

feeler gauge A tool comprising a series of blades of varying thickness that can be inserted into a gap to measure its size. The blade with the greatest thickness that can just be inserted into the gap corresponds to the clearance between the two surfaces. The gauge is used primarily to set the gap predetermined by design criteria between moving parts of machinery or to allow for expansion of materials in use.

feldspar (or felspar) A collective term for a group of minerals comprising predominantly aluminium silicate combined with various metallic ions. Feldspar generally appears as a white or red deposit in a range of rocks, including granite. It decomposes into clay.

See **also:** CHINA CLAY

felspar *see* FELDSPAR

felt need An action, intervention or treatment perceived by an individual as being something they require or desire. The term is usually used in relation to health status, the need relating to something that is perceived as improving or maintaining an existing health state. It may or may not be translated into some form of action. If it does lead to some form of positive action it is known as an EXPRESSED NEED.

fenestration The arrangement of windows in a building (from 'fenestra' – a perforation, window or other opening in a wall).

fermentation The chemical breakdown of CARBOHYDRATES by the action of ENZYMES produced by either YEASTS or bacteria (see BACTERIUM). Fermentation has three main applications. First, it is used in the manufacture of ALCOHOL (alcoholic fermentation), second, in the production of LACTIC ACID (lactic fermentation) and, third, in laboratories in the process of identifying certain bacterial species (e.g. *Clostridium* species).

fertiliser Any organic or inorganic material added to soil to improve its fertility and to provide a source of nutrients to plants. The three most important constituents of fertilisers are compounds containing nitrogen, potassium and/or phosphorous. The US spelling is fertilizer.

fetus The US spelling of FOETUS.

feudal Pertains to that in which the occupier of land or property holds an obligation of service (usually) to one of higher social class in return for permission or condition of occupation. A feudal social system was one in which the land occupier (vassal) had a duty of military service to the landowner (or 'land lord'). The landowner could in turn be called upon to raise such forces as demanded by the monarch, the latter usually being determined as the ultimate 'owner' of the land.

FEV The acronym for FORCED EXPIRATORY VOLUME.

fibre cement A boarding material composed of cement reinforced, potentially, with a range of fibrous materials such as cellulose fibres and FIBREGLASS.

fibreglass Properly a fibre made from molten glass that is then drawn out into very thin fibres. The name is used in practice to include fibreglass matting reinforced with plastic material (glass fibre-reinforced plastic or GRP) used as a construction or repair material and also (more incorrectly) to a range of plastic material used especially in DIY or repair applications for filling in holes or dents.

fibrin A substance formed from the FIBRINOGEN component of both blood and lymph such as when this is exposed to the air or as a result of the haemorrhage of a blood vessel. It is a thread-like material that promotes the formation of blood clots.

fibrinogen A fibrous PROTEIN that forms the basic matrix necessary for blood clotting. The process converts the protein into strands of FIBRIN.

fibrosarcoma A MALIGNANT tumour arising from or within the connective tissue of the body.

fibrosis The formation of FIBROUS TISSUE (such as scar tissue) within the living body. It can result in the replacement of existing tissue with that of a fibrous nature, especially after injury, trauma or loss of blood supply.

fibrous tissue That tissue which is composed essentially of fibres. It is one of the most abundant of all body tissues and can be sub-divided into two distinct types. The first of these, 'white fibrous tissue', is composed primarily of 'COLLAGEN' and is both strong and unyielding. It is found in muscles and ligaments, and is formed in response to injury (typically called scar tissue). The second type, 'yellow fibrous tissue', is comprised mainly of a substance known as 'ELASTIN'. Yellow fibrous tissue is elastic in nature and is found in elastic ligaments and in the walls of arteries. It is less abundant than white fibrous tissue.

field trial The practical testing of an artefact or system in a (usually limited or restricted) situation for which it was produced, the trial taking part prior to general introduction or adoption.

fine particle In relation to airborne contaminants, generally taken as one with an AERODYNAMIC DIAMETER of under 2.5 micrometres.

fingerprinting The name given to a number of techniques employed to characterise and differentiate various fugitive emissions from (primarily) industrial processes. The techniques can involve the use of a number of combinations of electron and normal microscopy together with various analytical methodologies in an effort to produce a clearly identifiable profile for a particular emission. The techniques are useful in identifying the source of particulate material, especially in situations where this is in dispute. The techniques are being used increasingly to identify fugitive emissions from sites such as those of OPEN-CAST MINING in efforts to identify and thence to minimise and control them.

finial A decorative or ornamental structure affixed to the top of a GABLE or canopy, or to the end of a curtain pole or similar.

fire damp A mixture of hydrocarbon gases, chiefly METHANE, formed in mines. It forms an explosive mixture with air.

See also: MARSH GAS

fire extinguishers These are colour coded to describe their intended use. In the UK a new standard was adopted in 1996 under the provisions of British Standard BS 7863. Colours should appear on the front of the extinguisher, immediately above, or in, section 1 of the operating instructions marking, and be visible in a 180° arc when correctly mounted. The information imparts, to the user, information on the contents and, thereby, on what type of fire the extinguisher should be used:

Extinguishing agent	*Colour used*
Water	Red
Foam	Pale cream
Powder (all types)	Blue
Carbon dioxide	Black
Vaporising liquid (including halons)	Green

fire hydrant *see* HYDRANT

fire point The temperature at which the vapour concentration of a liquid will sustain combustion.

See also: FLASH POINT

fixed-price contract An agreement between client and contractor in which the work to be done and the sum to be paid are agreed as legally binding before work starts. Neither party can alter the price unilaterally.

See also: ESTIMATE; QUOTATION

flagella (singular flagellum) Whip-like structures extending outwards from the cell wall of a (usually microscopic and aquatic) organism and used for the purpose of propulsion.

See also: DINOFLAGELLATE

flambé (or flambert) To cover a food (such as Christmas pudding) with an alcoholic spirit (such as brandy or whisky) that is then ignited to flame.

flammable In a condition of being ready to, or capable of being able to, catch fire. It is synonymous with inflammable.

flange A splayed or projecting edge, such as a disc-shaped rim, on the end of a pipe or machinery component. Its function is to aid attachment to another component or structure.

flash point The lowest temperature at which a liquid will give off sufficient vapour to ignite in the presence of a flame. The term is derived from the phenomenon in which such ignitions usually result in a violent and potentially explosive release of energy known as a 'flash'.

See also: FIRE POINT

flashing A material used to cover a joint between two units in the construction of a building in order to prevent the ingress of water. Flashings may be rigid (such as the cement flashing around the joint of a chimney pot to the chimney stack) or flexible (such as the lead or mineral felt flashing at the junction of a lean-to roof and the supporting wall).

flax One of several members of the *Linaceae* family of plants, some species of which are found wild in the British Isles. Common flax (*L. usitatissimum*) has blue flowers and is grown for its oil, known as linseed oil, which is obtained from the seed pods. Flax fibres are used in the production of linen cloth.

Flemish bond A type of BOND of a brick wall in which each course is made up of alternate HEADERS and STRETCHERS.

See also: ENGLISH BOND

fletton A type of brick in common use in England. It is manufactured primarily from Peterborough shale.

float finish A working of the surface of mortar, cement, concrete, plaster or similar with a hand tool known as a float in order to leave the surface smoother than before. The degree of smoothness achieved is largely dependent on the skill of the workman.

float glass That manufactured in a process of floating molten glass on a surface of molten metal to produce a product of a very high quality with virtually no imperfections.

See also: PLATE GLASS; SHEET GLASS

floating floor A construction in which the floor is isolated from the main structure by flexible or absorbent mountings in order to reduce, minimise or eliminate vibration either from or to the structure. Floating floors are used in manufacturing situations to reduce vibration transmission from machinery and in dance floors to minimise nuisance from transmitted sound and vibration.

flocculent A substance that, when added to a suspension in liquid, causes the suspended particles to aggregate together in clumps in a process known as flocculation. This phenomenon is widely used in water purification systems to aggregate suspended particles into clumps that are then large enough to be removed by filtration.

flow chart A representation in words, symbols and/or diagrams depicting the sequential stages in a process, usually of production or manufacture.

flow diversion valve A device within a controlled process handling liquids that comes into operation if sensors detect that the process is failing satisfactorily to achieve its purpose. The device usually redirects the flow of liquid back to the start to ensure its correct processing is achieved. In some operations in which reprocessing would potentially prejudice the integrity of the operation the flow diversion valve redirects the flow to waste.

flue The conduit used to convey the waste gases from combustion processes of a boiler, fire or similar. The term is often used differentially to denote a preformed structure used in the construction of a CHIMNEY.

fluidity A measure of the ability of a material or substance to flow freely.

See also: VISCOSITY

fluorescence A property possessed by some substances to emit radiation of a longer wavelength upon receiving radiation of a shorter wavelength.

See also: FLUORESCENT LIGHT

fluorescent light A device for providing illumination. It consists of a glass tube coated with a material that is capable of FLUORESCENCE together with a source of ULTRAVIOLET RADIATION. The ultraviolet radiation falling on the fluorescent material is converted into visible light for illumination. Fluorescent lights are more efficient in providing illumination per unit of energy than are incandescent lights. However, unlike the latter, they cannot operate under varying current and so are unsuitable for use with varying resistors such as dimmer switches.

fluoridation The term used to describe the addition of FLUORIDE to a drinking-water supply, usually to raise and maintain its level to around 1 part per million (1 milligram per litre) with the aim of reducing dental caries in the population. Although the fluoridation of water supplies is controversial it is a proven and effective public health measure (especially in relation to children's teeth). However, opponents claim that adverse health effects can be caused, notably a condition known as fluorosis, the main sign of which is a mottling of the teeth. The most comprehensive systematic review of the effectiveness of water fluoridation was conducted by the University of York (UK) and published in 2000. The review was supportive of fluoridation whilst recognising that the evidence of benefit and harm needed to be weighed alongside ethical, environmental, ecological, cost and legal issues.

fluoride A halide ION of FLUORINE. It is used predominantly as an additive in toothpaste as a protective against dental caries and is also added to public water supplies in some areas for the same purpose at concentrations of around 1 part per million. Its addition to water supplies is controversial in some regions although its efficacy in preventing tooth decay is well proven. The average cup of tea contains around 12 parts per million of fluoride. Fluoride is poisonous in large quantities, some sources suggesting that the fluoride content of a large tube of fluoridated toothpaste could kill a child.

fluorine A HALOGEN and a non-metallic element. Halogens are highly reactive and do not exist in nature, but the addition of another electron to their outer shell creates a halide ion and these do occur widely. FLUORIDE is a one such.

flush In constructional terms, flush refers to something that is level with the surrounding surface. The term is used in such applications as flush joint, flush switch, etc. The term is also used to describe an area of land on a sloping site through which water flows diffusely without producing defined channels.

flushing system One in which a (usually) measured or predetermined quantity of water or similar is discharged through a system for the purpose of cleaning or removing deposits. The system can be operated either manually or automatically, the choice depending primarily on the nature of the deposit and the frequency of cleaning required.

flux *see* SOLDER

fly ash Finely divided particulate matter emitted to atmosphere or deposited within chimneys and flue linings, being derived from combustion processes, especially those involving coal. Fly ash can contain portions of unburnt fuel as well as toxic chemicals emitted or created during the combustion process. Fly ash is nowadays usually removed and collected from flue gases before it can be discharged to atmosphere. Disposal of fly ash containing toxic material however needs careful control. Some fly ashes have POZZOLANIC properties.

flying buttress A type of BUTTRESS in which no part of the buttress itself is in contact with the ground. Usually a flying buttress is a curved (sometime straight) structure of stone or concrete composition in which both ends are attached to a wall or similar with the aim of transferring the load from one part of a building to another.

flyleaf A blank page (also known as a blank leaf or sheet) that appears at either the end or beginning of a book.

flying shore *see* SHORE

flywheel A heavy wheel attached to and revolving with a shaft in order to regulate its rotation within machinery.

foetotoxic That which has the potential to be toxic (see TOXIN) to the FOETUS, with possibly lethal effect.

foetus The unborn progeny of a mammal whilst still in the womb. In humans the term generally refers to the unborn child after 8 weeks of development. The US spelling is fetus.

fog Defined by the International Standards Organization (ISO) as a meteorological condition in which the presence of a suspension of water droplets in the air reduces visibility to less than 1 km.

See also: MIST

folates *see* FOLIC ACID

folding in A term used to describe the culinary practice of combining other ingredients with a mixture by continually turning over the ingredients with a spatula or similar so that the overall mixture retains a lightness of consistency.

folic acid A synthesised vitamin derived naturally from folates. It is often added to breakfast cereals and soft-grain white bread as a FOOD SUPPLEMENT. Its consumption by pregnant women can significantly reduce the risk of neural tube defects (such as spina bifida) in the foetus. It has also been suggested that folic acid can reduce the possibility of heart disease.

fomites Any articles sufficiently contaminated by infectious material to be capable of spreading the disease. They can include dust, clothing, equipment and household articles.

food allergy A form of FOOD INTOLERANCE that presents as an abnormal immunological reaction in response to a particular food or food type. Symptoms can be localised, e.g. being confined to the mouth or gastro-intestinal tract, or be more widely spread and include the production of a rash, flushing, conjunctivitis, vomiting, diarrhoea, abdominal pain and ASTHMA. Although around 20 per cent of people in the UK claim to have some sort of food allergy, medical practitioners suggest the true figure is more likely to be between 1–2 per cent.

See also: FOOD AVERSION

food aversion Classified as being mainly a psychological avoidance and psychological intolerance to food. The condition includes PSYCHOSOMATIC adverse food reactions, although, as these are thought to be primarily of mental rather than physical origin, it may not be possible to draw parallels with true allergy/intolerance in terms of diagnosis or treatment.

See also: FOOD ALLERGY; FOOD INTOLERANCE

food intolerance A reproducible and unpleasant adverse reaction (not being psychologically based) to a specific food or food ingredient. The term includes FOOD ALLERGY but is most commonly used to suggest generally a range of symptoms that are less severe than those of food allergy.

food poisoning Defined in 1984 by the World Health Organization as 'any disease of an infectious nature caused by or thought to be caused by the consumption of food or water'. The UK Government adopted the definition in 1992.

food supplement A concentrated source of nutrient or dietary requirement that is intended to be added to a food to enhance its dietary or nutritional value. The term is frequently applied to vitamins and minerals.

See also: fortified food

foot A traditional measure of length, the name deriving from the fact that it was based on the 'average' length of a man's foot. Not surprisingly the length was variable dependent on region or country. The standard British or US foot was standardised at 12 inches, or 30.48 cm. By comparison a Roman foot was equivalent to 29.56 cm.

Foot and Mouth Disease A viral disease particularly of cattle, pigs and sheep characterised by vesicles on the lips, tongue, mouth and feet of infected animals. Sheep in particular are capable of carrying infection with only minor clinical symptoms. Transmission is airborne, by direct contact with infected animals or by contact with contaminated materials. The disease is mildly zoonotic (see ZOONOSIS) exhibiting as a benign primary vesicle, possibly giving rise to secondary vesicles on the hands, mouth or feet, but it is thought that a massive dose is necessary to cause infection. Horses are immune to the infection.

foot-candle An obsolete British unit of luminance, one foot-candle being equivalent to 10.76 LUX.

footing Generally taken as the base portion of a building or other construction, especially one made of pre-formed units such as stone or similar. Some sources suggest that a footing is not a base structure cast IN SITU, such as one made from concrete, describing the latter as a FOUNDATION. Other sources describe the footing as any base structure, including those that are cast *in situ*.

forced circulation A term usually applied to fluids (especially liquids) in which the fluid is impelled by pumping around or through a circulatory system (such as a hot water system used for heating).

See also: GRAVITATIONAL CIRCULATION

Forced Expiratory Volume (FEV) Volume of air that can fully be expired by the lungs following a full INHALATION. The exhalation is usually conducted as quickly as possible and the volume measured at intervals of one half, one, two and three seconds. This information is used to contribute to an assessment of the respiratory capacity of the lungs.

forced ventilation Any system of ventilation in which air is circulated within a building by the means of artificially altering the air pressure so as to induce flow from an area of higher pressure to one of lower pressure. This can be accomplished either by forcing air into the building so that it passes outwards through spaces (such as vents and windows) in the fabric, or by extracting air so that the higher atmospheric pressure outside forces air in through similar spaces in the fabric.

See also: NATURAL VENTILATION

Fore Pronounced 'foray', a tribe of the Eastern Highlands in Papua New Guinea. Some members of the tribe were identified as suffering from a form of transmissible spongiform encephalopathy known as KURU, the recognition of which provided significant clues to the development of new variant Creutzfeldt-Jacob disease (see CJD) in humans.

fore stomach One of the two distinct regions comprising the stomach of a rodent.

See also: GLANDULAR STOMACH

foreshock A smaller tremor, sometimes occurring in a sequence, which precedes the occurrence of a major shock of an earthquake.

See also: AFTERSHOCK

forest degradation The reduction in the overall quality of a forest and/or its ecosystem. This could be due to the acidification of the soil due to ACID RAIN or to changes in rainfall, soil erosion or depletion of biodiversity. Forest degradation is distinct from DEFORESTATION, although the latter can lead to degradation of surrounding areas by promoting flooding or soil erosion, or by creating additional pressure on remaining habitats.

forge A workshop wherein metal (particularly iron and steel) is heated and shaped either by hammering or extrusion through rollers. As a verb, to forge means to shape heated metal by hammering or extrusion.

See also: WROUGHT IRON

formaldehyde Formaldehyde (H-CHO) is a powerful antiseptic and is widely used as a disinfectant. In weak solution it is used as a preparation for treating warts. Formaldehyde is also known as methanal.

See also: FORMALIN

formalin A solution of FORMALDEHYDE in water. It is usually stabilised with methanol. Classically formalin (also known as Formaldehyde Solution) contains between 34–8 per cent formaldehyde in water. Formalin is a powerful disinfectant and is also used as a solution to preserve animal specimens in the laboratory.

formwork A temporary restraining structure used to mould and contain a fluid material such as concrete in a predetermined shape for sufficient time to allow it to harden and then be incorporated into a building or similar construction. Formwork is also known as 'shuttering'.

fortified food A food to which additional nutrients, minerals or vitamins have been added to meet or supplement a (specified) dietary requirement.

See also: FOOD SUPPLEMENT

fossil fuel Any fuel derived from the ancient deposition of organic material and transformed by chemical action, heat and/or pressure into a usable fuel source. Fossil fuels include natural gas, PETROLEUM and coal.

foundation Generally taken as the base portion of a building or other construction. Some sources describe this portion as the FOOTING and define the foundation as the ground upon which the footing stands.

fractional distillation The primary processing stage in the refining of PETROLEUM. The process uses different temperatures to separate differentially the various components (known as fractions) from the raw material. The distribution of products obtained by the process running from lowest to highest temperature includes ethane, propane (as gases), petrol, kerosene, DIESEL, heating oil and heavy gas oil.

frankfurter A Continental sausage made from smoked meat, either comprising pork or a mixture of pork and beef.

franklin An Old English term used to describe a FREEHOLDER.

frappé In culinary practice, a drink or foodstuff that is frozen, chilled or iced (especially with crushed ice).

free field An environment within which there are no reflecting surfaces such as to affect the measurement of sound within a desired frequency band.

free house A public house not tied to a particular brewery or supplier and licensed for the sale of alcoholic beverages.

free radical A chemical molecule that has at least one unpaired electron. It has been suggested that they might contribute to the progression of many major degenerative processes in the human body and that these in turn can lead to serious conditions, including cancer. It has further been suggested that certain dietary constituents known as 'antioxidants' (such as vitamins C and E, and non-nutrients such as flavonoids) have the capacity to remove or reduce the levels of free radicals in the body and thereby reduce the risk of certain cancers and other chronic diseases.

This 'antioxidant hypothesis' builds on the fact that diets containing high intakes of fruit and vegetables can significantly reduce the risk of certain cancers and other chronic diseases. Although widely accepted and plausible, the antioxidant hypothesis has not been proven.

freehold A property or land held free from duty to any except the Crown.

freeholder Originally a person who was free from the duty of FEUDAL service. Nowadays the term is used to describe a person who is in possession of a FREEHOLD, i.e. someone who has absolute ownership of land or property, save for that owed to the Crown.

frequency The number of complete cycles in a wave's motion that pass a given point per unit of time. For the higher-frequency energy types the unit of measurement is the HERTZ (symbol Hz) with 1 Hz equating to 1 cycle per second.

See also: WAVELENGTH

frequency modulation One of the principal means of encoding a radio signal to transmit audio or visual images, differentiation being achieved by variations in the frequency, the amplitude remaining the same.

See also: AMPLITUDE MODULATION

friction The resistance exhibited between two surfaces when two bodies or substances move, or attempt to move, against one another. In order to overcome the resistance it is necessary to expend energy. This results in the generation of heat. The more resistance, the greater the friction and the greater the heat produced once the bodies are moving. This energy equation can be employed positively in, for example, friction brakes in which the KINETIC ENERGY of movement is dissipated into heat by added friction as it is applied. The creation of heat robs the body of kinetic power and its speed is reduced as a consequence. The coefficient of friction is the force required to move an object across a horizontal surface at a constant speed, divided by the mass of the object.

Friends of the Earth (International) An international federation comprising numerous autonomous organisations across the globe. The organisation exists to co-ordinate campaigns on environmental and social issues, in particular to further sustainable development. Originally formed in 1971 by organisations in France, Sweden, the USA and the UK, the federation boasted a presence in sixty-six countries by the turn of the millennium.

See also: EARTHWATCH INSTITUTE; GREENPEACE

frieze In classical architecture, the middle portion of an ENTABLATURE. The term is nowadays applied commonly to any (usually horizontal) band of sculpture, moulding or other decorative finish on a wall.

See also: BAS-RELIEF

frog A depression incorporated into one or both of the largest faces of a brick. The frog reduces the amount of clay needed to make the brick, makes it lighter and helps to form a key with the mortar.

front In the meteorological sense, a weather system created along the line where two dissimilar air masses meet.

fuel poverty A situation in which a household cannot afford properly to heat their home. This situation can arise as a combination of lack of sufficient income to buy or pay for fuel, and poor energy efficiency that results in a requirement to use more fuel to achieve and maintain a suitable temperature. Fuel poverty not only affects quality of life, but also can result in adverse health effects, especially to those in vulnerable groups such as the elderly, children, those with disabilities and people suffering long-term illness.

fulcrum A fixed point against which a lever pivots.

fumarole A natural hole through which gas or steam is vented into the atmosphere from the side of a volcano or a region of volcanic activity.

fume An AEROSOL that contains solid particulate matter.

See also: SMOKE

fume cupboard A protective working enclosure in which activities that might generate toxic or damaging fumes or gas are conducted, ensuring that any such are removed to safety by exhaust ventilation.

fumigation A process of disinfection achieved by destroying bacteria, insects and other pests (or of delivering pesticides across a controlled environment, such as an enclosed space) by using gases or vapours as the means of transportation. The success of the process is dependent on the type of active ingredient contained within the transport medium, but the nature of the process ensures relatively uniform distribution across the desired environment. It is important to consider any undesirable effects before carrying out fumigation as the process will deliver potentially toxic substances to all points. The active ingredient and transport medium can have deleterious effects on plants, materials and surfaces. Both pests and domestic animals can pick up deposited material, and contamination of food and water is also possible. Operatives will need full training, and consideration will have to be given to the provision of personal protective equipment and safe systems of work.

functional claim A term used to designate a stated beneficial effect on the human body resulting from the ingestion of a type of food.

See also: NUTRITIONAL CLAIM

fungicide A chemical substance that kills fungi.

fungus (plural fungi) Simple organisms, usually considered as members of the plant kingdom, but they do not possess CHLOROPHYLL and are unable therefore to generate energy by PHOTOSYNTHESIS. They obtain their nutrients directly from their environment (HETEROTROPHIC). There are many thousand different fungi known, ranging from microscopic YEASTS to the more familiar mushrooms and toadstools. Many fungi are essential to the processes of decay, especially of vegetable material.

funicular railway A transportation system whereby a vehicle is propelled along rails by means of a cable. Funicular railways were a common way of transporting people, goods and materials along steep gradients in situations where normal friction wheels would cease to operate.

furan A general term used to describe a group of chlorinated organic chemicals. They are generally described as dibenzofurans and linked together with polychlorinated dibenzo-p-dioxins as 'DIOXINS'.

furring The description applied to the condition of the inside of pipework, utensils and equipment caused by the deposition of calcium and magnesium salts.

See also: HARD WATER

fuse A safety device incorporated into an electrical circuit consisting of a wire with a known melting point that will melt (i.e. it will 'fuse') if that temperature is reached. Such conditions will occur if the current in the circuit exceeds design parameters.

G

G7 The appellation used to describe collectively the Group of Seven major industrialised countries. The Group comprises Canada, France, Germany, Italy, Japan, the UK and the USA.

gable The triangular portion of an external wall defined at its upper edges by the slopes of a pitched roof. A gable roof is one that has two sloping portions, each falling away from a RIDGE. A gable wall is any external wall, usually at the side of a building or at the end of a terrace or semi-detached property, in which a gable forms a part.

Gaia hypothesis Originally proposed by James Lovelock in the early 1970s. The theory was published in 1979 in *Gaia, a New Look at Life on Earth*. The hypothesis holds that the earth was colonised around 1 billion years after its formation by a meta-life-form that began the process of transforming the planet and incorporating it into its own substance. All life forms on earth are, according to the hypothesis, constituent parts of Gaia, acting together to sustain the life of the overall organism. All systems should therefore be in balance with the waste products from one life form providing the nutrients for another. The key component of the hypothesis is that the earth is the organism, comprising the rocks, waters, life environment and climate, in other words – everything. The theory propounds therefore that the environment is determined by life and not, as Darwin would have held, that life is a product of the environment. Gaia would eliminate anything that threatens the integrity of the system, such as humans, as being a threat to its own life.

gale A moderately strong wind defined by the World Meteorological Organization as registering force 8 or above on the BEAUFORT SCALE, i.e. having a wind speed equal to or in excess of 34 knots (around 64 km/hr).

galvanise To coat a material (especially iron or steel) with a protective coating (usually zinc or zinc with aluminium) by any of several means including the use of an electric current, immersion in the molten coating material or cold deposition as part of a chemical process.

galvanometer A sensitive scientific instrument used to measure electric currents, especially small electric currents.

gambrel roof A variation of a HIP ROOF in which a small GABLE is added to each of the upper portions of two opposing sloping sides.

gamma radiation High-energy electromagnetic radiation with a wavelength within the range 1×10^{-10} to 1×10^{-13} metres. It has more penetrative power than X-RAYS and can pass through human tissue or even demonstrate limited penetration of concrete or lead. It is an IONISING RADIATION, its penetrating power being capable of access to even deep-seated organs of the body and thereby potentially generating CANCER.

ganglion (plural ganglia) In anatomy, an aggregation of nerve cells along the route of the nerve. In disease the term refers to a fluid-filled swelling or enlargement of the protective sheath of a tendon.

See also: REPETITIVE STRAIN INJURY

gangrene The condition in which sufficient portion of the outside of the living body has died as to be visible to the naked eye. Gangrene is distinguished as one of two types. Dry gangrene occurs when circulation to that area ceases and the area consequently simply withers away. Wet gangrene is usually the product of infection and is characterised by inflammation and putrefaction.

Gantt chart A representation of the scheduling of individual tasks necessary to complete a whole process. It uses bars of varying length to represent the time taken to complete each task, the start and ends of each bar corresponding to the start and finish times depicted on the scale; the relative positioning of the bars indicates when in the process each task needs to take place.

See also: BAR CHART

gargoyle A waterspout that projects outwards from the guttering of a building such as to discharge the water into the air and being directed away from the walls of the building itself. Many gargoyles were originally carved to represent human heads or grotesque animals, especially on churches and other religious buildings.

garret A room formed just under the roof of the house, especially one in which the slope of a pitched roof is incorporated into the structure of the room.

gas chromatography *see* CHROMATOGRAPHY

gas mantle A gauze covering for the jet of a gas-powered lamp. When heated the mantle becomes incandescent and thereby significantly improves the performance of the lamp. The original gas mantle was a cotton gauze covering impregnated with thorium and cerium oxide, and was invented by Welsbach in 1885. It is therefore sometimes known as the Welsbach mantle.

gas poker A device with a gas-powered flame at its end inserted into flammable material for the purpose of generating fire.

gasahol A fuel comprising petrol and alcohol, usually in a ratio of between 9:1 to 8:2 respectively.

gasification The name given to the process of producing gaseous fuel by heating organic material in a vessel in the presence of air, steam or oxygen. The process also produces ash and tar. Some see the process as leading, once more fully developed, to a more environmentally friendly way (than direct incineration) of disposing of waste, especially municipal solid waste.

See also: ENERGY FROM WASTE; PYROLYSIS

gasket A flat packing material of compressible nature that fits between two surfaces such as to form a leak-proof joint upon compression when the two surfaces are fastened together.

gasoline The term used in the USA to describe that which in the UK is known as PETROL.

gastro-intestinal Refers to the totality of the alimentary tract, i.e. the mouth, oesophagus, stomach and intestines.

gastroschisis A CONGENITAL ABNORMALITY in which there is inadequate development of the abdominal wall of the foetus creating a fissure. Sometimes this is accompanied by protrusion of the intestines.

gate valve A device for limiting or restricting the flow of liquids or gases in a system by means of a flat plate lowered or raised across the line of flow.

See **also:** GLOBE VALVE

gauged arch A semi-circular arch constructed from GAUGED BRICK.

gauged brick A brick of tapering width that, when assembled with others of similar size, produces a curved aspect of predetermined radius. They can be used in the construction of a GAUGED ARCH.

Gaussian distribution That which when represented graphically produces an equal smooth apportionment on either side of the MEAN, i.e. the graph appears as if the cross-section of a bell. Gaussian distribution is also known as 'normal' distribution.

gavage The administration of a liquid directly into the stomach via a tube inserted through the mouth.

gazebo Also known as a belvedere, a turret, pavilion or summerhouse with open sides to admit breeze or to facilitate viewing the surroundings.

GDP The acronym for GROSS DOMESTIC PRODUCT.

gear A machine part that transmits the energy from one moving part of a machine to another by means of a toothed wheel intermeshing with another. Gears are usually of different diameters ensuring that the number of rotations of the driving shaft is transformed into a different number of rotations on the driven shaft. The smaller of two intermeshing toothed wheels is known as the pinion, the larger as the gear. The relationship of the rotations of one shaft to another is known as the gearing.

Geiger counter Also known as a Geiger–Müller counter, an electronic instrument used to detect the presence of high-energy radiation particles. The device comprises a single wire held at a high positive voltage surrounded by a gas such as argon, the whole being contained within a cylindrical electrode. A particle such as an ion or electron penetrates the wall of the cylinder directly, and gamma and X-RAYS generate IONS from the gas. These ions or particles are attracted by the high voltage in the wire, gain energy, collide with other atoms of the gas and release more electrons; the process ceases due to limiting factors in the gas itself or in the electronic circuitry. This process creates a clear pulse that is detected (i.e. 'counted') by the instrument. In this way the Geiger counter is a highly accurate instrument for the detection of radiation and quantifying the number of disruptions of the electronic field. The instrument is however not capable of differentiating between radiation types, and nor does it describe the energy of the particles (save to identify that they are capable of penetrating the cylinder wall).

gelatin Sometimes gelatine, a transparent or colourless substance derived from the COLLAGEN component of connective tissue on boiling. It is soluble in boiling water and sets as a jelly-like material at room temperature.

gene One of a series of units contained within a cell that together comprise the CHROMOSOME. Each gene is self-reproducible when cells divide and each carries a specific encoded instruction for the development of a particular characteristic of the body, such as for eye colour.

See also: GENOME

gene expression The total process whereby the instructions encoded on a gene are translated into the production of a living entity.

See also: PHENOTYPE

genetic drift The process in which random change or variation in the occurrence of a particular gene occurs within a defined population. It is thought that this phenomenon is responsible for, or contributes to, the process known as SPECIATION.

genetic engineering (GE) The common term used to describe the process more correctly known as GENETIC MODIFICATION.

genetic modification The deliberate and artificial (or non-natural) alteration of the genetic material of animals or plants that results in changes to the progeny or resultant life. The practice is controversial. Proponents claim that the benefits include, for example, improved disease resistance, better livestock or crop material, improved resistance to herbicides, improved keeping quality and improved drought resistance. Opponents claim that we do not know enough about the technique to avoid creating unwanted effects or side effects, either in the target organism or in passing on such effects to progeny or transmitting them to other species.

genetically modified organism (GMO) Any living organism whose natural DEOXYRIBONUCLEIC ACID (DNA) sequence has been altered by the insertion of an alternative portion of DNA from another source. The release of genetically modified organisms into the environment (especially in industrialised countries) is carefully controlled through systems of detailed risk assessment. However, their introduction is controversial and the subject of much debate.

See also: FARM SCALE EVALUATION; GENETIC MODIFICATION

genome The totality of the genetically encoded instructions contained within a cell for the self-reproducible development of life. Humans have around 10 million individual genes contained within the genome.

See also: GENE

genotoxic The term used to describe a CARCINOGEN that can adversely affect the genetic material of the cell. Genotoxic carcinogens can either act directly or as a result of being transformed (e.g. metabolised) within the body.

genotype An organism differentiated from others of the same species by virtue of differences in their genetic make-up. The genetic component can be differentiated even if it is not manifest as a difference in the physical characteristics of the organism.

genus A classification of organisms defined together as a group by broad similarities. A genus can be sub-divided into SPECIES.

geodesy The term applied to the surveying or mapping of the earth's surface. It is undertaken on such a large scale as to necessitate making allowance for the earth's curvature.

See also: GEODETIC LINE

geodetic line An imaginary line of the shortest possible length, used in GEODESY, linking two points on the surface of the earth and allowing for its curvature. The term is also applied to any line of the shortest possible length linking two points on a curved surface.

geographic information system A sophisticated computer-based MODEL that allows the analysis of interrelated data to be conducted with the results being presented as an image relative to a real geographical location.

See also: VIRTUAL REALITY

geological fault A fracture in a once-whole rock stratum, usually in which the rocks on either side of the fracture are displaced from each other as a result of movement in the earth's crust. The line along which the fault runs is known as the 'fault line' and is an area of weakness in the ground structure. Fault lines can also act as conduits for sub-soil water and gases.

geometric mean A point in a series of asymmetrical data derived by converting the data into logarithms, establishing the MEAN and taking the anti-logarithm of that point.

geostationary orbit Another name for a SYNCHRONOUS ORBIT.

geostrophic wind One that is not affected by the frictional effects of passing over surfaces, such as those of the earth. They form at altitudes of around 500–1,000 m and, because they are largely free from attenuation by friction, they flow parallel to ISOBARS.

geotaxis A response of a living organism to gravity.

See also: GEOTROPISM

geothermal energy That which is derived from the earth's core. It is obtained either by extracting heat from steam, superheated water or mud welling up through the earth's crust, or from boreholes drilled for the purpose.

geotropism A response of a living organism to the stimulus of gravity. A positive response is one in which the organism moves or grows downwards (i.e. in the direction of the gravitational pull), while a negative response is one in which the organism moves or grows upwards (i.e. in the opposite direction to the gravitational pull).

See also: GEOTAXIS

germ An imprecise, all-embracing term to describe a (usually potentially pathogenic) micro-organism. The term is most frequently applied to bacteria or viruses, but can also be used to describe protozoans or rickettsiae (see RICKETTSIA).

German measles *see* RUBELLA

germ cell The generic term used to describe any reproductive cell, such as the sperm and the egg cells and their precursors. Germ cells carry half the normal complement of chromosomes of a normal cell, e.g. in humans a germ cell has 23 chromosomes instead of the normal 46 chromosomes.

See also: HAPLOID NUMBER

Gertsmann–Straussler–Scheinker syndrome (GSS) *see* TRANSMISSIBLE SPONGIFORM ENCEPHALOPATHY

gestation The period of time from conception to birth in a normal pregnancy.

giblets Comprise the neck, gizzard, heart and liver of a fowl. In some eviscerated poultry giblets are returned to the abdominal cavity prior to sale for use in cooking for making stock.

GIS *see* GEOGRAPHIC INFORMATION SYSTEM

glacé In culinary practice, refers to food that is either iced or frozen. The term is alternatively applied to any foodstuff that has a glazed or glossy surface giving the appearance of having been iced or frozen.

glandular stomach The second of the two distinct regions that together comprise the stomach of a rodent, the other region being known as the FORE STOMACH. The glandular stomach is analogous to the human stomach.

glare Either direct or reflected light of sufficient intensity so as to adversely affect the normal vision of an observer.

glass fibre *see* FIBREGLASS

glass wool A MINERAL WOOL made from spun glass. It is used primarily as an insulation or packing material.

glaze In culinary practice, anything that is used to give a glossy or reflective surface to a foodstuff.

glazing bar A dividing section in a window, shaped to accept the glass or other glazing material and to which the latter is fastened (usually) by pins and/or PUTTY, by pre-formed clips or by a GLAZING BEAD.

glazing bead A strip of wood, plastic metal or similar used to fasten a glazing panel to the GLAZING BAR.

glia A type of nerve cell that functions as a means of fighting infections. It also provides certain regulatory functions within the body.

global commons The term used to describe common environmental assets of the earth that are not the property of any single country. Included in the term are the atmosphere and the oceans, together with the life that these contain.

global warming The common term applied to the phenomenon more properly known as 'global climate change'. The phenomenon is by no means universally accepted as a fact but the basic concept is that GREENHOUSE GASES in the earth's atmosphere trap heat from the sun, creating a 'greenhouse effect'. This causes the sometimes catastrophic disruption of weather and climactic conditions, and could result in melting of the polar ice caps causing flooding to lower-lying coastal regions of the world with commensurate knock-on effects for the planet's ecology.

See also: CARBON SINK

globalisation A term that came into common usage in the early 2000s used to describe the end product of an ill-defined set of activities allegedly practised by international conglomerates or corporations to harness workforces and promote brand loyalty. The term is usually applied as derogative implying that the corporations distort employment patterns, especially in poorer countries, to the extent that they become monopoly providers in some areas. It is also suggested that the promotion of brand loyalty can be taken to extremes, sometimes to the extent of subtle brainwashing, in order to produce loyal customers for life.

Some see globalisation as an inevitable consequence of free-market expansion by business and as a way of speeding the distribution of wealth and opportunity across the globe. Opponents feel that the global corporations have become so powerful that they can effectively dominate local economies, sometimes even distorting those of smaller (especially developing) countries, in a way that is ultimately detrimental.

globe valve A device for limiting or restricting the flow of liquids or gases in a system. This is done by means of a spherical or globular insert lowered or raised along the line of flow by means of a screwing or leverage mechanism.

See also: GATE VALVE

globulin A SALINE-soluble PROTEIN that is insoluble in water.

glottis The opening portion of the larynx or windpipe.

See also: EPIGLOTTIS

glucose Glucose ($C_6H_{12}O_6$) (also known as DEXTROSE) is one of the MONOSACCHARIDE types of SUGAR. It is produced in plants by the process of PHOTOSYNTHESIS and is the basic source of energy in both plant and animal METABOLISM. In such a metabolism, carbohydrates (including other sugars) are converted by the organism into glucose before use.

glutelin A PROTEIN that is soluble in dilute acids or bases but insoluble in neutral solutions.

gluten An adhesive protein substance that is a constituent of wheat flour. It aids the process of bread 'rising'. It has been associated with varieties of 'malabsorption syndrome' in which the intestines are unable properly to absorb food. Those persons affected are required to pursue a gluten-free diet.

glycogen A storage POLYSACCHARIDE found in animals. It is readily broken down into GLUCOSE. In the body it is deposited mainly in muscle tissue and the liver.

See also: STARCH

glyphosphate Glyphosphate($C_3H_8NO_5P$) is a water-soluble broad-spectrum herbicide. It is widely used in agriculture and by domestic gardeners. It breaks down on contact with soil and has the advantage therefore that it is not environmentally persistent. It acts as an enzyme-inhibitor in plants, preventing growth and causing them to yellow, wither and (providing they do not recover) die from around 5 days to several weeks following application. Glyphosphate is the active ingredient in the commercial herbicide preparation known as 'Roundup'.

GMO *see* GENETICALLY MODIFIED ORGANISM

gnomon The part of a sundial that casts the shadow, thereby indicating the time.

GNP The acronym for GROSS NATIONAL PRODUCT.

going The horizontal distance in a staircase travelled between each successive step. It can alternatively be described as the horizontal distance between each successive RISER. The distance does not include the NOSING.

gonorrhoea An infection of the genitourinary tract by the gonococcal bacterium *Neisseria gonorrhoeae*. Gonorrhoea is also known colloquially as 'clap' or 'a dose'. The bacterium is capable of infecting other sites such as the pharynx and the anorectal area. It is also capable of infecting the conjunctiva, especially of the newly born, in which case the infection is known as 'gonococcal conjunctivitis (neonatorum)'. Infections of sites other than the genitourinary tract are sometimes grouped more accurately within the term 'gonococcal infections'. Gonorrhoea in males presents primarily as a purulent discharge from the urethra; in females it presents mainly as inflammation of the genitourinary area. If not treated, more serious symptoms can develop, including skin lesions, arthritis and, rarely, septicaemia. Death is rare, but is not unknown, especially in patients who develop endocarditis.

good manufacturing practice (GMP) The term used to describe a process within (particularly) the food industry whereby goods are produced to a specified quality within a time frame when all necessary controls for the safe production of the product are in place. In effect the term GMP is convenient shorthand to describe the integrity of a process in its entirety without having to specify each individual control in detail.

See also: HAZARD ANALYSIS CRITICAL CONTROL POINT

goosebump *see* HORRIPILATION

gout A constitutional disorder caused by an excess of uric acid in the blood. As a result depositions of uric sodium monourate arise in the joints causing inflammation and pain. The condition may be acute or chronic and appears to have a strong hereditary component. Rich diet and excess intake of alcohol also appear to influence its occurrence but this is not a universal requirement as vegetarians and teetotallers may also be affected.

See also: CAFFEINE; DIURETIC

Government In the UNITED KINGDOM, the body responsible for the actual running of the country (as opposed to its legislature). It comprises the Prime Minister, formally appointed by the monarch (usually, but not necessarily), being the leader of the majority party in the HOUSE OF COMMONS, together with those ministers who have been appointed (by the Prime Minister) with departmental responsibilities.

The term Government is often confused with the term PARLIAMENT. The distinction is best explained by considering that Government proposes (or in some instances, provided legislative provisions allow, determines) policy, Parliament determines (in the absence of specific permissive or PRIMARY LEGISLATION) whether or not the policy becomes law.

governor A mechanical device used to regulate automatically the speed of a machine or appliance by applying variable frictional resistance to rotating or other moving parts.

Gram negative *see* GRAM'S STAIN

Gram positive *see* GRAM'S STAIN

Gram's stain An important microbiological tool first described in 1884 by the eminent bacteriologist H.C.J. Gram. The technique involves the application of a dye to bacteria and the subsequent application of iodine. When treated in this way some bacteria retain (or 'fix') the dye whilst others do not. Those that retain the dye are known as 'Gram positive', while those that do not are termed 'Gram negative'. A long-held hypothesis was that the ability to fix the dye was dependent on the permeability of the outer shell membrane of the bacterium and therefore predicted how susceptible the organism would be to ANTIBIOTICS. Although this has been shown not to be universally infallible, the technique is still valuable for differentiating bacteria and for providing an indicator to assist the early diagnosis and treatment of infections.

granite A generic term applied to a number of igneous PLUTONIC rocks of coarse grain. It is relatively dense and, under normal circumstances, highly durable, although it has limited durability against fire (see IGNEOUS ROCK). Many granite rocks have been found to yield RADON but this is neither an exclusive nor inevitable property.

granuloma A TUMOUR comprising granulation tissue (the tissue comprising masses of new cells suffused with small blood vessels that form, for example, over wounds). A granuloma may be initiated by both infectious and non-infectious agents.

graphite A form of elemental carbon. It occurs naturally or can be made synthetically. It is used in pencils, electrodes and paints, and is incorporated as an additive into specialist lubricants.

gravid Means pregnant or in a condition such as to be able to produce new life.

gravimetric analysis That which seeks to characterise articles or the constituents of compounds by weight.

gravitational circulation A system whereby a fluid circulates within or through a system as a result of different temperatures of the fluid. This creates differential density in the fluid at various parts of the system. The method is used in some heat transfer systems (such as some domestic indirect hot water systems) in which a boiler situated at the lowest part of the system heats water that then expands, thereby reducing its density. This hot water is displaced in the system by more dense water, flowing downwards by gravity, as it loses its heat within the system, creating the circulatory flow.

See also: FORCED CIRCULATION

gray A unit of absorbed energy imparted by IONISING RADIATION. One gray is equivalent to the absorption of 1 joule of energy per kilogram.

Great Britain The name given collectively to those countries of mainland UK comprising England, Scotland and Wales.

See also: BRITAIN; UNITED KINGDOM

green belt A swathe or broad strip of land surrounding an urban area that has been designated as being preserved for either natural or agricultural use and is thereby protected against urban or industrial encroachment.

green chemistry The term used to describe a system for the chemical industry that maximises the environmental considerations of waste elimination/minimisation, sustainability and emission reduction or elimination by the use of alternative technologies and the use of energy efficient processes and technologies. It was defined by the authors P.T. Anastas and J.C. Warner in their book *Green Chemistry – Theory and Practice* (Oxford University Press, 1998) as 'The utilisation of a set of principles that reduces or eliminates the use or generation of hazardous substances in the design, manufacture and application of chemical products.'

Green Data Book A publication of the WORLD BANK that presents the key indicators of the environment and its relationship with mankind under the headings of agriculture, biodiversity, energy, emissions and pollution, water and sanitation, and 'green' activities.

See also: RED DATA BOOKS

green field site An area of undeveloped farm or natural land destined or intended for new buildings for domestic, commercial or industrial usage. Such sites are often both commercially and aesthetically attractive as they do not incur the expense of clearance of existing buildings or contamination, they allow a freer hand for design and layout, and they are frequently surrounded by other green field areas. Once used however such sites are effectively lost to wildlife and open-space amenity in perpetuity.

See also: BROWN FIELD SITE; SUSTAINABLE DEVELOPMENT

green revolution The term applied to the phenomenon of rapid increases in the grain and cereal crop yield experienced by (especially) developing countries in the 1950s and 1960s. The revolution was instigated by the introduction of improved varieties of plants, many of which required significant quantities of additional fertilisers to fulfil their potential. Some environmental campaigners have criticised the green revolution for damaging the environment by the introduction of large quantities of expensive fertilisers, mass irrigation schemes and the promotion of MONOCULTURE, all of which can be seen as potentially damaging to BIODIVERSITY and SUSTAINABLE DEVELOPMENT.

green wood Timber that contains all or most of the moisture that it did as a piece of the living tree. Green timber is liable to distortion or developing cracks upon drying out and should not be used for constructional purposes until it has been properly seasoned.

See also: SEASONING (OF TIMBER)

greenhouse gases A group of around twenty gases in the atmosphere that trap heat from the sun and contribute to GLOBAL WARMING. CARBON DIOXIDE is perhaps most important in this context because, although not the strongest greenhouse gas, volume for volume, the amount in the atmosphere contributes most to the phenomenon. Estimates vary, but most accept that this is over 50 per cent. It is postulated that the increasing burning of fossil fuels (such as COAL, natural gas and oil) and the incineration of BIOMASS are a major contributor to the 'greenhouse' effect. Other gases that have a similar or, in some instances, stronger 'greenhouse' potential include water vapour, methane, CFC-11, CFC-12 (CFC stands for chlorofluorocarbon) and nitrous oxide.

Greenpeace An international organisation that grew from a protest group set up in 1971 to stop nuclear testing by the USA at Amchitka Island, Alaska, through a process of direct action. Since then they have grown into probably the major environmental campaign organisation on earth. They are perhaps most famous for their association with campaigning to stop commercial whaling and harvesting seal pups. They continue to campaign on a whole range of environmental issues including pollution, SUSTAINABLE DEVELOPMENT and disarmament.

See also: EARTHWATCH INSTITUTE; FRIENDS OF THE EARTH (INTERNATIONAL)

grey water A term used principally in the USA to describe predominantly (but not necessarily exclusively) domestic wastewater that does not contain bodily waste. Grey waste is therefore predominantly composed of water from bathing, laundry and dishwashing.

grilse A stage in the life history of the SALMON.

Gross Domestic Product (GDP) The totality of a country's economic output as generated by the resident population per given year.

See also: GROSS NATIONAL PRODUCT

Gross National Product (GNP) A similar concept to GROSS DOMESTIC PRODUCT, the principal difference being that it comprises all the income earned by the citizens of a country, whether or not they reside in that country. It thus includes income earned abroad, provided such income is used solely for the benefit of the residents of the country.

gross output/human capital approach The value of a particular human life expressed in terms of the economic value of future output that would be lost to society if that person were to die.

groundwater Water that has percolated into the earth and flowed along impermeable sub-soil strata either to collect in the ground or to emerge as springs. The permeable strata through which it flows will act as a filter and (generally) reduce the levels of organic material and microflora present. As a consequence groundwater is preferred to SURFACE WATER as a source for obtaining POTABLE water.

See also: CRYPTOSPORIDIUM; DEEP WELL; SHALLOW WELL

grout A generic term given to a range of liquid CEMENT, CONCRETE or MORTAR mixtures that are sufficiently fluid to be able to penetrate existing structures.

Grout is used principally for filling gaps, weatherproofing and decorative purposes.

growth promoter An ANTIBIOTIC used in veterinary medicine to prevent infection with a bacterial agent in order to assist the growth rate of the animal. In other words the antibiotic is administered more as a PROPHYLACTIC than as a treatment for a specific infection. Growth promoters are widely administered to pigs and to poultry, but not to RUMINANTS such as cattle and sheep because of the negative effect they would have on the vital bacterial flora in the latter's digestive systems.

groyne A wall, jetty or other barrier built outwards towards the sea at right angles to the shore for the purpose of reducing erosion.

GSS (Gertsmann–Straussler–Scheinker syndrome) *see* TRANSMISSIBLE SPONGIFORM ENCEPHALOPATHY

gully trap A device in a drainage system in which a vertical 'U' bend or similar is incorporated into a drainage channel such as to fill with water to provide a seal between the pipework on one side of the trap and the other.

gum arabic A sticky exudate from certain types of acacia trees, especially *Acacia senegal* and used in the manufacture of a range of emulsifiers, food thickeners, inks, paints and medications.

gumtree Any tree that yields the sticky substance known as gum. The term is especially, but not exclusively, applied to trees of the eucalyptus family.

gunmetal An alloy containing copper, tin and zinc in proportions of approximately 90:8:2 respectively.

gutting A synonym for evisceration or drawing (see DRAW). It is especially applied to this type of operation in relation to fish.

guy (or guy rope) A rope used to brace, support or steady something else, or to hold it in position.

gybe A term used in sailing to describe the moving of a sail from one side of the boat to another and in doing so to cause the boat to move forwards on a zigzag path.

gymnosperm A seed-bearing plant of the conifer family.

See also: ANGIOSPERM; HARDWOOD; SOFTWOOD

gypsum Hydrated calcium sulphate (chemical symbol $CaSO_4.2H_2O$) used in the production of cement, certain paints and plaster of Paris.

H

HACCP *see* HAZARD ANALYSIS CRITICAL CONTROL POINT

haché In culinary practice refers to food that has been chopped or minced.

haemodialysis The process of DIALYSIS using an artificial kidney to remove nitrogenous waste directly from the blood.

See also: PERITONEAL DIALYSIS

haemoglobin The portion of the blood that produces its red colour. It is made up of a pigment containing iron (haemin) and a protein (globin). Each human red blood cell contains around 600 million molecules of haemoglobin. Its function is to carry oxygen around the bloodstream. When combined with oxygen the haemoglobin molecule is known as oxyhaemoglobin. Haemoglobin will also combine with CARBON MONOXIDE, creating a molecule known as carboxyhaemoglobin. This process is practically irreversible; excess absorption of carbon monoxide therefore reduces the oxygen-carrying capacity of the blood and, in extreme cases, can be fatal.

See also: ANAEMIA

haemolytic That which causes the catastrophic breakdown of blood corpuscles through the action of a chemical poison or toxin, these latter usually being proteinaceous in nature.

haemolytic-uraemic syndrome (HUS) A serious illness that arises from kidney failure associated with problems with the clotting mechanism of the blood. The illness is most commonly reported in children under the age of 4 years. HUS is one of the more serious outcomes associated with infection with ESCHERICHIA COLI 0157.

haemorrhage Any loss of blood from the blood vessels. Haemorrhage can be caused by rupture (for example as a result of HYPERTENSION) or by physical damage and may be external or internal. Haemorrhages are usually classified by the blood vessel involved, including arterial haemorrhage, venous haemorrhage of capillary haemorrhage.

See also: ARTERY; CAPILLARY; VEINS

haemorrhagic colitis A medical term used to describe an inflammation and bleeding of the large intestine. It is usually caused by the invasion of the digestive tract by a pathogenic micro-organism, such as ESCHERICHIA COLI 0157.

ha-ha A landscape feature in which a ditch or sudden drop in levels is introduced to prevent livestock from entering the gardens of a country house without the feature being visible from the main house. It thereby gives the impression that the gardens extend uninterrupted into the countryside.

half bat Half a brick produced by cutting the parent brick across its length.

See also: KING CLOSER; QUEEN CLOSER

half-life The time taken by a radioactive substance for its radioactivity to decay to half its level of its original activity. It is designated by the symbol t.

See also: BIOLOGICAL HALF-LIFE

Halisbury's Laws of England An authoritative reference encyclopaedia of all the legislation pertaining to the United Kingdom.

See also: HANSARD

halocarbon Any one of a group of ANTHROPOGENIC chemicals containing carbon together with one of the HALOGEN group of chemicals. The group includes chemicals such as chlorofluorocarbon and HALONS that have reacted in the atmosphere to destroy OZONE.

halogen Any one of a group of five non-metallic elements comprising astatine, bromine, chlorine, fluorine and iodine. They all have seven electrons in their outer shells, are powerful oxidising agents and all are toxic. They are highly electronegative and reactive, and consequently do not exist in elemental form in nature.

halons Generally defined as being stable gaseous compounds that contain bromine and have long atmospheric lifetimes. They are alternatively defined as any compound produced from a member of the HALOGEN group. They are used in fire extinguishers and contribute to the destruction of OZONE in the STRATOSPHERE.

handicap Often used as a synonym for DISABILITY. More precisely it is used to describe the effect of a disability or similar in reducing an individual's ability to participate in social interactions.

Hansard The name given to the publications detailing the verbatim proceedings of the UNITED KINGDOM Government's sittings.

See also: HALISBURY'S LAWS OF ENGLAND

haploid number The number of chromosomes in the sex cell. It equates to half the normal chromosome complement (the DIPLOID NUMBER) of a normal cell.

hard water That containing a high level of dissolved salts. These are derived from the sub-soil strata through which it has passed; limestone and dolomite are notable for producing hardness in water. The more salts the water contains the 'harder' the water is said to be. As a general rule the more acidic the water, the harder it is likely to become in a given geological location. The harder the water, the more difficult it is to get soap to lather and the greater the risk of causing the FURRING of pipes and utensils.

There are two types of hardness, 'temporary' and 'permanent'. Temporary hardness is caused by the bicarbonates of calcium and magnesium. Permanent hardness is caused by sulphates of calcium and magnesium. Hardness of whatever cause is generally expressed as the equivalent of milligrams per litre of calcium carbonate (mg/l $CaCO_3$). Water with less than 100 mg/l is regarded as soft, while more than this is associated with increasing degrees of hardness. Very hard water would contain more than 300 mg/l $CaCO_3$.

hardboard A construction material formed under pressure and made from sawdust or wood fibre held together with an adhesive resin or similar. It is usually manufactured in thin sheets with a smooth surface on one side and a rough one on the other. Specialist hardboards can be made with decorative finishes, and they can also be produced to be water resistant.

hardness *see* HARD WATER

hardware The physical equipment of a computing system that is required to run the programs.

See also: SOFTWARE

hardwood Wood from any tree other than a CONIFEROUS one. The wood so designated is generally harder than a SOFTWOOD, but this is not invariably so. BALSA WOOD, for example, is one of the physically softest of woods, but is technically designated as a hardwood.

hazard Essentially anything with the potential to cause harm. For environmental health purposes it is most frequently used in relation to health and safety or to food. In terms of the latter the definition was ascribed by the Codex Alimentarius Commission in 1998, and is most frequently quoted as being 'a biological, chemical or physical agent in, or condition of, food with the potential to cause an adverse health effect'.

See also: RISK

Hazard Analysis Critical Control Point (HACCP) A system of QUALITY ASSURANCE that relies on the identification and control of HAZARD and RISK; specifically at those points in the system at which they can be eliminated or reduced to acceptable levels (known as CRITICAL CONTROL POINTS). The system was originally developed in the USA in the 1960s as a way of ensuring the safety of food used by astronauts in the US space programme.

The system has been adapted to many applications over the years, including engineering and manufacturing. It is most commonly encountered as 'HACCP' in relation to food safety, in which it is applied as a highly structured system demanding precise analysis and control mechanisms for implementation and operation. In 1997 the Campden and Chorleywood Research Association in the UK proposed seven principles whereby HACCP could be derived in practice; in 1998 the Codex Alimentarius Commission endorsed these principles, which are reproduced below:

Principle 1	Conduct a hazard analysis. Prepare a flow diagram of the steps in the process. Identify and list the hazards and specify control measures.
Principle 2	Determine the critical control points. A decision tree can be used.
Principle 3	Establish critical limit(s) that must be met to ensure that each CCP is under control.
Principle 4	Establish a system to monitor control of the CCP by schedule testing or observations.
Principle 5	Establish the corrective action to be taken when monitoring indicates that a particular CCP is not under control or is moving out of control.
Principle 6	Establish procedures for verification to confirm that HACCP is working effectively, which may include appropriate supplementary tests, together with a review.
Principle 7	Establish documentation concerning all procedures and records appropriate to these principles and their application.

Adherence to HACCP principles should ensure that a product is safe and suitable for use, provided the hazards and risks have been identified correctly and action taken to control them. HACCP should not be considered a stand-alone system as producers and manufacturers need to implement GOOD MANUFACTURING PRACTICE as well.

hazard assessment An evaluation of the intrinsic potential of something to cause harm to humans, animals or the environment. It is not an assessment of the likelihood that harm will be caused.

See also: RISK

hazard profile The cumulative description of the physical, chemical and interactive properties of a substance or phenomenon in relation to its capacity to cause harm to humans, animals or the environment. A complete profile will include details of toxicity, bioaccumulation and environmental persistence.

hazardous waste Considered by the European Union to be that waste which is the most harmful to either humans or to the environment.

haze A suspension of microscopic particles in air.

head When referring to a specified point in a liquid, means the depth of liquid bearing down on that particular point.

header A unit material of construction such as a brick, block or stone incorporated into a wall in which the length of the unit runs at right angles to the wall surface, thereby exposing only the end of the unit on the external surface. Headers are usually incorporated into solid wall constructions in conjunction with STRETCHERS in order to improve structural strength. Headers are not incorporated into CAVITY WALL construction, as they would bridge the cavity; instead their binding together role is achieved by using WALL TIES in this type of construction.

headspace Volume within a closed container between the top surface of the contents and the inner surface above it of the container. The term is used in the canning industry to describe the 'airspace' between the top level of the product and the inside of the can lid.

health Defined by the World Health Organization as follows: 'Health is a state of complete physical, mental and social well-being and not merely the absence of disease or infirmity.'

Health Impact Assessment An evaluation of the impact of a (proposed) development on health (especially in respect of Government policy), usually but not necessarily exclusively confined to that of the local population. Ideally the evaluation should include economic and social impacts as well as direct health links. However, for many variables the impact is likely to be unknown and the interrelationships between various factors may well be imperfect. This makes health impact assessment less than precise in many instances, but the simple fact that health should be considered as a factor at all is often an improvement over previous arrangements.

health inequality Generally taken as referring to any differential state of health between two or more identifiable individuals or groups. The term is usually used to describe differential health states in identifiable and defined populations that occur as a result of DEPRIVATION. In this context deprivation relates to the differential status between the groups in respect of many factors including economic, social and financial capability and capacity together with psychosocial components such as stress.

health outcome The resultant health status accruing from an INTERVENTION.

heart attack *see* CORONARY HEART DISEASE; MYOCARDIAL INFARCTION

heavy metal An imprecise term used usually to describe metallic elements whose ions are environmentally persistent and potentially toxic. Some sources define the term as referring to any metal that has a density in excess of 5 grams/cm^3.

heavy water The name applied to water containing DEUTERIUM rather than hydrogen in combination with oxygen.

hecatomb Originally an ancient Roman or Greek feast in which 100 oxen were sacrificed. The term has come to mean any great sacrifice.

hectare A metric unit of area. It is equivalent to 100 ares, 10,000 m^2 or 2.471 ACRES.

hedge laying A country craft in which the upright branches in a hedge are partially cut through, then laid and fastened almost horizontally along the line of the hedge in order to create a thicker, stronger and more impregnable barrier.

helminths Members of a class of parasitic worms. The classification includes tapeworms, roundworms and flukes.

hemicellulose A compound that comprises those polysaccharides that are associated with CELLULOSE and LIGNIN in the cell walls of green plants. Hemicellulose differs from true cellulose in that it is soluble in alkalis.

hepatic Means of, or pertaining to, the liver.

hepatitis Literally any inflammation of the liver. More commonly the term is used to describe the disease 'hepatitis' as being one caused by one of several viruses; more correctly the disease should therefore be referred to as 'viral hepatitis' to distinguish it from any general, non-specific, causes of inflammation. Hepatitis viruses can be divided into two distinct groups, being those that are enterically transmitted viruses (hepatitis A virus and hepatitis E virus) and those that are blood-borne (hepatitis B, C, D and G). It should be noted that there is no virus known as hepatitis F – the original identification of such a virus is now considered to be flawed, but the designation 'F, having already been used, is effectively deleted from the nomenclature.

Hepatitis A virus is the causative agent of the disease now known as hepatitis A (previously known as infectious jaundice). Globally hepatitis B virus affects more people, and generally leads to more deaths, than any other hepatitis strain. It is present in the blood, semen, vaginal secretions, saliva and skin lesions of people who are infected. It is estimated that over 300 million people worldwide are chronically infected with hepatitis B. The virus is not exclusively sexually transmitted and blood transfusion, contaminated needles and maternal transmission have all been recorded.

Hepatitis C virus is thought to be exclusively a human virus, although chimpanzees have been infected experimentally. A specific test for HCV only became available in 1989. Transmission from infected people is of major public health concern and infection rates in health care workers are especially worrying. Hepatitis D virus is also known as delta agent. It occurs only in the presence of hepatitis B infection. Hepatitis E virus is enterically transmitted. The virus usually occurs in those areas of the world with poor sanitation.

hepatotoxic Means having the potential to cause toxic (see TOXIN) damage to the liver.

herbaceous Refers to those plants that do not form wood. The term is also sometimes applied to the non-woody tissues of plants.

herbicide A chemical substance that kills plants and vegetable life. Herbicides may be general or specific, natural or artificial.

herbivorous Means deriving sustenance from a diet consisting exclusively (or predominantly) of plant or vegetable material.

See also: CARNIVOROUS; OMNIVOROUS

herd immunity The phrase used to describe the overall level of resistance in a population (human or animal) to a specified infectious agent based on the proportion of individuals resistant to the infection. The level of herd immunity is an important determinant of the possibility of an EPIDEMIC occurring should the infection gain access to the community. Such a level is not the same for every disease as it depends on factors such as transmission characteristics and environmental conditions.

hermaphrodite An animal or plant that possesses both male and female reproductive organs.

hermetic Means airtight. A hermetic seal is one that will therefore prevent ingress or egress of air. It will also, importantly in some applications, exclude microorganisms.

herpes simplex An acute infectious viral disease characterised by the manifestation of vesicles on the skin. The vesicles can appear anywhere on the skin or mucous membranes but if they appear on the mouth or lips they are commonly referred to as 'cold sores'. Sexual transmission is possible. Infection is for life and although recurrent the vesicles in adults usually resolve within around 10 days. In very young infants a generalised infection can prove to be fatal. Medication is available to prevent or reduce the effect of the vesicles provided it is applied to the potentially affected area of skin before they erupt.

herpes zoster Also known as 'shingles', an acute eruption of minute yellow vesicles on the skin caused by the virus responsible for chickenpox. It derives its name from the Greek for a girdle in recognition of the symptoms that characteristically appear as a circlet covering half the area of the chest, although other areas of the body can also be affected.

herpetofauna A collective term referring to amphibians and reptiles.

herringbone drain A wastewater disposal system in which a number of blank-ended side drains with open joints run at an angle to a main spine in the fashion of the main bones of a fish. The object is to distribute the water over as wide an area and as speedily as possible.

hertz A unit of measuring FREQUENCY, it is designated by the symbol Hz. One hertz is equivalent to one cycle per second.

heterologous protection A type of IMMUNITY in which an individual is resistant to one strain of a particular PATHOGEN but not to other strains of the same pathogen.

heterotrophic In respect of living organisms, means one that derives sustenance or nutrients directly from the environment.

heuristic The term applied to describe a method of problem solving through the exploration of a set of possibilities rather than by following a specific set of rules. Teaching methods that are heuristic are those that allow the pupil or student to discover or learn things for themselves rather than through being specifically taught.

See also: ALGORITHM

hibernation The phenomenon in which some creatures reduce body activity and enter a dormant state in order to survive the winter months.

See also: AESTIVATION

high-acid food Generally accepted as describing a food product with a PH of 4.5 or above.

See also: LOW-ACID FOOD

High Court A more senior court (i.e. above the COUNTY COURT) for hearing civil matters in the UNITED KINGDOM. The High Court is divided into three Divisions, being the Queen's Bench Division, Family Division and Chancery Division. The High Court also hears appeals from the County Court. Appeals on points of law are referred to the Queen's Bench Division. Appeals against conviction or sentencing are referred to the Court of Appeal (Criminal Division).

See also: COURTS

high production volume (HPV) A term used in relation to chemicals to describe one that is produced or imported into a particular country in quantities over a specific level. Different regimes use different quantities as the defining quantity. The OECD defines this quantity as at, or in excess of, 1,000 tonnes per year, while the US Environmental Protection Agency defines the quantity as in excess of 1 million lbs. per year (i.e. around 444 tonnes).

high seas The term generally applied to describe the totality of fishable waters lying outside all countries' territorial fishing limits.

hip roof One with four sloping portions falling away from each other, normally with one slope facing each quarter of the compass. The hips are those non-horizontal junctions of a hip roof at which two of the sloping portions meet.

See also: GABLE; GAMBREL ROOF; RIDGE

histamine Histamine (chemical symbol $C_5H_9N_3$) is an amine formed in the body from HISTIDINE. Its release in response to the presence of a foreign substance can lead to an allergic reaction.

See also: SCOMBROTOXIN POISONING

histidine A non-essential amino acid that occurs in many proteins.

histogram A graphical representation of statistical variations of frequency and quantity by portraying these as rectangles of size appropriate to the variations in each of a number of defined bandwidths.

See also: BAR CHART; GANTT CHART

histopathological Refers or pertains to those changes in the structure of a tissue that are brought about by the action of disease or physical damage.

HIV or Human Immunodeficiency Virus A ribonucleic acid (RNA) retrovirus responsible for the condition known as AIDS.

homeopathic In relation to a treatment for disease, refers to one that would induce the same symptoms in a healthy person as the disease is producing in the person affected by the disease. For example a medicine that raised body temperature and was administered in response to a fever could be considered to be homeopathic. The philosophy of the treatment is to induce or promote the body's own immune system to fight the infection or produce a cure. Dr Samuel Hahnemann developed the system of homeopathy in the 18th century.

See also: ALLOPATHIC; ALTERNATIVE TREATMENT

homeworker A person who is employed to use their home as their place of work. Traditionally many homeworkers have undertaken relatively simple manual tasks of manufacture or assembly or carried out clerical tasks such as addressing envelopes. They were generally paid on a piecework basis at the rate of a fixed sum per number of completed items. Although not all homeworkers fit into this category the absence of an easily identifiable or registered place of work has made identification by regulatory authorities difficult and resulted in considerable exploitation. This is especially so since homework often appeals to parents with young children whose opportunity for other employment is limited.

homogenous Also homogeneous; means having the same properties or substance throughout the whole. Homogenised milk is that in which the cream and the milk are mixed together such as to ensure that the two parts do not separate out again and that the composition of the bulked whole is therefore the same from wherever a sample may be taken.

homologous In biology refers to having or possessing similar structures or locations, but performing different functions. In genetics the term is usually used to describe two genes that are evolutionary related.

See also: ANALOGOUS

horizontal Means level. It is applied in medical practice as horizontal transmission and means the transmission of a disease between people, either directly or indirectly. In legislature it is often applied to Directives from the EUROPEAN UNION (EU) as in the term horizontal (food) Directive, in which case it means a Directive applicable across a range of foodstuffs, but at a particular stage in the food chain (e.g. at production, wholesale or retail level).

See also: VERTICAL

horizontal gene transmission (HGT) The process, similar in operation to that of an infection, whereby genetic material is transferred from one individual to another. The term distinguishes the process from the normal transfer of genetic material from parent to offspring (in a process known as 'vertical gene transmission'). Horizontal gene transmission occurs naturally between micro-organisms such as bacteria (see BACTERIUM) (either using viruses, TRANSPOSONS or PLASMIDS), such as in the process of transmitting ANTIBIOTIC RESISTANCE. The technique is also used in genetic modification to transfer genetic material between two individuals.

hormone A chemical substance in animals secreted by an ENDOCRINE GLAND. In plants the term refers to a chemical substance associated with plant growth. Hormones are used to regulate metabolic activity.

See also: METABOLISM

hormone replacement therapy (HRT) The medical application of synthetic versions of the female hormones progesterone and oestrogen that are produced naturally. The treatment is used principally to relieve symptoms associated with the menopause, after certain surgical procedures or for post-menopausal women who are considered to be at risk of developing osteoporosis.

horripilation The contraction of the cutaneous muscles such as to raise the hairs of the skin and to cause goosebumps. This is an automatic response to cold and to fear, probably in early evolution to trap air next to the skin to act as insulation, in the case of the former, or to make one appear bigger to an aggressor, in the case of the latter.

hors d'œuvres Dishes of savouries, olives, sardines, etc. used to whet the appetite before the main meal.

horsepower A measure of the power output of an engine. Unlike BRAKE HORSE-

POWER it is usually measured directly from the engine before any transmission losses occur. One horsepower is equivalent to 550 foot-pounds per second (33,000 foot-pounds per minute), i.e. 745.7 watts.

host An organism upon or within which another organism lives as a PARASITE being dependent upon it for sustenance, habitation or protection. The host can remain in this condition for the life of the parasite, can perform this function for a part of the parasite's life or can act as a VECTOR for passing the parasite or its offspring on to others.

See also: DEFINITIVE HOST; INTERMEDIATE HOST

hot desk working A system of maximising the utilisation of office space by discontinuing the allocation of available desk space to individuals and sharing that which is available amongst a number of workers. The system can either allocate individual desks to specified individuals (most common in shift systems) or allow any worker to use any desk if it is free (more common where workers are transitory in the workplace, such as for travelling representatives). The system can have disadvantages in that perceived 'ownership' of space and equipment is reduced or eliminated and damage and maintenance charges per desk can consequently increase.

House of Commons The legislative assembly of the UNITED KINGDOM. At the time of writing (AD 2002) it consists of 659 Members of Parliament (MPs) elected by simple majority vote in a general election that must occur at a minimum of every five years.

House of Lords A part of the UNITED KINGDOM legislature and the most senior of the United Kingdom's appeal courts. The legal appeal function is conducted entirely separately from its legislative role – cases of appeal are heard by up to thirteen senior judges, who together comprise the LAW LORDS.

See also: HOUSE OF COMMONS

household waste That which would normally be expected to be produced by a household as a result of the normal occupation of a dwelling and would include kitchen waste, packaging, discarded household items, garden waste and waste DIY materials.

hub height wind speed The air velocity as measured at the height of the hub (i.e. the centre of the rotor) of a wind turbine generator. If the generator is in operation an adjustment to the measured air velocity is necessary to obtain the true velocity, as an operational rotor will slow the air down.

Human Immunodeficiency Virus *see* HIV

humectant A substance that is capable of absorbing moisture.

See also: DELIQUESCENT

humidifier A piece of equipment designed to release additional water vapour into an atmosphere and thereby raise the level of HUMIDITY.

See also: SONIFICATION

humidistat A sensing device incorporated into an air conditioning or management system to measure the RELATIVE HUMIDITY and to signal the need for remedial action to be taken if the reading extends beyond defined limits.

humidity A measure of the amount of water vapour contained within a given volume of air at any particular time.

See also: ABSOLUTE HUMIDITY; AIR CONDITIONING; RELATIVE HUMIDITY

humus Partially decomposed organic (usually vegetable) material of dark brown or black appearance. It is used to improve fertility and water retention in soil.

hurricane A strong wind defined by the World Meteorological Organization as registering force 12 on the BEAUFORT SCALE, i.e. having a wind speed equal to or in excess of 64 knots (120 km/hr).

HUS *see* HAEMOLYTIC-URAEMIC SYNDROME

hybrid The term used to describe the first-generation progeny resulting from the artificial and controlled cross-fertilisation between two parent organisms that differ in one or more genes.

hydatid disease Also known a hydatidosis. The disease is caused by infection with the cestode helminth parasite *Echinococcus granulosus* (the dog tapeworm). The adult tapeworm is around 3 to 9 mm in length and lives in the intestines. Its body consists of segments called proglotids (three to four in number), the terminal segment of which becomes gravid and is voided in the faeces. In the natural life cycle, the eggs pass on to pastureland where they infect sheep, cattle, horses and pigs. Eating contaminated vegetables has infected humans. It is also possible that flies may also carry the eggs to uncovered food or they could be inhaled in dust.

Following ingestion by the INTERMEDIATE HOST (i.e. the host in which the immature cysts are formed) the egg hatches in the intestine and the oncosphere burrows through the intestinal wall to enter the hepatic portal system where it is transported to the liver. The oncospheres may lodge in the liver or pass through to lodge in the lungs, brain or other internal organ, at which point the oncosphere encysts to become what is known as a hydatid cyst. The cyst is essentially a bag of fluid consisting of a germinal layer within which brood capsules develop. The hydatid cyst is capable of producing infection in the DEFINITIVE HOST (i.e. the host in which the adult tapeworm lives) after around five or six months. When the intermediate host dies it may be scavenged by dogs, and if the viable hydatid cyst is eaten then the life cycle begins again. If the eggs of the tapeworm are eaten by other species the eggs will hatch and create cysts in the alternative intermediate host; humans are capable of filling this role.

hydrant A supply point in a main supply that can be used to extract or tap the contents. The term is used especially in relation to a mains water supply in which a vertical pipe equipped with a valve can be used to provide water in emergency situations (such as in fire fighting – then described as a 'fire hydrant') or as a purging outlet for flushing the system.

hydrated Means having undergone the addition of water such as to bind the water molecules to the chemical compound to which it is added.

hydrated lime Hydrated lime (chemical symbol $Ca(OH)_2$) is made by adding water to QUICKLIME (chemical symbol CaO). It was used in the construction industry in the manufacture of lime mortar. Hydrated lime is also known as 'slaked lime'.

hydraulic cement CEMENT that will harden under water.

See also: POZZOLANIC

hydrocarbon An ORGANIC compound containing carbon and hydrogen atoms bound together in either linear or ring formation. Many hydrocarbons are used as fuels. When combusted in the presence of oxygen they release CARBON DIOXIDE and water vapour.

See also: GREENHOUSE GASES

hydrocephalus An increase in the size of the upper portion of the head caused by an abnormal accumulation of fluid within the cranial cavity. The causes of hydrocephalus are various and many are congenital in origin.

hydroelectric power Electricity produced by the action of water falling under the power of gravity to drive a generating turbine or similar. The electricity thus generated is known as hydroelectricity. Hydroelectricity is produced either directly for use in an electricity system or produced at times of high demand from water pumped to higher levels by excess electricity generated at times of low demand.

hydrogen swell The distortion or bursting of a can due to the build-up of hydrogen gas from within generated by the activity of the acid contents on the (usually mild steel) material of the can itself. It is usual for such cans to be coated with a lacquer (previously tin, hence the name); if the lacquer is inadequately applied or damaged by physical action the lacquer coating can become breached allowing the contents directly to attack the can structure. Hydrogen swell is indicative of food contents that are unfit for human consumption. Dependent on can composition hydrogen swell could also indicate contamination of the contents with metallic salts.

hydrogenous Means containing or relating to hydrogen.

hydrolysis A chemical reaction in which a compound undergoes chemical change in the presence of water to produce other compounds, such as the change in a salt to form an acid and a base.

hydrometer A scientific instrument used to measure the density of a liquid.

See also: HYGROMETER

hydrophilic Means having an affinity for water.

hydrophobic Means having an aversion to water. Hydrophobia is an old name for RABIES, the name deriving from the aversion to water exhibited by sufferers.

hydroscopic Means capable of taking pictures or forming images underwater. A hydroscope is an instrument used for making observations underwater.

hydrostatic pressure The specific pressure at any given point in a liquid that is at rest. It is a function of the density of the liquid and the HEAD of liquid at that particular point.

hydrotropic Refers to the directional growth of plants (usually of the roots) towards water.

hydrous Means containing water or, in respect of a chemical compound, having water combined or incorporated within the molecular structure.

See also: HYDRATED

hygrometer Also PSYCHROMETER, a scientific instrument used to measure RELATIVE HUMIDITY.

See also: HYDROMETER

hygroscopic Means capable of absorbing, or tending to absorb, moisture from the air.

See also: HYDROSCOPIC

hyperplasia An increase in the normal size of a living organ or tissue attributable to an increase in the total number of cells that the structure contains.

See also: HYPERTROPHY

hypersensitivity An abnormal allergic reaction produced by the body's immune system to re-exposure to a particular ANTIGEN.

See also: ALLERGY

hypertension The medical term for high blood pressure.

See also: HYPOTENSION

hypertrophy An increase in the size of an organ or tissue, or the cells of which they are comprised. Hypertrophy is usually caused as a result of an increased demand placed upon the affected tissue, such as the thickening of the heart manifest as a result of thickening of the arteries necessitating greater pressure to force blood through the circulatory system.

See also: HYPERPLASIA

hypocaust An ancient Roman system for heating buildings in which hot air from a fire was directed under the floor and between double walls.

hypochondria Or hypochondriasis, a chronic medical condition in which the sufferer exhibits a heightened and abnormal anxiety about their health, especially in the absence of any clinical symptoms.

hypodermic Means of, or pertaining to, that layer of skin below the EPIDERMIS or to the region immediately below the skin.

hypodermic syringe A medical instrument comprising a cylinder and plunger to which a fine hollow needle is attached and through which drugs or therapeutic medication is administered by way of an injection below the skin. Any breach of the skin is considered to carry a risk. Injections are therefore usually restricted to circumstances in which the medication via oral administration would be inactivated in the stomach (e.g. insulin), where the patient might vomit up any medication or where it is necessary to get the medication speedily into the body.

hypodermis The layer of thick-walled cells found in some plants below the outer epidermis. The cells have the function of adding support to the plant and/or carrying water.

hypotension The medical term for low blood pressure.

See also: HYPERTENSION

hypotenuse The longest side of a right-angled triangle, neither of whose angles forms a right angle (i.e. 90°).

hypothesis A statement made at the start of an epidemiological study about which evidence will be gathered to test (i.e. attempt to prove or disprove) the contention that is the hypothesis. The use of a hypothesis narrows the focus of the study, enabling better study design and more focused research.

hypoxia The medical term used to describe a low blood concentration of oxygen. It may be caused by disease of the heart or lungs, or by abnormal respiration.

Hz The symbol for HERTZ.

I

IARC The acronym for the International Agency for Research on Cancer. It is a World Health Organization body commissioned to conduct and co-ordinate research on human cancers, including carcinogenicity and the development of strategies for cancer control.

iatrogenic disease One in which the physician (accidentally or inadvertently) has introduced an infectious or causal agent to the patient. The term iatrogenic comes from the Greek and a literal translation means 'physician born'. The definition includes disease caused by drugs and agents responsible for, for example, hepatitis, HIV and vCJD.

See also: CJD

ICD(N) *see* INTERNATIONAL CLASSIFICATION OF DISEASE

Ice Age A description of the periodic fluctuations in the earth's climate in which glaciers extend from the polar regions to cover large areas of the planet's surface. There is no precise definition of the term since, even within a period of high glaciation, there are fluctuations in their extent. The most recent Ice Age was really a series of events that began 2.5 million years ago and ended around 10,000 years ago. The triggers for the commencement or termination of an Ice Age are subject to debate and many theories are controversial. Hypotheses include fluctuations in sunspot and volcanic activity.

igneous A term used to describe something derived from fire or of a fiery nature.

See also: IGNEOUS ROCK

igneous rock That which is generated as a result of the solidification of volcanic magma.

See also: METAMORPHIC ROCK; SEDIMENTARY ROCK

illuviation The process of the deposition of fine particles such as those of clay and organic matter originally leached (see LEACHATE) from the upper layers of soil by the action of water. The process of removal is known as eluviation.

imago The adult stage in the life cycle of an insect.

See also: LARVA; PUPA

immiscible Refers to two liquids and means that they are incapable of being mixed together to form a HOMOGENOUS whole. Oil and water are usually quoted as being the classical immiscible substances. Those liquids that are capable of being mixed together are known as MISCIBLE.

immunisation The process whereby the body develops a resistance or protection against certain diseases or toxic agents. There are two main classifications of immunity – natural and artificial. Natural immunity can be sub-divided into two categories, inherent and acquired. Inherent immunity is that which is intrinsic within a species; in other words the entire species is not susceptible to infections or poisons that affect other species. Acquired immunity is that which derives from the body surviving an attack by a certain pathogenic organism. Immunity derived in this way may last for varying times; sometimes the immunity is acquired for life.

Artificial immunity can also be sub-divided into two categories, active and passive. Active immunity is achieved by the artificial introduction into the body of a non-lethal dose of a pathogenic organism or toxic substance sufficient for the body to produce an immune reaction. Such doses can be administered orally or by VACCINATION. For some organisms or substances it is necessary for several introductions to be made. For other organisms a single administration might be sufficient for immunity to be acquired.

The dose administered to confer active artificial immunity must be carefully controlled to avoid serious harm. This can be achieved in one of several ways. In the first case the dose might either be of limited quantity or be diluted. Alternatively one type of organism can be used to develop immunity to another, or the organism can be attenuated to reduce its pathogenicity. Finally artificial immunity can be acquired in some cases by the administration of deactivated organism or toxin.

Passive immunity is achieved by the artificial administration of blood serum within which the agents of immunity are present. Resistance is conferred simply by the presence of resistance agents from elsewhere. Immunity in this case will only last so long as these agents are present in the bloodstream; the body does not develop a capacity to produce these agents for itself.

immunity The phenomenon whereby an individual is protected from the effects of infectious agents or poisonous substances that would otherwise result in illness. Immunity may be inherent within the individual or be developed or acquired during life in response to infection or artificial IMMUNISATION. Immunity may be partial or complete and may last for a short period or for life.

immunoassay Any form of analysis that uses the specific reaction between an ANTIBODY and an ANTIGEN for the purposes of differentiation.

immunocompetent Refers to an individual whose body is able to mount a normal immune response.

See also: IMMUNOCOMPROMISED

immunocompromised A condition in which an individual's immune system is unable to mount a normal response to the threat of infection. The condition can be a natural phenomenon, brought about by infection or induced during surgical procedures (such as transplant) in order to suppress the possibility of rejection of the transplanted organ.

See also: HIV; IMMUNOCOMPETENT

immunoglobin One of a group of proteins that are ANTIBODIES. Their action is very specific and usually limited to a single ANTIGEN.

immunological test An analytical test that is based on the reaction between an ANTIBODY and an ANTIGEN.

See also: IMMUNOASSAY

immunomagnetic separation An analytical technique used for isolating a specific micro-organism. It works through the use of magnetic beads coated with the ANTIBODY to the organism being sought, against which the organisms form clumps.

immunosuppression The term usually used to describe the intentional medical intervention, especially through the use of drugs, of preventing the normal operation of the body's immune system in order to aid the acceptance of transplanted material.

immunotherapy A medical intervention intended to benefit a patient by the attenuation of the body's immune response mechanisms. Immunotherapy includes stimulating the immune system such as to attack pathogenic microorganisms and suppressing it such as to aid the acceptance of transplanted organs or other material.

imprest A cash fund or advance held by an individual or department for paying incidental expenses and that, following expenditure, is topped up periodically to a predetermined level.

in situ A Latin term meaning 'in the original situation'. It generally refers to an act of examination, maintenance or similar conducted without removing the object of the exercise from its location or environment.

in vitro A Latin term meaning literally 'in a glass'. It is applied to experimentation on living cells conducted outside the body, for example in petri dishes etc.

See also: CELL CULTURE; IN VIVO

in vivo A Latin term describing activity or observation that takes place within the living body.

See also: IN VITRO

incidence The number of new cases of a disease, usually expressed as a RATE, occurring in a defined population within a specified time frame.

See also: PREVALENCE

incidence rate *see* ATTACK RATE

incineration The process of burning waste under controlled conditions to either reduce its bulk or denature toxic or hazardous characteristics. The term usually refers to the process of direct incineration, in which the calorific value of the waste itself is utilised during burning. Direct incineration is often associated with production of heat or power. Variations of the incineration process are being developed in which the objective of heating the waste is not its direct incineration.

See also: ENERGY FROM WASTE; GASIFICATION; PYROLYSIS

income elasticity The proportionate change in an individual's lifestyle and/or health status produced by a proportionate change in that person's income.

incubation period That time between the initial invasion of a susceptible host by a PATHOGENIC organism and the first appearance of the clinical symptoms associated with the infection. In the majority of cases the incubation period is fairly consistent within a defined time scale, and knowledge of the organism, together with knowledge of the onset of symptoms, can be used as a retrospective indicator as to when the initial infection occurred. This is particularly useful in CONTACT tracing and in ascertaining when potentially contaminated food was ingested in cases of suspected food poisoning.

See also: LATENCY PERIOD

index case The first case identified in an outbreak of infectious disease.

See also: MARKER ORGANISM

index organism *see* MARKER ORGANISM

Indian club A bottle-shaped club, usually around 2 feet in length, used by jugglers and in fitness training. For the latter, they are held in either hand and rotated or swung around the body in limbering-up and suppleness exercises.

indicator organism In microbiology, a (generally) non-pathogenic organism. The detection of this can be used to suggest that there has, at some stage, been the potential for that which is being examined to have been contaminated by a pathogenic organism. The presence of *ESCHERICHIA COLI* in foodstuffs, for example, is often used as an indicator of potential faecal or sewage contamination.

See also: MARKER ORGANISM

indigenous Means of, from or produced by a defined and discrete geographical area.

industrial waste A classification of waste used in the UK to describe that waste which arises solely as a result of industrial activity (possibly excluding waste from mines and quarries). It is distinct from commercial or domestic waste. Its principal distinguishing feature is that it is more likely to contain substances that are hazardous to wildlife, humans and/or the environment, and might need special processing or handling before disposal.

inert In the chemical sense means unreactive or having only limited capacity to react.

inert gases Also called 'noble gases', any of the gaseous elements, being argon, helium, krypton, neon, radon and xenon. Gases such as CARBON DIOXIDE and nitrogen are sometimes called noble gases since they are non-oxidising and do not support combustion.

infant An individual under the age of 1 year.

infant botulism An infection, first described in 1976 in the USA, caused by the spore-forming bacterium *Clostridium botulinum*. It differs from true BOTULISM in that illness follows direct ingestion of the spores rather than the pre-formed toxin. The spores are unable to germinate in those over the age of around 12 months, so the illness is unable to develop in this group. Although infant botulism is generally a mild infection, with symptoms typically of constipation, weakness and neurological disorders, it can result in lengthy hospitalisation.

infant mortality The number of deaths of children under 1 year of age in a defined population expressed in terms of the number of live births in that population during the same period. It is usually expressed as a rate per 1,000 live births and is taken as a measure of the overall status and health capacity of a population. Infant mortality ranges from below 5 per 1,000 in developed countries to over 150 per 1,000 in deprived populations in the least developed countries.

infantile paralysis An old name for POLIOMYELITIS (also known as 'polio').

infarct A localised area of necrotic (dead) tissue.

See also: INFARCTION

infarction A change in an organ of the body brought about as a result of a blockage of an ARTERY. The reduced blood supply causes the affected area of tissue to become dense and (as a result of the blood supply pattern) wedge shaped.

See also: ARTERIOSCLEROSIS; CORONARY HEART DISEASE; STROKE

infection The process whereby a body becomes infected by a PATHOGENIC organism. Diseases transmitted in this way are known as infectious or communicable diseases. The descriptors 'infectious' and 'communicable' when applied to disease are effectively interchangeable in common parlance, although since the 1950s the scientific establishment has fluctuated in preference from one to the other over time.

infectious agent Any agent, such as a bacterium or virus, which, when introduced to a host, is capable of multiplication within that host. It is not necessary for an agent to be described as infectious for it to produce symptoms of a disease.

See also: ASYMPTOMATIC

infectious disease Any disease that is caused by an INFECTIOUS AGENT.

infectious dose The number of bacteria, viruses or similar that is necessary to produce an INFECTION.

inflammable Means in a condition of being ready to, or capable of being able to, catch fire. It is synonymous with flammable.

influenza Commonly known as 'flu'. It is an acute viral disease of the respiratory tract. Typical symptoms include fever, headache, myalgia (muscular pain), sore throat, cough and malaise. Gastrointestinal symptoms, such as diarrhoea and vomiting, may also be present. Influenza is often confused (especially by sufferers) with the common cold. The common cold is generally less severe than true influenza, although laboratory diagnosis may be necessary in sporadic cases to provide a differential diagnosis. The major significance of influenza as a human pathogen lies in its ability to produce new forms of the virus and thereby to exploit new situations in which immunity to a particular strain has not yet been developed. If the new strain is particularly virulent there is the possibility of an EPIDEMIC or even PANDEMIC. The elderly and immunocompromised are especially vulnerable.

information bias A form of SYSTEMATIC ERROR in an epidemiological study or similar in which incorrect data has been recorded and input to the calculations. There are several sorts of information BIAS. The most common are misclassification due to incorrect diagnosis (identification bias), errors in data entry (transcription bias) and inaccurate recall of events or symptoms (recall bias).

informed consent The voluntary agreement of an individual to the performance of an action or undertaking, or their participation in a programme, on the basis that they fully appreciate any risks or benefits that may accrue together with any other consequences that might follow as a result of their agreement. These latter might include publicity, financial implications and time commitments.

infrasound The sound of a frequency below the audible range of the human ear. This is generally taken as around 20 Hz (see HERTZ).

infrared radiation ELECTROMAGNETIC RADIATION with a WAVELENGTH between that of light and RADIOFREQUENCY RADIATION. It is capable of causing the sensation of heat in a receiving person. There are three sub-bands of the broader infrared frequency designated as A, B and C respectively.

infusing The extraction of a flavour from a spice or other food ingredient by steeping it in a liquid, such as the extraction of flavour from the dried leaves of the tea plant by immersing it in boiling water.

inhalation The stage in the respiration cycle in which air is taken in to the lungs. In a normal respiratory cycle inhalation takes longer than the exhalation phase; if this pattern is reversed it might suggest the individual is suffering from a pathological condition such as asthma.

injunction A legally binding order from a court directing a person, group or body to a particular course of action. Injunctions may either prohibit an action or activity or be a requirement to undertake an action or activity.

inoculation A procedure whereby infective or toxic material is introduced into the body by way of a small wound or puncture made in the skin or mucous membrane. The procedure is undertaken artificially to facilitate the introduction of a VACCINE to the body.

insecticide A chemical substance that kills insects.

insulin A HORMONE secreted by the pancreas. Its function is in enabling the muscles of the body to take up sugar (GLUCOSE) from the blood. It is nowadays produced artificially, predominantly for the treatment of DIABETES.

integer A whole number, i.e. it is not a fraction.

integrated pest management (IPM) An internationally recognised system that is being developed to reduce the use of broad-spectrum herbicides and pesticides by trying to create ecologically integrated crop and livestock management systems. The system does not seek to ban the use of chemicals in agriculture but does try to find alternatives – the philosophy being that ideally chemicals should only be used if more environmentally friendly control and management techniques fail to keep pests below levels that create unacceptable damage.

See also: ORGANIC FOOD; SUSTAINABLE DEVELOPMENT

Integrated Pollution Control (IPC) A system established in Great Britain to control pollution from the most potentially polluting or complex industrial processes. The objectives of the system are to use BATNEEC to prevent or minimise emissions, to render harmless any potentially damaging substances that are released and to consider the effects of release on the environment as a whole.

See also: LOCAL AIR POLLUTION CONTROL

Integrated Pollution Prevention and Control (IPPC) A system for considering, at the planning stage, how a proposed development might affect the environment and how it might impact on human health. The system requires regulatory authorities, in consultation as deemed necessary, to ensure, by the imposition of conditions, that a high level of protection of the environment as a whole is achieved and that appropriate preventive measures are taken to ensure that no significant pollution from either routine or accidental releases is caused. Regulators must ensure that they also set any conditions deemed to be necessary for the protection of human health.

The system is based on the concept of the prevention of emissions or waste production, or, where this is not practicable, reducing these to a minimum. The system applies only to defined developments (these are deemed 'prescribed activities' and are generally larger-scale industrial developments and food production facilities or intensive livestock-rearing establishments) and conditions must be based on the concept of 'best available techniques' (BAT). IPPC was introduced into European legislation following publication of a Directive in October 1996.

See also: INTEGRATED POLLUTION CONTROL

integrated waste management A system of waste disposal in which all key parties in the waste production process are able to have an input to the decision-making process to produce solutions to disposal problems. Such systems are likely to use a range of waste disposal options and recycling strategies especially focusing on the needs identified by local sustainable waste management systems.

intellectual property An imprecise term generally used to describe a right of ownership or use over an intangible item. The term includes patents, trademarks and copyright.

inter alia Means 'amongst other things'.

inter partes A term used to describe legal proceedings or actions between two parties, especially in respect of disputes or adversarial actions.

See also: EX PARTE

intermediate host In relation to a PARASITE, the host within which the immature cyst stage of the life cycle is formed.

See also: DEFINITIVE HOST

International Agency for Research on Cancer *see* IARC

International Bank for Reconstruction and Development Also known as the WORLD BANK, a financial institute of the United Nations. It was originally formed in 1945 with the intent of providing funding to rebuild Europe after the Second World War. Its remit has now been extended to cover global issues of trade and development. It is particularly active in pursuing initiatives to reduce poverty and improve living standards through sustainable growth and investment in people.

International Classification of Disease (ICD) An internationally recognised systematic classification into which individual disease diagnoses can be recorded by allocating them to specified categories, each designated by an identifying number (the ICD Number, or ICDN). The principal advantage of the system is in removing uncertainty of language or description by close definition of the specified categories. The classifications are reviewed periodically – by 2002, for example, edition 10 (i.e. ICD10) was in use.

interstitial condensation *see* CONDENSATION

inter-tidal zone Refers to the area between high and low tides that occurs at the edge of (tidal) seas and oceans. It is a highly sensitive habitat and places great demands on life forms that live there, due predominantly to the variability in water and saline content of the area and the physical pounding received by tidal action and storms. It is especially vulnerable to marine pollution such as MARITIME oil spillages.

See also: LITTORAL

interval scale One that apportions data into categories with a natural interval between them, such intervals being equally spaced across the range, e.g. categories with intervals such as 10–20 years, 20–30 years and 30–40 years.

See also: NOMINAL SCALE; ORDINAL SCALE

intervention Any action in a natural process of events that will cause, or has the potential to result in, a differential outcome than would have happened if the intervention had not been made. Many medical procedures such as the administration of medicine or undertaking surgery are broadly considered as coming under the umbrella term of 'intervention'. In practice the use of the term is more accurate in this context than the use of the word 'treatment' as the latter carries an implication or expectation that the outcome will necessarily be an improvement or that the patient will inevitably be cured.

intestinal flora A term that refers to the normal complement of micro-organisms normally found living within the intestinal tract. It would be expected that the complement would comprise essentially COMMENSAL organisms. This latter distinction is sometimes emphasised by describing the complement as the 'normal' intestinal flora.

intramuscular Means being inside a muscle, such as an intramuscular injection.

intraperitoneal Means being within the abdominal cavity.

intrasternal Means into or through the sternum, such as an intrasternal injection.

intravenous Means inside a vein, such as an intravenous injection.

intubation The insertion of a tube through the mouth to the larynx with the intention of keeping the air passage to the lungs open.

inversion layer A meteorological phenomenon of the TROPOSPHERE in which the normal TEMPERATURE GRADIENT in a body of air becomes inverted and an increase in height is accompanied by an increase in temperature rather than a decrease. This can happen for example in a valley in still conditions in which night-time cooling of the air by the land reduces the temperature of the air to below that of the upper layers, effectively trapping the cool air below the warm. The temperature gradient becomes distorted as heat from the warm layer above gradually diffuses to the cooler lower layers, but the normal pattern of convection currents does not appear.

An inversion layer is stable over a short period of time, say for a few days, and can trap emissions to the atmosphere from houses and industry, and cause a build-up of potentially harmful pollutants. Inversion layers can also be created by the warming of descending air in an ANTICYCLONE.

invert level The lowest internal point of a system (such as a drain) used to convey liquids.

invertebrate An animal that does not possess a backbone or spinal column. Such creatures as insects and crustaceans are invertebrates.

See also: EXOSKELETON; VERTEBRA

ion An electrically charged atom or group of atoms. This effect is usually occasioned by the loss of one or more electrons, the process being known as ionisation.

See also: ANION; CATION

ionic theory Developed around 1884 by Svante Arrhenuius to explain the behaviour of acids, bases and salts when dissolved in water. The theory holds that the constituent units of atoms and molecules are held together by the mutual attraction of positively or negatively charged units of electricity (the charged units being known as 'ions'). The creation of a solution greatly weakens these bonds, potentially leading to the separation of the units within the solution and thereby providing a mechanism whereby these units are readily attracted to the electrodes upon the passage of an electric current.

ionising radiation Any form of radiation that is capable of removing an electron from an atom and thereby creating an ION. ALPHA PARTICLES, BETA PARTICLES and ELECTROMAGNETIC RADIATION such as X-RAYS and GAMMA RADIATION are all forms of ionising radiation.

IPC *see* INTEGRATED POLLUTION CONTROL

IPCS The acronym for the International Programme on Chemical Safety, a programme of the WORLD HEALTH ORGANIZATION.

IPM The acronym for INTEGRATED PEST MANAGEMENT.

IPPC *see* INTEGRATED POLLUTION PREVENTION AND CONTROL

iris The muscular diaphragm with an open hole (the 'pupil') that forms the coloured portion of the eye when viewed *IN SITU*. The iris varies the size of the pupil by contraction or relaxation, thereby altering the amount of light that can enter the eye.

See also: CORNEA; RETINA; AQUEOUS HUMOUR

irradiation of food Involves exposing food to ionising radiation usually generated by Cobalt 60. Radioactive waste is entirely unsuitable for use as an ionising source since the neutrons that are emitted would induce radioactivity in the target foodstuff. The energy of the electrons and photons in the accepted irradiation process is too low to induce radioactivity. Expert opinion appears to be consistent in accepting that neither mutagens, carcinogens, polyploidy (effect on chromosomes) nor botulism arise from food irradiation. In addition there is no exceptional loss of nutrients as a result of the process. The irradiation of food is permitted in the UK under strict licensing conditions. Irradiated food must be correctly labelled to indicate that it has been so treated. The permitted labelling is either the designation 'irradiated' or 'treated with ionising radiation'. Although there are thought to be no health implications in the use of irradiated food, there is considerable consumer resistance to the practice, probably allied to misunderstandings and wider concerns in relation to radiation in other contexts. Irradiated food is not widely available in the UK.

ischaemia A reduction or loss of blood flow to a part of the body. Ischaemia may be caused by constriction of the blood vessels due to a spasm or constriction, or it could be due to a physical blockage of the blood vessels.

See also: ARTERIOSCLEROSIS; CORONARY HEART DISEASE; STROKE

ischaemic heart disease (IHD) *see* CORONARY HEART DISEASE; ISCHAEMIA

isobar A contour on a map delineating lines of equal atmospheric pressure.

isolate In microbiological terms, a single species of a micro-organism retrieved from an individual sample or from a defined environment.

isomer One of two or more substances whose molecules possess the same atoms (i.e. they have the same molecular weight) but these arranged in a different structure – the isomers consequently have at least one difference in their physical or chemical properties from each other.

isopleth A contour on a map delineating lines of equal concentrations of airborne contamination or pollution.

isosceles Means having two sides of equal length.

See also: EQUILATERAL

isotherm A contour on a map delineating lines of equal temperature.

isotope An atomic variant of the same element in which the atoms of the different isotopes have the same number of protons but a different number of neutrons. Thus the isotopes of a particular element have the same atomic number, but a different atomic weight.

isotropic A term used to describe substances or entities that have the same physical properties or structures.

See also: ANISOTROPIC

IV An abbreviation of the term *intra venous*, literally meaning 'into the veins'.

J

jackhammer A hand-held tool driven by compressed air and used to drill holes in masonry, rocks, etc.

jamb One of the vertical side members used to frame a door or window.

See also: REVEAL

Jarman index A comparative score used to measure relative deprivation in, or of, a community. It is based on data obtained from the Census and reflects a range of parameters including perceptions of GP workload together with socio-demographic and economic factors.

Jeroboam A bottle equivalent in size to a double MAGNUM (i.e. four bottles).

See also: METHUSELAH

jet stream A narrow band of air moving at speeds of around 60 miles per hour (i.e. around 95 kilometres per hour) in the areas of a strong temperature gradient in the upper atmosphere, such as those between the cold polar regions and the warmer regions. The jet stream is usually located in the TROPOSPHERE at an altitude of between 5.5–11 miles (i.e. around 9–18 kilometres). Although aircraft travelling in the direction of a jet stream can significantly improve journey times and reduce fuel consumption, such areas can be turbulent and reduce passenger comfort.

jib The triangular sail at the front of the foremast of a sailing ship. A jib-crane is one in which an inclined arm (sometimes referred to as the jib) is attached to the base of a rotating vertical support post, the upper portion being attached by GUY ropes.

joist A small horizontal structural beam incorporated into a building, usually to carry the floor or the ceiling.

See also: RAFTER

joule A unit measure of energy, work and heat in the SI system. One joule is the energy dissipated in one second by a current of one AMPERE across a potential difference of one VOLT. In terms of work one joule is that work done when a force of one NEWTON is advanced from its point of application by one metre. One joule is equivalent to 1×10^7 ERGS, or 0.239 CALORIES.

Judicial Precedent A legislative standard applied in the United Kingdom following the interpretation of a higher court of a particular STATUTE. The interpretation is generally deemed to be correct for that particular piece of legislation unless overturned by appeal or by subsequent amendment of the legislation.

junket A sweet dessert made from curds and is produced by treating milk with RENNET. Junket was once a popular food for children and invalids as the rennet acted on the milk proteins to make them more easily digestible.

jurat The final statement, signed by the person authorised to act as witness to a signature on an AFFIDAVIT or similar, to declare when such documentation was sworn, affirmed or declared.

K

k-value A measure of the thermal conductivity of a particular material. It is measured as the amount of heat energy per hour transmitted through a unit thickness of the material for each degree of temperature difference. The summation of k-values and surface transfer resistance of individual materials in a structure such as a wall together provide the U-VALUE for the structure.

kaolin Another word for china clay, a pure white form of hydrated aluminium silicate produced by the decomposition of FELDSPAR in IGNEOUS rock. It is used in the manufacture of fine porcelain ('china') and in some medical preparations, notably for the treatment of diarrhoea.

keds Wingless flies of the species *Melophagus ovinus* that live on the wool and skin of sheep. They have tripartite bodies, are dark brown in colour and are around one-quarter of an inch long with a sharp biting proboscis when mature. If left unchecked they cause the condition on sheep fleeces known as COCKLE. They can be controlled by shearing or by dipping with approved insecticides.

kentledge Ballast attached to a crane, hoist or similar in order to act as a counterbalance to the load it is expected to bear.

keratin A fibrous PROTEIN found in hair, wool and the skin.

kerbside collection A system of regular or routine waste collection from domestic or commercial premises in which the items to be disposed of are collected from the premises within which they are generated.

kerosene *see* DIESEL

keystone The stone at the apex or crown of an arch. In a VOUSSOIR arch the keystone is essential to the capacity of the arch to be self-supporting as it transfers the loading to the stones comprising the lower portion of the arch.

kibbling In culinary practice, the coarse chopping or grinding of an ingredient.

kilometre A metric unit of length comprising 1,000 METRES. It is equivalent to 0.62137 of a MILE.

kinetic energy An expression of the energy a body has as a result of its being in motion. The amount of kinetic energy it holds is equal to the amount of work the body would do if it were brought to rest.

king closer A brick of three-quarters dimension created by cutting off one corner from the full sized brick. They are used near the corners of the wall with the smallest end facing the outside and the main bulk of the brick running along the inner side of the rebate in a double skin wall. Their function is to ensure that the bond of the brickwork is regular when the design dimensions for the wall cannot be achieved precisely, given the size of the bricks from which it is constructed.

See also: HALF BAT; QUEEN CLOSER

knock In reference to an internal COMBUSTION engine, the sound that the engine makes if the fuel ignites prematurely in the cylinder. This phenomenon is caused when part of the fuel ignites (usually) prior to the advent of the ignition produced by the spark plug, and is due to excessive heat or to pressure during compression. The pre-ignition results in excessive pressure in the cylinder as the piston is rising, and the 'knock' is the sound produced as the engine effectively starts working against itself. The production of 'knock' is both inefficient and, because of the strain produced, potentially damaging to the engine.

See also: ANTIKNOCK COMPOUND

knot In timber, that portion of the wood in which the structural fibres of a portion of the whole run across the direction of the majority. It is caused by cutting through the trunk of the tree at that point at which a branch or limb is embedded in the main growth. The knot can adversely affect the structural strength of the timber, especially if it is loose or decayed. Conversely the appearance of knots might be considered aesthetically appealing, especially in decorative timber or cladding.

Koch's postulates Four rules devised in 1884 by the German physicist Robert Koch to ascertain whether or not the cause of an infection had been established. The rules relate to four stages – the first is to find evidence that the infection is caused by a microbe, the second is to isolate the microbe, the third is to reproduce the disease by inoculating the microbe into a healthy individual and the final stage is to re-isolate the microbe from the second subject. Although they represented a milestone in understanding human pathogens, the postulates are unacceptable for use in modern times on ethical grounds alone.

kuru Pronounced 'koo-roo', a progressive and ultimately fatal brain disease detected in a tribe in Papua New Guinea. The disease is spread by cannibalism, notably amongst women and children who ate the infected brain tissues of deceased people. The disease was almost exclusively confined to a tribe known as the FORE and research into its occurrence has aided consideration of the development of the new variant form of Creutzfeldt-Jacob disease (see CJD) in humans.

L

La Niña *see* EL NIÑO

labile A general term referring to that which is prone to change or movement. 'Heat labile' means unstable in the presence of heat.

lactase An ENZYME that converts LACTOSE into galactose and glucose.

lactic acid Lactic acid ($CH_3CHOHCO_2H$) is an organic ACID (also known as 2-hydroxypropanoic acid). It is found in human and cow's milk, and it is produced when GLUCOSE breaks down under anaerobic conditions and during some FERMENTATION processes.

lactose A sugar. It is found in milk.

lampblack A very fine and almost pure form of carbon used as a pigment or, when formed into a solid block, to form the carbon electrodes used in electrical mechanisms.

lamprey Any of the primitive fish of the family *Petromyzondidae*. They are characterised predominantly by their eel-like appearance and in having a rounded mouth supplied with rasping teeth. They are parasitic, attaching themselves to the flesh of other fish with their sucker-like mouths and eating their flesh or drinking their blood. In the UK the largest lamprey is the sea lamprey (*Petromyzon marinus*), being capable of attaining 3 feet in length and weighing up to 5 lbs. The Lampern (*Lampetra fluviatilis*) is a freshwater species attaining perhaps 16 inches in length and the Brook Lamprey (*Lampetra planeri*) is the third species found in the UK – it rarely exceeds 6 inches in length. Henry I of England is reputed to have died from eating a 'surfeit of lampreys'; these are thought to have been Brook Lampreys. In some parts of the UK lampreys are also known as Lampern eels.

landfill The process whereby solid waste material is disposed of to land, either in terms of landscaping, land reclamation or simply as a means of getting rid of the waste. The latter has historically been practised as being the cheapest or easiest option. The term is especially applicable to the disposal of municipal solid waste to land. The limited land now available for landfill, the loss of possible recyclable materials by simply throwing them away and the potential for pollution have all contributed to making landfill less of a sustainable option for waste disposal in the future. The increased use of recycling is seen as a key strategy for reducing the amount of waste that goes to landfill.

landfill tax A tax levied on those who dispose of waste to LANDFILL. It is seen as a way of discouraging such disposal in favour of waste reduction and recycling strategies.

landspreading The disposal of certain types of BIODEGRADABLE waste material (such as sewage sludge and paper pulp) by spreading it on agricultural land. The system needs careful monitoring and control to ensure that any risks to humans, wildlife and the environment are minimised, as well as ensuring that such disposal does not lead to the accumulation of hazardous or potentially toxic material.

Langrangian point A point in the space between two celestial bodies in which an object can be held in a constant position by their joint gravitational attraction. In theory the sun/earth system has five such points.

laparoscopy An examination of the PERITONEAL cavity of the body with a laparoscope – an instrument that allows visual examination of the abdominal cavity of the body through a small incision in the abdominal wall.

LAPC *see* LOCAL AIR POLLUTION CONTROL

larva The immature stage in the life cycle of an insect between the egg and the PUPA.

See also: IMAGO

laser A device for creating a very intense and focused non-divergent beam of monochromatic light from a mixed-frequency source, or the beam emitted therefrom. The beams are used at various power strengths in a range of applications, including communications, light shows, surveying, weapons guidance systems, surgery and (potentially) offensive weapons. The name derives from the acronym for Light Amplification by Stimulated Emission of Radiation.

latency period The time interval between exposure to a substance and the manifestation of symptoms arising as a result of that exposure.

See also: INCUBATION PERIOD

latent heat The heat energy necessary to change the physical state of a solid to a liquid, or liquid to a vapour, without changing its temperature. The energy is absorbed on translation of the substance from solid to liquid or from liquid to vapour and released when the process is reversed. At standard ATMOSPHERIC PRESSURE the latent heat of vaporisation of water (i.e. the amount of heat needed completely to convert 1 gram of water at 100°C to steam at the same temperature) is 2,256 Kj/kg.

See also: SPECIFIC HEAT (CAPACITY)

latent period *see* LATENCY PERIOD

latex Also known as 'natural rubber latex', a milky-white fluid produced by a number of different plants. The latex from the rubber tree *Hevea brasiliensis* is used in the manufacture of RUBBER. Synthetic latex is used in the manufacture of synthetic rubber.

See also: LATEX ALLERGY

latex agglutination An analytical technique used for isolating a specific microorganism through the use of latex, coated with the ANTIBODY to the organism being sought, to which the organisms adhere and form clumps.

latex allergy A heightened sensitivity to natural rubber latex and presents as a skin irritation. It can be developed from two distinct reactions. Immediate hypersensitivity is produced predominantly as a result of contact with the natural proteins derived from the raw product. Delayed hypersensitivity (also known as allergic contact dermatitis) is produced predominantly as a result of contact with the chemical agents used in the manufacture of the product.

See also: ALLERGY

latitude The distance of a place on the surface of the earth measured in degrees from the equator along a line of latitude. A line of latitude is a (imaginary) circle running parallel to the equator; the sequence of circles terminates at the North or South Pole.

See also: LONGITUDE; MERIDIAN

laver Any one of a number of types of edible seaweed, especially of the genus *Porphyra*. Laver is a popular regional dish in several areas of the British Isles and is usually boiled before eating. Laver bread is a dish of fried laver, being eaten mainly as a breakfast dish, especially in areas of Wales.

Law Lords The most senior judges in the UNITED KINGDOM. They are all members of the House of Lords and a selection from their membership comprises the ultimate court of appeal in the United Kingdom legislature.

LC50 A measure of the toxicity of a substance on inhalation. It expresses the concentration at which 50 per cent of the population of an animal species (the 50 of the LC50 expression) in experimentation will die – i.e. they will have received a lethal concentration (the LC). The LC50 is usually expressed as parts per million by volume of the substance per each million parts of air over a given period of exposure.

See also: LD50

LD50 A general measure of the toxicity of a substance as administered to a population of animals under laboratory conditions. The substance can be administered orally, through the skin or by injection. The LD50 is the dose that will result in the deaths of 50 per cent of the animals. It is usually expressed as milligrams of substance per kilogram of body weight of the animal species to which the substance is administered.

leachate Dissolved material washed out of a deposit (usually in the ground) by the action of water. The term is especially applied to environmentally toxic or potentially toxic soluble chemicals washed out of landfill sites, mines, quarries, etc.

leaching The process whereby soluble material is washed out of a medium (such as soil) by the passage of water (such as rainwater or groundwater). The material washed out is known as the LEACHATE.

lead Lead (Pb) is a dense, relatively inert, metallic element. It is one of the oldest known materials for which occupational health concerns have been expressed. It is used in a wide range of industries, notably in the production of lead-acid batteries, the manufacture of secondary chemical compounds (especially pigments) and in the ceramics industry. Exposure can produce a range of clinical symptoms including anaemia, central and peripheral nervous system damage, aching in bones and muscles, kidney damage and digestive disorders. There is also concern that it reduces both male and female fertility. Developing foetuses and breast-feeding babies are thought to be particularly at risk from secondary exposure, especially as it is thought that it impairs their mental development.

leaded light *see* STAINED GLASS

lean-burn Refers to a type of internal COMBUSTION engine that has specifically been designed to run on a weaker air/fuel mixture than engines of similar size/design. The use of a weaker mixture is intended to reduce fuel (usually petrol) consumption and minimise exhaust emissions.

learning curve A visualisation of new knowledge or experience gained over unit time. It recognises that the amount of knowledge gained per unit of time at the start of the process is likely to be more than later on. A 'steep learning curve' is one in which it is necessary that a great deal of knowledge, experience or information needs to be gained in a short time period.

lease A contract whereby the owner of a property (the lessor) confers certain rights or privileges to another (the lessee) for a specified period (with or without conditions) and usually upon payment of rent.

leaven A substance that produces FERMENTATION.

See also: BREAD

lectin Any one of a group of proteins of plant origin that have the capacity to bind to certain carbohydrates and can be used as an AGGLUTININ.

leeward The side facing away from the direction of the wind. It is the opposite of WINDWARD.

Legionnaires' disease A potentially fatal respiratory disease, characterised by pneumonia, resulting from the inhalation of water droplets contaminated by the bacillus *Legionella pneumophila*. There are around fourteen sero-groups of the organism and, whilst all are capable causing disease in humans, the most common cause of infection is those organisms in sero-group 1. The name derives from its first identification in a group of US ex-servicemen (known as Legionnaires) attending a conference in Philadelphia in 1976.

See also: PONTIAC FEVER

lessee The person who takes on a property or land leased to them by the LESSOR.

See also: LEASE

lessor The person (usually the owner) who leases a property or land to the LESSEE.

See also: LEASE

leucocyte Another term for the white blood cell of the body's defensive mechanism. They are a component of BLOOD and, since they contain no HAEMOGLOBIN, they are colourless. They vary in size between around 8–15 micrometres diameter and there are around 8,000 per cubic millimetre of blood. There are three main classes of leucocyte, being granulocytes, lymphocytes and monocytes.

Granulocytes are formed in the red bone marrow and are the most numerous of the leucocytes, comprising up to around 70 per cent of their total number. Granulocytes themselves may be further sub-divided into neutrophils (which stain with neutral dyes), basophils (which stain with base dyes) and eosinophils (which stain with the acid dye eosin).
Lymphocytes are produced predominantly in LYMPHOID TISSUE and comprise around 25–30 per cent of all white blood cells. They may be sub-divided into small lymphocytes and large lymphocytes.
Monocytes comprise around 5 per cent of all white blood cells.

See also: ABSCESS; BLOOD; ERYTHROCYTE; LYMPHOCYTE; PUS

leucoderma A condition in which there is a depigmentation of the skin caused by systemic reaction to certain chemicals such as hydroquinones and phenols or their metabolites. The loss of skin colour is caused by the destruction of MELANOCYTES in the skin. The condition may be permanent.

leukaemia A form of cancer of the blood in which the number of white blood cells is increased. There are a number of different types of leukaemia, both acute and chronic. The various types are chiefly differentiated dependent upon the type of white blood cells affected.

level of significance *see* SIGNIFICANCE LEVEL

lewis A device for lifting stone blocks. It consists of a number of curved metal grabs each fitting into a dovetailed groove carved into the block for the purpose. The action of lifting increasingly secures the grabs in place.

lichen A compound organism comprised of algae and fungi living together in a SYMBIOTIC relationship. They appear as vegetative growths on the surfaces of rocks, buildings and tree bark. They are often highly sensitive to chemical atmospheric pollution and can be used as indicators to this effect. Some lichens are very slow growing and through examination of bioaccumulation (see BIOACCUMULATE) of chemicals can sometimes be used as calendars of historic pollution episodes.

life expectancy The theoretical maximum average age to which individuals in a specified group would be expected to live.

life table A tabulation showing the life expectation at different ages for a given population.

ligand A molecule that is bound to a RECEPTOR.

light A form of ELECTROMAGNETIC RADIATION with a WAVELENGTH between INFRARED RADIATION and ULTRAVIOLET RADIATION.

lignin The material that provides the structural rigidity in the cell walls of thick-celled and woody plants. It has a complex polymer structure. The term 'lignum' means 'wood'. Lignification is the process whereby wood is produced as the plant grows.

limestone A porous SEDIMENTARY ROCK composed primarily of calcium carbonate ($CaCO_3$). It is also known as 'chalk'. Limestone sometimes acts as a reservoir for PETROLEUM.

See also: HARD WATER

linen *see* FLAX

linseed oil A vegetable oil produced from the seed pods of the common FLAX plant. It is used as an addition in paints and as a lubricant for PUTTY.

lipid One of the substances that comprise the four major classes of compounds that constitute the living body (the others are CARBOHYDRATES, NUCLEIC ACIDS and PROTEINS). They include FAT, STEROIDS, terpenes and waxes. Lipids contain fatty acids and are soluble in alcohol and ether but are insoluble in water.

lipopolysaccharide A part of the outer membrane of a Gram negative (see GRAM'S STAIN) bacterium that performs the function of a somatic (also known as the 'O') ANTIGEN.

liquefied petroleum gas *see* LPG

liquid chromatography *see* CHROMATOGRAPHY

listeria A Gram positive (see GRAM'S STAIN), non-sporing bacterium. There are six species, being respectively *Listeria monocytogenes*, *Listeria ivanovii*, *Listeria seeligeri*, *Listeria innocua*, *Listeria welshimeri* and *Listeria greyi*. In humans the most significant pathogenic species is *L. monocytogenes*. This bacterium causes the disease in humans known as listeriosis. It is widespread in the environment and can be found in soil, dust, mud, vegetation, silage and sewage. It has been detected in most of the animal species that have been tested. It has been found in up to 5 per cent or more of normal healthy people, usually in the gut. Exposure to the bacterium is thought to be unavoidable. It has the unusual characteristic of being able to grow, albeit slowly, at temperatures as low as 5–6°C.

Listeria monocytogenes can cause a variety of diseases (including infections in pregnancy). Symptoms of infection vary from a mild chill to a severe illness that may precipitate premature birth or miscarriage, and meningitis in new-born children. The infection can also be serious in adults whose immunity to infection is impaired. Infection is extremely rare in healthy adults and children. The infection may be treated with antibiotics but in about one-third of cases the full-blown disease is fatal.

lithosphere The outermost and solid portion of the earth. It comprises the earth's crust and the upper part of the mantle.

See also: TECTONIC PLATE

littoral Relates to the edge of a body of water and is often used to denote a particular type of habitat. It is usually applied to the edges of lakes or, in respect of oceans and (tidal) seas, to the INTERTIDAL ZONE.

See also: BENTHIC

loam A type of soil of medium texture much favoured as a growing medium. It comprises roughly equal proportions of clay, sand and grit, and usually contains adequate quantities of HUMUS and trace minerals to support plant growth.

Local Agenda 21 A development of the principles introduced in AGENDA 21 to promote local action or activity for SUSTAINABLE DEVELOPMENT. It is crucial to the development of the 'think globally; act locally' principle.

Local Air Pollution Control (LAPC) A system in Great Britain for the regulation, by local authorities, of atmospheric emissions from processes that are generally less polluting than those covered by the INTEGRATED POLLUTION CONTROL (IPC) regime.

See also: POLLUTION PREVENTION AND CONTROL

lockjaw *see* TETANUS

long-wave radiation Generally taken as electromagnetic radiation with a wavelength in excess of 0.3 microns. In terms of light long-wave radiation is differently described as being that which extends beyond the infrared spectrum and with a wavelength in excess of 0.7 microns.

See also: SHORT-WAVE RADIATION

longitude The distance east or west on the surface of the earth of a line of longitude measured from Greenwich, London. A line of longitude is one of the imaginary great circles of the earth, each of which passes through north and south poles.

See also: LATITUDE; MERIDIAN

longitudinal study Prospective examination of a defined group of people (the COHORT) who have been selected from the population on the basis of their exposure to certain defined parameters. The purpose of the study is to ascertain whether or not there are discernible differences in the manifestation of disease or health status between various groups or cohorts over time, and to suggest what risk factors might be associated with the difference.

Lord Advocate The principal law officer of the Crown in Scotland. The equivalent in the rest of the UK is the ATTORNEY GENERAL.

Lord Chancellor A cabinet member of the UK government who heads the judiciary in England and Wales, and officiates as Speaker of the House of Lords.

Lord Lieutenant In the UK, the representative of the Crown at the county level.

louping ill A paralytic disease of sheep. It is caused by a virus and transmitted by the tick *Ixodes ricinus*. The symptoms are broadly classified into two forms. In the acute form the sheep have a raised temperature and may stagger or exhibit partial paralysis but the distinguishing feature is a tendency to leap or to take sporadic jumps. The sub-acute form exhibits similarly but with a tendency to take pronounced high steps with the fore legs.

Love Canal A neighbourhood in Niagara Falls, New York, USA. The name derives from William Love, an entrepreneur who began digging a canal in 1896 in an attempt to connect Lake Ontario and Lake Erie. The canal was never completed but the section that was excavated was subsequently used as a chemical by-products dump between 1942 and 1955. Once full the land was sold for redevelopment and a school and houses were built on the site. Local residents complained about a range of adverse health events such as birth defects, spontaneous abortions, low birth weight, cancers and respiratory diseases. Eventually, in April 1978, the site was declared to be a threat to health, ultimately leading to the closure of the site. Although not the first or worst incident of its kind, Love Canal received considerable media attention and was a significant milestone in the understanding of the potential threats that can be occasioned to communities from environmental pollution.

Lovibond comparitor Also known as a Lovibond tintometer, an instrument used to differentiate the concentrations of chemical in liquid form. It does so by the comparison of a sample to which an indicator dyestuff has been added against the colours on a standard colour chart.

low-acid food Generally accepted as being a food with a PH of below 4.5.

See also: HIGH-ACID FOOD

LPG An acronym for liquefied petroleum gas. It is so named because it is stored in liquid form under pressure and can consequently be supplied in pressurised cylinders and tanks. It is a popular fuel source in areas where access to mains supply is unavailable. The term refers generally to either butane or propane, the latter having a higher calorific value. LPG is sometimes known as bottled gas.

lubricant Any material (usually liquid but sometimes solid) that acts to reduce FRICTION between two surfaces.

See also: LUBRICITY

lubricity A measure of the capacity of a substance (usually oil or grease) to act as a LUBRICANT.

lukewarm The imprecise term used to describe the temperature of something that is moderately warm – usually applied to something at a temperature of around 35–40°C.

lumbar puncture A surgical penetration of the spine in the region of the lower back undertaken with the aim of procuring a sample of the spinal fluid for the purposes of examination or analysis.

lumen The international standard measure of the total power emanating from a light source.

See also: LUX

lung function test Any of a series of tests designed to measure physical characteristics of the lung functions, such as maximum capacity by volume or air flow rate into or out of the lungs.

lux The international standard unit for measuring luminance per unit area. One lux is equivalent to one LUMEN per square metre.

See also: FOOT-CANDLE

Lyme disease The name of a disease caused by infection with the bacterium *Borrelia burgdorferi*. The name Lyme disease is derived from the first outbreak that was described in Lyme County, Connecticut, in 1975. The organism is transmitted by ticks (*Ixodes ricinus*) which parasitise deer and wild rodents. The bacterium is transmitted to humans if the tick passes the bacterium from infected animals that have previously been bitten. Lyme disease can affect many body functions, although the skin is the primary site of infection, the bones, joints, cardiovascular and neurological systems can be secondarily affected if the disease enters a chronic phase. The tick is generally the size of a pinhead, but they can reach the size of a coffee bean when they are full of blood. The tick feeds once at each stage of its life cycle, as larva, nymph and adult. Larvae and nymphs are too small usually to feed on animals much larger than small mammals, although the adults will feed on larger mammals such as sheep, deer, horses or humans. They are found in grassy areas in both moor and woodland.

lymph A colourless watery fluid, derived from BLOOD, which circulates in the LYMPHATIC SYSTEM of the body. It comprises salts, proteins and defence cells such as lymphocytes as well as fats derived from the alimentary canal.

See also: LEUCOCYTE; LYMPHATIC SYSTEM

lymphatic system A circulatory system of the body that parallels to a greater extent the capillary system of the BLOOD. The lymphatic system acts as a transportation mechanism for fat derived from food in the alimentary canal and for certain salts, fluids and bodily defence mechanisms. The system has no direct pump mechanism (such as the heart in the blood system) and lymph is driven round the body predominantly by the secondary action of the pulse.

See also: LYMPH; LYMPHOID TISSUE

lymphocyte A white blood cell that produces antibodies or chemical messengers used to promote a body's immune response.

See also: LEUCOCYTE

lymphoid tissue That tissue which is involved in the generation of LYMPH.

lymphoma A MALIGNANT tumour that arises in the lymphatic tissue of the body. Lymphomas usually occur at multiple sites in the body, especially the lymph nodes, spleen and thymus, but also potentially affecting other sites such as the bone marrow.

M

macro-analysis A technique whereby the results of a range of different studies into an occurrence or phenomenon are considered together to establish whether or not there is any agreement in the findings. The use of macro-analysis increases the sample base upon which the evaluation is made and increases the POWER of the study.

macrocyte An abnormally large red blood cell. They are a characteristic of PERNICIOUS ANAEMIA.

macrophage A form of PHAGOCYTE found in many organs and tissues of the body as opposed to those found in the blood.

mad cow disease The popular term for the cattle disease more properly known as BOVINE SPONGIFORM ENCEPHALOPATHY.

maggot The larval stage in the life history of the fly. Maggots are frequently torpedo-shaped and usually have no distinctively discernible head. Maggots do not possess legs.

Magistrates' Court The lowest of the criminal courts in the UNITED KINGDOM court system. Hearings are conducted in front of either lay (i.e. unqualified) magistrates (usually three in number) or in front of a (usually singular) qualified Stipendiary (i.e. in receipt of an income) Magistrate.

See also: COURTS

maglev A variety of high-speed train that is propelled by magnets and is supported by a magnetic field generated around the track.

magma Hot molten rock formed in the upper part of the earth's MANTLE and that sometimes is expelled into the upper crust of the earth or onto the surface where it cools to produce IGNEOUS ROCK. Magma can reach temperatures as high as 1,000°C.

See also: GRANITE

magnetopause The outermost boundary of the MAGNETOSPHERE separating the plasma of the earth from that originating from elsewhere in the galaxy and impelled by the solar wind.

magnetosphere The outermost area of the earth bounded at its outer limit by the MAGNETOPAUSE and dominated by the earth's magnetic field.

magnum A large wine bottle equivalent in volume to two standard bottles.

See also: JEROBOAM

magnox reactor A type of thermal, gas-cooled NUCLEAR REACTOR that uses uranium metal housed in magnesium alloy cladding as the fuel. Carbon dioxide is used as the coolant.

See also: ADVANCED GAS-COOLED REACTOR

maisonette (or maisonnette) Originally the term applied to describe a self-contained apartment, usually over two floors, forming part of a larger residence and having its own outside entrance. Nowadays, in the UK, the term refers generally to any apartment on two levels, especially those with the bedrooms on the upper level.

See also: DUPLEX

malaria An infectious disease characterised by recurrent bouts of fever and chills. The name derives from the Italian '*mala aria*', or 'bad air'. Malaria is also known as ague, jungle fever, marsh fever or swamp fever. Malaria is caused by the transmission of parasitic protozoans of the genus *Plasmodium* introduced into the bloodstream by the bite of an infected anopheles mosquito. There are three types of Plasmodium, being *P. falciparum*, *P. malariae* and *P. vivax*.

The disease has been known from the earliest of times and has been recorded at least as long ago as 1500 BC. Malaria has a worldwide presence in the tropics, being associated with the swampy habitat favoured by the mosquito. It occurs throughout Africa, India, South-Eastern Asia and in Central America. Although the mosquito that generally acts as the VECTOR has been detected in several areas of the UK over recent years (possibly due to GLOBAL WARMING), no home-generated case of malaria has yet been associated with this occurrence.

malignant (tumour) An abnormal growth of body tissue, frequently resistant to therapy, which has the potential to become life threatening.

See also: BENIGN (TUMOUR); CANCER

malleable Refers to a characteristic of a solid indicating that it is capable of being shaped or worked by pressure (or by blows) without breaking.

See also: DUCTILE

malpigian layer *see* EPIDERMIS

manifold A pipe or chamber attached to several separate inlets or outlets and used for conveying liquids or gases.

mandamus An order from a superior court to an inferior one to carry out a public duty. The document so ordering such activities and signed on behalf of the higher court is known as a 'writ of mandamus'.

manometer An instrument used for measuring the difference in gaseous pressure between two environments, one of which is usually the AMBIENT atmospheric pressure. The instrument usually has the configuration of a transparent 'U' tube containing a measuring fluid (usually mercury) with a graduated scale, against which difference in pressure is measured as being the difference between the levels in either side of the 'U' tube.

mansard roof A pitched roof in which the area on each of the two sides of the RIDGE comprises two sloping planes, the lower slope being steeper and generally longer than the upper slope.

mantle The layer in the cross-section of the earth that lies between the crust and the core. The upper part of the mantle where it meets the crust is relatively rigid and together these two components are described as the lithosphere.

See also: TECTONIC PLATE

manual handling The physical transportation or support of a load either by hand or bodily force. In the UK it is estimated that over 30 per cent of all serious accidents are caused by incorrect manual handling.

marble Either a METAMORPHIC ROCK formed from limestone or an industry name for much non-metamorphic limestone capable of being polished to a durable and aesthetically pleasing finish. This latter classification is highly variable in finish and colour, dependent upon where the stone was originally quarried.

marbling The creation on (or appearance of) a surface of a pattern that resembles MARBLE.

marginal land Generally land of poor agricultural quality due to deficiencies in soil structure or composition, irrigation, site or climate. It is usually found on the margins of developed land, having escaped development because of its poor economic potential.

marginal utility of wealth The extra utility capability accruing to an individual for a small change in wealth.

mariculture The term sometimes used to distinguish the cultivation of the plants or animals of the natural marine (i.e. saltwater) environment from that of organisms of the freshwater environment.

See also: AQUACULTURE

marine ply (or plywood) *see* PLYWOOD

maritime Means relating to or of the sea. It also applies to the edges of the sea or TERRESTRIAL regions bordering the sea, such as in reference to a 'maritime climate'.

marker organism In food technology, a type of (usually) bacteria that responds to food safety processing techniques in a way that allows for an assessment of the effectiveness of the techniques to be made. Marker organisms are normally subdefined as either indicator organisms or index organisms. The detection of index organisms is used to imply the presence of ecologically related pathogens. The detection of indicator organisms is used to assess the effectiveness of processing by comparing results to predetermined indicator values. There are practical limitations in the use of markers in that at best they are only ever substitutes for more detailed assessment or evaluation methods.

marquenching A form of QUENCHING in which the cooling process is extended to minimise distortion or cracking.

marsh fever Another name for MALARIA.

marsh gas A gas, principally comprising METHANE, emitted by the decay of organic material in the absence of air.

See also: ANAEROBE; FIRE DAMP

masking The attenuation of an effect by the imposition or introduction of some form of impedance. Masking occurs in acoustics when the threshold of audibility of one frequency is raised by the introduction of audible sound of another frequency. Masking occurs in drug therapy when the effect of one chemical is impeded by the action of another chemical. The recognition of the possibility of masking is important in taking measurements as a true effect cannot be ascertained if masking is occurring.

masonry Refers to constructions of stone, block or brickwork laid in MORTAR.

Mass Median Aerodynamic Diameter *see* MMAD

mass number The number of PROTONS and NEUTRONS in the NUCLEUS of an ATOM.

See also: ATOMIC MASS

mass spectrometer A scientific instrument used to provide both QUALITATIVE and QUANTITATIVE analysis of chemicals. The analysis is performed by accelerating molecules in a circular path within an electric field. The molecules in the sample are separated by CENTRIPETAL FORCE, larger molecules being forced towards the outer periphery of the circular path. Analysis may be undertaken photographically or electronically.

mastic Originally an aromatic resin obtained from the Mediterranean mastic tree *Pistacia lentiscus*. The term is now also applied to many jointing or sealing compounds, especially those that dry or harden on the surface but remain plastic underneath. Alternatively the term is applied to an adhesive constructional bedding material for wood, timber and block work or similar or to sound-absorption treatments for a variety of surfaces.

mastitis An inflammation of the mammary gland(s).

maternal immunity Another name for VERTICAL immunity, in which the offspring of a mammalian species derives immunity by acquiring ANTIBODIES from the mother.

maximum residue level (MRL) That level of residual substance considered to be acceptable to allow food to be used for human consumption. The level may be mandatory (i.e. STATUTORY) or specified for guidance. The level is determined by expert opinion and usually incorporates a safety factor.

MBM (meat and bone meal) Protein derived from an animal by the process of RENDERING. MBM was used in animal feed as a source of protein.

See also: RUMINANT FEED BAN

MDF Stands for MEDIUM-DENSITY FIBREBOARD.

mean Also known as the arithmetic mean, is the average of a set of numbers derived by adding all the numbers together and dividing this by the number of individual numbers that were added together to give the total.

See also: GEOMETRIC MEAN; MEDIAN; MODE

measles An acute viral disease and one of the most highly communicable of all human diseases. It has a worldwide distribution, causing around 1.5 to 2 million deaths per year in children and crippling many more through blindness and lung disease. It is responsible for around half of all those deaths in developing countries that could be prevented by vaccination. The infectious agent is a member of the Morbillivirus group, of the family Paramyxoviridae. Measles is characterised by a high fever and a blotchy red rash that occurs between the third to seventh day following infection. Around one child in fifteen is at risk of complications that include chest infections, fits and brain damage. The disease is frequently more severe in infants and adults than it is in children.

The reservoir of the infection is man. The infection is spread by airborne droplets or by direct contact with secretions from the nose or throat of infected persons. More rarely infection can take place through contact with articles that have been freshly contaminated with secretions from the nose or throat. The incubation period varies between 7–18 days, typically being around 10 days, with the rash appearing normally around day 14. The disease is communicable from just before the presentation of symptoms to 4 days after the appearance of the rash, although communicability is considered to be minimal after the second day of the rash. Immunisation by live attenuated vaccine is frequently conducted in association with live virus vaccines for mumps and rubella. The combination is commonly called MMR VACCINATION.

measles, mumps and rubella vaccination *see* MMR VACCINATION

meat and bone meal *see* MBM

mechanically recovered meat (MRM) A product derived from that meat which is left on the bones of animal carcasses after the normal de-boning process has been carried out. This meat is stripped from the bones by either mechanical or hydrostatic high pressure, creating a slurry, which can subsequently be dried or reconstituted and used in a variety of ways. The most common uses in the human food chain include in those products that use meat pieces as part of the product, such as meat pies, pasties or chicken nuggets, or as a binding agent. EU legislation, however, prohibits the use of MRM in minced meat. Cattle, lamb, pig and poultry bones are all used to produce MRM. Production essentially breaks down the muscle structure of the meat and thereby creates the potential for the dissemination and growth of bacteria. It is critical that the manufacturing process is properly controlled and supervised. Because the product is denatured there is potential for fraud or misrepresentation. The importance of securing authorised sources for, and acceptable quality of, the raw materials should not be underestimated.

mechanisation The replacement of manpower by machinery.

median The number that occurs in the middle of a range of numbers.

See also: MEAN; MODE

mediator A chemical produced within a living body and which is used to provoke or initiate a specific bodily response.

medical device A term whose use in the UK describes any product, other than a MEDICINE, which is used in connection with a health care situation for the diagnosis, prevention, monitoring or treatment of illness or injury. The term is deliberately very wide in order to ensure that the Medical Devices Agency can consider any failure of such item. The term includes such as contact lenses, condoms, dressings, heart valves, hospital beds, resuscitators, radiotherapy machines, surgical instruments, syringes, wheelchairs and walking frames.

medicine A substance given (especially orally) to a person suffering from a DISEASE with the intent of effecting a cure or relieving the symptoms.

medium-density fibreboard (MDF) A construction material, usually in the form of sheeting, in which wood fibres in an adhesive binder are compressed into a density of around 450–800 kg/m^3. It has no grain and is used in the manufacture of furniture and artefacts. It is easy to work and has found much use as a DIY material. It has some structural strength and has (limited, usually internal) use as a material in the construction of buildings. The dust generated by (particularly powered) sawing can adversely affect health if inhaled and suitable personal protective equipment should be used when working with it.

Medline An electronic database containing medical and biomedical literature. It is a highly useful resource to aid the identification of published research on a variety of medical topics, including those that have an environmental component.

meiosis A special division of the cell in organisms that reproduce sexually. The nucleus of the cell divides to produce four daughter nuclei, each with half the complement of the parent nucleus. The female sex chromosomes each give rise to a sex cell (egg) containing an X chromosome whilst the male sex cell divides to provide one sperm with an X-chromosome and another with a Y-chromosome. On successful mating the egg and the sperm combine to produce a complete viable cell with a full complement of chromosomes that will then grow to become the offspring.

See also: MITOSIS

melanin A brown pigment that produces the colouration of the skin, hair and eyes in animals, including man. It is produced by cells known as melanocytes. Melanin is one of two pigments found in human skin and hair, the other being CAROTENE. Melanin is capable of absorbing ultraviolet light and therefore plays a role in the protection of the individual from ultraviolet radiation. In certain diseases and cancers the production of melanin is excessive.

See also: MELANOMA

melanocyte A type of cell in the skin responsible for pigmentation.

See also: MELANIN

melanoma A TUMOUR arising in the MELANIN-producing cells of the skin. In its MALIGNANT form it is potentially fatal although it responds well to surgery if caught early enough in its development. It is thought to be caused by, or has been associated with, excessive exposure to ultraviolet-B radiation. It has been speculated that depletion in the OZONE LAYER will result in increased UV radiation and commensurately increased incidence of melanoma. There is epidemiological evidence to support this contention.

meltdown The result of the catastrophic overheating of the core of a nuclear reactor. A total meltdown would result in the core of the reactor becoming so overheated as to melt through any protective containment structure and (at least theoretically) descend into the earth. The results of such an occurrence can only be a cause for speculation as no such event has been recorded. Partial meltdowns have occurred, one of the most notorious being at THREE MILE ISLAND in the USA in 1979.

See also: CHINA SYNDROME

meninges The collective term used to describe the layer of three membranes surrounding the brain and the spinal cord. The outermost and toughest membrane is the dura mater, the middle membrane is known as the arachnoid and the innermost membrane is known as the pia mater.

meningitis An inflammation of the membranes (MENINGES) surrounding either the brain (cerebral meningitis) or the spinal cord (spinal meningitis). It is usually caused by an infection with a VIRUS or a BACTERIUM. Viral meningitis is usually mild, but bacterial meningitis can be life threatening.

See also: ENCEPHALITIS

meniscus The curved surface of a liquid created when the edge of the liquid comes into contact with a solid. The phenomenon is most noticeable when a glass tube (especially a CAPILLARY TUBE) is inserted into the liquid with part of its length projecting above the surface of the liquid. The curvature is created by SURFACE TENSION.

Mercalli scale A comparative method of describing the effects of an earthquake. The scale runs from 1 to 12, with 1 equating to no damage and 12 equating to near total destruction. Point 6 on the scale is reached when light structural damage occurs. Unlike the RICHTER SCALE it is not solely dependent on the amount of energy released as location of the EPICENTRE is an important consideration.

mercaptan A generic term used to describe one of a number of toxic substances containing sulphur that have an unpleasant odour, sometimes described as that of rotting cabbage. They are produced naturally as part of the process of the decay of organic matter and are produced as by-products of some industrial processes. Mercaptans are a contaminant of PETROLEUM but they are usually removed by the refining process. Mercaptan is sometimes added at very low concentrations to both natural gas and to LPG to provide a warning mechanism in case of leakage. These substances are also sometimes known as thiols.

mercury Mercury (Hg) is a very dense silvery coloured metallic element. It is liquid at room temperature. It occurs naturally as mercury sulphide in many rocks, albeit at very low concentrations. The combustion of fossil fuels is the largest source of atmospheric emissions in the UK. Mercury is used in the manufacture of temperature and pressure-monitoring instruments, the manufacture of fluorescent and mercury discharge lamps and in dentistry (as an amalgam with other raw materials for filling teeth cavities). Its principal industrial use is as a liquid cathode in the electrolysis of brine for the production of chlorine and alkali metal amalgam (the amalgam is subsequently hydrolysed to produce hydroxide and the mercury is recycled).

Elemental mercury is easily absorbed by the respiratory tract; absorption by the gastrointestinal tract is less so for elemental mercury and even less so for mercury compounds. The skin also, to a small extent, absorbs elemental mercury vapour. Once absorbed, mercury is widely distributed around the body but it accumulates particularly in the kidneys. Passage of mercury from the mother to the foetus is apparent, especially in respect of elemental mercury. Excretion via urine, faeces or exhalation is relatively slow, and bioaccumulation will therefore occur with repeated exposure. Mercury has been linked with human toxicity for many years. The chronic toxicity exhibited by workers in the hat industry due to the use of inorganic mercury salts for shaping felt used in that manufacture is probably one of the most famous examples of industrial poisoning. The effects were immortalised in the character of the Mad Hatter in the children's book *Alice in Wonderland* by Lewis Carroll.

See also: ACRODYNIA

mere A synonym for lake.

meridian One of the (imaginary) great circles of the earth running along its surface and passing through both North and South Poles. To an observer on the earth's surface the meridian for that location is that which passes directly overhead. The meridian at Greenwich, London, is ascribed as 0° LONGITUDE. To an observer at a fixed point the greatest altitude (i.e. the maximum degrees of elevation from the horizon) the sun reaches in the sky in a day signifies the 'solar noon'.

mesopause That layer of the earth's atmosphere which forms a junction between the MESOSPHERE and the THERMOSPHERE at around 80 km from the earth's surface.

mesosphere That layer of the earth's atmosphere extending between the STRATOSPHERE and the THERMOSPHERE. It ranges in temperature from around 0°C at the STRATOPAUSE to around –80°C at the MESOPAUSE.

mesophylic organism One that has an optimum growth temperature between 35–7°C.

mesothelioma A malignant tumour of the lung. It is primarily associated with exposure to ASBESTOS fibres.

meta-analysis A statistical technique that uses the data from several studies of a particular phenomenon or occurrence in order to produce a composite picture of that which has been studied. The advantage of a meta-analysis is that it produces a balanced view indicating where the strength of the available evidence lies.

See also: BRADFORD-HILL CRITERIA

metabolic activation The conversion of a chemical from one state to another by the action of an ENZYME.

metabolic disease One that affects the body's METABOLISM. It is usually an inherited characteristic and is manifest as a result of ENZYME abnormality.

metabolism The sum totality of physical and biochemical processes whereby life is maintained within an organism itself. 'Basal metabolism' is that minimum level of metabolism (heartbeat, breathing, energy generation, etc.) essential to maintain life.

metabolite Any chemical produced as a result of the action of a living organism's METABOLISM on another substance it has absorbed. It is important to recognise that bodily processes can act on many substances in this way. If one is looking for a particular substance in living tissue it may be more appropriate to look for the metabolite rather than the original substance as the latter could have undergone complete change since it was first introduced into the tissue.

metallo-enzymes Those ENZYMES that contain at least one metal atom in their structure. Zinc is one of the more common metals found in enzymes.

metamorphic rock Any rock that has been changed by physical or chemical action in such a way as fundamentally to alter its original characteristics since it was first deposited in the earth's crust. The term essentially refers to what has happened to it rather than its original manner of generation. SLATE and MARBLE are examples of metamorphic rocks, slate deriving *INTER ALIA* from clay or shale and marble from limestone.

See also: IGNEOUS ROCK; SEDIMENTARY ROCK

metamorphosis The process of physical change in the life cycle of an insect from egg to LARVA through PUPA to IMAGO (adult).

metaphase A stage in the division of the living cell in which the chromosomes are arranged along the central line (equator) of the nucleus (known as the nuclear spindle). The chromosomes are most visible at this stage of division.

See also: MEIOSIS; MITOSIS

metastasis The process whereby a MALIGNANT cell is detached from a primary TUMOUR and is disseminated in the body to set up secondary tumours elsewhere. The process is not random as the migrating malignant cells of many cancers have a predilection for certain sites.

methaemoglobinaemia A condition in which the ferrous component of the HAEMOGLOBIN has been converted to the ferric form, reducing the oxygen-carrying capacity of the blood. The condition exhibits with a typical blue tinge appearing around the lips and on the extremities. It can be either hereditary or toxic. Infantile methaemoglobinaemia (also known as 'blue baby syndrome') has been associated with excess levels of nitrate in drinking water although not all sources accept the association as causal.

See also: CYANOSIS; NITRITE

methanal *see* FORMALDEHYDE

methane Methane (chemical symbol CH_4) is the principal gas found in natural gas. It is an odourless, flammable gas of low density. It is produced on the decomposition of many organic molecules. It is of environmental health importance predominantly because it can escape from landfill sites, enter buildings constructed on or near (especially) decommissioned sites and accumulate to form a potentially explosive mixture in air. It is advisable for buildings located near to potentially methane-producing sites to be equipped with suitable monitoring and alarm sensors to give advance warning of gas accumulating towards dangerous concentrations. Methane can be produced in commercial quantities from decomposing organic material and used as a fuel

See also: ENERGY FROM WASTE; FIRE DAMP; MARSH GAS

methanol Methanol (chemical symbol CH_3OH), also known as 'wood alcohol' or 'methyl alcohol', is the simplest ALCOHOL and the one with the lowest molecular weight. It can be produced by the anaerobic decomposition of wood. It can be dangerous if consumed and may be a contaminant of illegally distilled (and therefore more likely to be inexpertly distilled) spirits. In some countries it is sometimes added to petrol to reduce exhaust gas emissions.

Methuselah A large wine bottle, equivalent in size to eight standard bottles.

See also: JEROBOAM; NEBUCHADNEZZAR

methyl tertiary-butyl ester (MTBE) A volatile, colourless liquid primarily used (particularly in the USA) as an additive for petrol or gasoline to improve the oxygenation or OCTANE capacity of the fuel and to reduce harmful exhaust emissions.

See also: ANTIKNOCK COMPOUND

metre A standardised unit of length in the metric system originally introduced by the French Academy in 1791. Although originally defined as being 1×10^{-7} the distance from the North Pole to the equator, it has now been redefined in terms of the wavelength of light at a specified frequency. The metre is equivalent to the imperial measure of 39.37 inches.

See also: SI

metric system *see* SI

mezzanine (floor) An intermediate floor between two main floors in a building. The term is especially (but not exclusively) applied to a floor intermediate between ground and first-floor levels.

microaerophilic Refers to an atmosphere in which the amount of oxygen by volume is considerably lower than that in air. A micro-organism that is described as microaerophilic is one that can only survive in such an atmosphere.

microcoulometry An electrochemical analytic technique used, for example in the petrochemical industry, to measure the chlorine and sulphur content in a sample. The technique is highly sensitive and accurate.

microfiche A record-storing system using reduced-size, photographically reproduced documents that can then be read through a magnifying reader.

microflora A collective description for a population of micro-organisms.

micronucleus (plural micronuclei) A portion of a fragmented chromosome that has atypically remained in the body rather than being expelled. The presence of micronuclei following exposure to a chemical agent can be used to assess the capacity of the chemical to induce ANEUPLOIDY.

microwave An ultra-high-frequency electromagnetic radiation with a free-space wavelength of between 0.001–0.3 m (i.e. a frequency of around 1–100 gigahertz – a gigahertz is equivalent to a billion cycles per second). Microwaves propagate in straight lines and are not reflected by ionised regions in the upper atmosphere. They do not diffract readily around barriers (such as buildings) but do exhibit some attenuation when passing through trees or wooden structures. They are used for cooking, WIRELESS transmission and RADAR.

mignon In culinary practice refers to a foodstuff that is small or of dainty appearance, such as small fillets of beef.

mild steel STEEL with a carbon content of between 0.1–0.2 per cent.

mile A measure of distance or length. There are various definitions, but the term is usually applied to the standard British or US mile, a distance of 1,760 yards, equivalent to 1.609 KILOMETRES (km). The Admiralty nautical mile is equivalent to 1.853 km; an international nautical mile is equivalent to 1.852 km.

Minamata disease A disease of the central nervous system identified in residents around the town of Minamata, Japan, in the 1950s. It took a further 20 years to resolve the problem and satisfactorily to identify the cause of the disease as being the consumption of locally caught fish and shellfish that had been contaminated by discharges of methyl mercury from a local plastics factory.

mineral Theoretically any chemical that occurs naturally. In general usage the term has come to apply particularly to any inorganic material found naturally within the earth. Thus all the rocks of the earth are composed of minerals.

See also: ORE

mineral oil Oil derived from PETROLEUM. The term distinguishes such oil from that of animal or vegetable origin.

mineral wool A generic term applied to any of several artificial fine fibrous materials manufactured from inorganic materials such as glass, rock or SLAG. The manufacturing process requires that air or steam jets are blown through the molten MINERAL to generate the strands or fibres of the mineral wool.

minimum infective dose The fewest number of PATHOGENIC organisms that are necessary to produce symptoms of a disease or infection.

Minister of State A member of the HOUSE OF COMMONS who has been appointed by the Prime Minister to the GOVERNMENT and is thereby in charge of a Government department. The Ministers of State together form the CABINET.

Minister without Portfolio A Cabinet Minister of the Government who does not have responsibility for a specific department.

See also: PORTFOLIO

miscarriage *see* SPONTANEOUS ABORTION

miscible Refers to two or more liquids that are capable of being mixed together without separating out afterwards into the original constituents. The term IMMISCIBLE refers to those liquids that are not capable of being mixed together in this way.

mist Defined by the International Standards Organization (ISO) as a meteorological condition in which the presence of a suspension of water droplets in the air reduces visibility to more than 1 km but less than 2 km.

See also: FOG

mitosis Simple cell division such as to further the process of growth or cell replacement. The chromosomes of the cell divide into two thereby doubling their number. The parent cell then divides to form two new cells, each with its own full complement of chromosomes.

See also: MEIOSIS

MMAD The acronym for Mass Median Aerodynamic Diameter and is used to characterise the distribution of particle sizes in a given sample of particles. A sample with a MMAD of (say) 0.5 mm will have 50 per cent of the total mass (i.e. not the total number) of particles with a diameter of more than 0.5 mm and 50 per cent with a diameter of less than 0.5 mm.

See also: AERODYNAMIC DIAMETER

MMR vaccination A single-dose combined vaccine against measles, mumps and rubella (German measles), usually offered to children in their second year of life. It was introduced, for routine immunisation, into the UK in 1988. Following published research (much of which was said to have been misreported) there was considerable concern expressed in the media regarding alleged associations between MMR vaccine with Crohn's disease, autism and other complications. In spite of reassurances by the medical authorities many parents are seeking to have their children immunised using individual rather than combined vaccinations, or avoiding having their children immunised at all. The non-vaccination of children, especially in relation to measles, is predicted by many experts as leading unavoidably to a measles epidemic at some time in the future.

mode That number which occurs most frequently in a set of numbers.

See also: MEAN; MEDIAN

model An ALGORITHM that allows analysis of a problem to be undertaken using data as a proxy for real life. Ideally the model will contain all the information that is relevant to the problem but exclude that which is irrelevant. In practice models are often generated in line with economic or practical constraints; it can, for example, be more expensive and time consuming to produce sophisticated computer-generated models than to conduct investigations in real life.

See also: VIRTUAL REALITY

modem A device that connects two computers by telephone line allowing them to send signals to one another by converting the digital signal from one into an analogue signal for transmission and, at the other end, converting this back to a digital signal again. The name derives from MOdulate (converting digital to analogue) and DEModulate (converting analogue to digital).

moderator In a NUCLEAR REACTOR, a material used to control the speed (and thereby the energy) of NEUTRONS emitted during the process of NUCLEAR FISSION.

modified-atmosphere packing (MAP) A food industry technique in which the air in a food pack is replaced with one or more alternative gases, the pack then being sealed to maintain its integrity. MAP is used predominantly to increase the shelf life of the product by excluding air (principally the oxygen component of the air) and thereby reducing the rate of microbial growth and commensurate spoilage.

modulation A periodic fluctuation in the amplitude, frequency or phase of a noise.

moisture barrier Any impermeable component incorporated into a structure for the purpose of excluding moisture.

See also: DAMP-PROOF COURSE; VAPOUR BARRIER

moisture content The amount of water held in a solid, usually expressed as a percentage by dividing the weight of water by the total combined weight of the solid plus the water.

moisture gradient The variation in the water content of a given material between two specified points. In timber this is usually between the inside and the outside; in building materials it may be between one side and the other.

molecular fingerprinting A technique in which an organism is distinguished from others of the same species by virtue of characteristic differences in its genetic makeup.

molecular weight The total mass of a molecule's constituent atoms.

molecule The smallest portion of a substance that can exist in the free state and retain the physical and chemical properties of the substance.

molybdenum disulphide Molybdenum disulphide (MoS_2) is a dry-film lubricant used as both a lubricant and as an additive to lubricants. It has good high-temperature performance characteristics. It is sometimes referred to solely as 'moly' or as 'molysulphide'.

monoclinous Refers to the sexual characteristics of any flowering plant that has both male and female reproductive organs on the same flower.

See also: DICLINOUS

monoculture An agricultural practice in which only crops of one specific type are grown in a defined area (such can be a field or larger area). This is usually accomplished by using herbicides to kill all other plants (such as weeds or indigenous plants) leaving the selected crop to grow in the absence of competitive species. The use of genetic modification to provide crops with selective herbicide resistance is likely to increase the use of monoculture as crops can then be sprayed *IN SITU* during the growing period without risk of damaging them. The use of monoculture however reduces biodiversity by removing natural food species, especially for insects and micro-species, thereby removing the food source for animals further up the food chain. This can lead to a complete change in an area's flora and fauna.

See also: ORGANIC FOOD; SUSTAINABLE DEVELOPMENT

monolith A large block of stone, or anything, such as a block of concrete, which resembles one.

monolithic construction Any structure created as a single unit or made by assembling MONOLITHS.

monomer A single unit building block in organic chemistry that may link in chain formation with other similar units or identical blocks to form more complex compounds. These latter are known as POLYMERS.

monosaccharide A simple SUGAR containing either five or six carbon atoms. GLUCOSE ($C_6H_{12}O_6$) is an example of a monosaccharide.

See also: POLYSACCHARIDE; SUCROSE

monoxenous The adjective used to describe a parasite that can develop fully in a single host.

Mooney viscosity A measure of the resistance to deformation of raw rubber. The measurement is taken using equipment known as a Mooney viscometer by embedding a steel disc in a sample of the rubber to be tested. Energy is then applied to the disc to make it rotate; the point at which the resistance is overcome is known as the Mooney viscosity value. Continuing to rotate the disc increases the viscosity of the rubber, and the time necessary to raise the viscosity by a specified amount is known as the Mooney scorch value. This latter is used as an indication of the tendency of the sample to undergo VULCANISATION.

morbid Means of or pertaining to disease.

See also: MORBIDITY

morbidity The state or condition of being affected by disease. The term in epidemiology is often used to denote the ratio of those in a population with a disease to the totality of the population being considered.

See also: MORTALITY

mortality Refers to death. The term in epidemiology is often used to denote the death rate – i.e. the ratio of those who have died from a (usually specified) condition or cause compared to the totality of the population under consideration.

mortar In construction terminology, a material, nowadays usually consisting primarily of a sand/CEMENT mixture, used to bind bricks or stones together in the form of walls and other structures. Previously lime was commonly incorporated into the mixture as an alternative to cement.

See also: GROUT

mortise A rectangular slot cut into a piece of timber with the aim of receiving a TENON or projection from an interlocking piece. A 'mortise and tenon' joint in woodworking is one in which the tenon of one piece fits closely at a right angle into the mortise of another piece. The US spelling is 'mortice'.

mortise lock A fastening device for a door or similar in which the tongue (or 'TENON') of the lock enters, upon locking, into a rectangular hole (the 'mortise') in the door JAMB.

See also: RIM LOCK

mosaic A decorative surface of a building or artefact comprising small pieces of glass, pottery, tile and marble or similar inlaid to an adhesive surface. The mosaic is usually constructed to form a picture or geometric pattern. The small pieces from which the mosaic is made are known collectively as tesserae (singular 'tessera').

motile Means capable of independent movement.

mould Is either a forming for producing or casting an article of production or a type of fungi. In relation to the latter the term in use is relatively imprecise. It is generally applied to smaller saprophytic (see SAPROPHYTE) fungi that appear on the surface of foods, buildings and other articles and which, to the naked eye, usually appear to have an indistinct structure. Mould grows in the presence of moisture and, especially on inorganic material, often stains the surface. The growth of mould on living organic material is sometimes known as mildew although common parlance frequently uses the terms as interchangeable.

See also: ERGOT

moulding A shaped piece of construction material (such as wood, stone or plasterwork) used as a decorative finish or to cover and disguise a joint or similar.

mouse spot test A laboratory test to assess the MUTAGENIC effects of a foreign substance. The test is conducted *IN VIVO* by dosing mice with the test compound and subsequent examination for the development of mutations (i.e. 'spots') in the skin pigment cells (known as melanocytes).

MRL *see* MAXIMUM RESIDUE LEVEL

MRM The acronym for MECHANICALLY RECOVERED MEAT

MRSA An abbreviation of Methycillin-Resistant *STAPHYLOCOCCUS AUREUS*. This is a type of the BACTERIUM known as *Staphylococcus aureus* that has developed a resistance to the antibiotic methycillin. MRSA is carried normally in the nose and throat of a small percentage of the population without ill effect. However, its potential to cause illness, particularly to people in hospitals, is a major concern.

MRSA does not behave any differently from other strains of *S. aureus* but if it infects patients with severely impaired resistance or infects wounds or other breaches of the skin it can cause septicaemia, pneumonia and other serious infections. Although it is a potential risk to those in hospital MRSA poses no particular risk to those in the community, including to those in residential and most nursing homes. In the event of a flu epidemic, however, because of its antibiotic resistance, secondary infection with MRSA could potentially kill more patients than the primary infection with the influenza virus itself.

See also: CARRIER; NOSOCOMIAL

MTBE *see* METHYL TERTIARY-BUTYL ESTER

mucosa *see* MUCOUS MEMBRANE

mucous membrane The generic name applied to any of the membranes that line many of the hollow organs (such as the lungs, the intestine, etc.) of the body. Mucous membranes may also be referred to as 'mucosa'. In healthy individuals the mucous membranes are only covered by a thin layer of mucus; excessive production or discharge is a sign of inflammation of the relevant bodily tissues.

mucus The generic name given to the slimy secretions produced by the MUCOUS MEMBRANES. Mucus varies in composition dependent on the body site from which it is derived, but it is mainly comprised of a substance known as 'mucin'.

mulling The process of warming an alcoholic beverage such as cider or wine, usually with the attendant addition of spices or sweeteners.

mullion A vertical dividing rail of a window. The mullion differs from a GLAZING BAR in that it is usually a structural member and consequently possesses the capacity to be load bearing.

See also: TRANSOM

multiple chemical sensitivity A condition whereby exposure to certain chemicals (including notably pesticides, solvents, fragrances and food allergens) is thought to generate or increase the likelihood of the production of an allergic-like response with commensurate symptoms such as irritation to eyes, respiratory tract and skin sensitivity following exposure to other chemicals.

multiple sclerosis A chronic, progressive, degenerative disease of the brain and the spinal cord. Initially the disease develops as a result of the degeneration of the nerve sheaths but this can be followed by the development of random scattered nodes of connective tissue throughout the central nervous system. The disease is characterised by many symptoms, especially by stiffness and paralysis, the degree or extent of which is dependent on the portion of the central nervous system affected.

mumps An infectious viral disease predominantly of childhood or young people. It is characterised by an increase in temperature, fever, sore throat and swelling of the parotid glands (located below and slightly in front of the ear). The disease is not usually fatal although some sources report fatality of around 1.4 per cent. Meningitis can be a complication. Mumps is also known as 'epidemic parotitis'.

See also: MMR VACCINATION

muntin An alternative name for a SASH BAR.

muntz metal An alloy of copper and zinc mixed in the ration 3:2.

Murphy's Law A humorous expression denoting the perception of perversity usually summarised as 'anything that can go wrong, will go wrong'. It is alternatively known as 'Sod's Law'.

mutagenic Means having the capacity to increase the rate of MUTATION in cells. The capacity of being mutagenic is not normally applied as referring to the capacity to increase the range of mutations. Chemicals or physical agents that are mutagenic are known as mutagens; the occurrence or induction of mutation is known as mutagenesis.

See also: CARCINOGENICITY; GENOTOXIC; TERATOGENIC

mutation A permanent change in the genetic material of the cell potentially leading to a change in the physical characterisation of the progeny of that organism on reproduction. Upon mutation the resultant DAUGHTER cells, following subsequent cell division, will be different from those or the precursors. Mutated cells might either be destroyed by the body's own defence systems or develop into TUMOURS.

mycoplasmas Micro-organisms that do not possess a cell wall. Mycoplasmas can form colonies on media enriched with bodily fluids such as serum. The group includes organisms that are PATHOGENIC to animals and man, but not all possess this quality.

mycotoxins A diverse group of toxic compounds produced by some fungal mould species. Consumers are most likely to be exposed to mycotoxins through contaminated foods, including cereal products, fats and oils, nuts and nut products, seeds, spices and herbs, pickles, sauces, some varieties of canned vegetable and pickle products, chilli powder, curry powder and ginger.

See also: AFLATOXIN

myocardial infarction A blockage (INFARCTION) of a blood vessel supplying the muscular tissue of the heart (the myocardium). The blockage reduces or prevents a supply of oxygen to the heart. An acute myocardial infarction is known as a 'heart attack'.

myocardium The muscular tissue of the heart.

myoglobin A protein that contains iron and is found in muscle cells. It has the function of storing oxygen.

myriapod A member of the classification of animals defined by their having multiple legs. The classification includes centipedes and millipedes.

myxomatosis An endemic viral disease of rabbits in South America. It was deliberately introduced into Australia in the 1950s in an effort to control the population explosion in rabbits originally introduced there for sport and food. It was subsequently introduced to Britain with devastating effect, not only reducing rabbit numbers but also removing them from the food chain for larger predators.

N

nacelle The enclosure located at the top of the tower of a wind turbine generator, usually housing the gearbox and generator, and from which the shaft of the rotor protrudes.

nanometre A unit of length, one nanometre being equivalent to one-millionth of a millimetre.

naphthene A generic industry name given to a number of hydrocarbon compounds containing saturated carbon atoms in a ring structure. Naphthenes have the general formula C_nH_{2n}. Naphthenes are also known as cycloalkanes and cycloparaffins. Naphthenes are used in lubricating oils and have good solvency properties.

See also: PETROLEUM

nasal septum The central wall of tissue dividing the nostrils. It is vulnerable to CHRONIC damage by inhaled pollutants and chemicals, and its damage or destruction is one of the classical physical manifestations of long-term cocaine inhalation.

national formulary A publication containing the key details and information in relation to the prescribing, dispensing and administration of medicines. The formulary contains advice on the therapeutic use of medicine and on the choice of drugs available. Many countries publish their own formularies; in the UK this is the British National Formulary.

National Institute for Clinical Excellence (NICE) A Special Health Authority set up as part of the National Health Service (NHS) in England and Wales on 1 April 1999. Its primary function is to provide advice to patients, health professionals and the public in respect of best practice in relation to both specific diseases (or conditions) and health technologies. NICE also provides advice in relation to the methods for conducting clinical audits.

National Radiological Protection Board (NRPB) An independent public body established in 1970 in the UK to advise on the protection of the public and employees in relation to all aspects of exposure to radiation. The Government's Chief Medical Officer proposed in 2002 that the NRPB would become part of a National Infection Control and Health Protection Agency.

National Service Framework (NSF) A national standard for the UK for the provision of health care services specified for individual services or care groups. The NSFs establish performance measures against which improvements can be measured or tracked.

natural gas The term used to describe any gas that occurs naturally and includes gas that is produced by the decomposition of oil or coal. Natural gas is often found in association with deposits of CRUDE OIL in layers of limestone or sandstone capped by impermeable shales or clay. It is used widely as a fuel source but its composition is variable dependent on source. Natural gas generally contains between 80–95 per cent methane with around 3–8 per cent ethane and lesser amounts of propane, butane and pentane. It may also contains varying amounts of carbon dioxide, nitrogen and helium.

natural rubber A naturally occurring ELASTOMER obtained commercially primarily from natural rubber latex produced by the rubber tree *Hevea brasiliensis*. It is used unblended in the production of tyres for larger vehicles such as aircraft and heavy agricultural machinery, and as an additive to synthetic rubber in the manufacture of car tyres.

natural selection Also known as 'survival of the fittest', natural selection is the keystone of the theory of evolution proposed by Charles Darwin. In essence the theory proposes that the development of life arose from the mechanism that the life form which can best survive or exploit a given set of conditions would be more likely to survive and multiply than competing life. The term 'survival of the fittest' is often misinterpreted as survival of those in the best physical condition – it should properly be interpreted as survival of those 'most fitted to survive in given conditions'.

See also: GAIA HYPOTHESIS

natural ventilation Any system whereby air is circulated through and within a building by allowing natural variations in air pressure to move it from one area to another.

See also: FORCED VENTILATION

Nature Conservancy Council (NCC) Established in Britain in 1973 to deal with all aspects of nature conservation, education and research. It was subsequently reorganised on a regional basis to create English Nature, the National Countryside Council for Wales and Scottish Natural Heritage.

nausea The sensation that one is about to vomit. It need not necessarily be followed by actual vomiting.

nautical mile *see* MILE

Nebuchadnezzar A large wine bottle, equivalent in volume to twenty standard wine bottles.

See also: METHUSELAH

necrosis The death of a cell or portion of a tissue or organ of a plant or an animal at a time when the remainder of the organism is still alive. The term in animals is usually applied to the death of internal body parts whilst the death of cells on the outside of the body is normally referred to, dependent on extent, as either GANGRENE or ULCERATION.

needle A strong metal or timber horizontal beam that fastens a SHORE to the wall.

nematode Any worm of the class *Nematoda* or unsegmented worms. The group includes both free-living and parasitic worms.

neonatal Literally means 'new born'. The term is applied in humans to describe events or occurrences relating to the period encompassed by the first 28 days (i.e. the first 4 weeks) of life.

neonatal sepsis A condition of disease or infection in a NEONATE associated with microbial infection.

neonate A new-born baby within its first 28 days of life.

neoplasm A synonym for TUMOUR.

neoplastic An adjective used to describe cells, the growth of which is more rapid than that of other cells.

neoprene A synthetic rubber compound. It is highly durable and has good resistance to water, oil, fire and many chemicals. It has many applications including providing insulation for electric cables, the manufacture of flexible hoses and as an additive to specialist adhesives and paint.

nested case control study A CASE CONTROL STUDY in which both cases and controls are selected from an identified cohort defined as those with similar exposure to the risk being examined. Choosing people in this way reduces the effect of CONFOUNDING FACTORS.

neurobehavioural That behaviour determined by the operation of the nervous system.

neurological shellfish poisoning (NSP) An illness caused by the consumption of shellfish (primarily bivalve molluscs) that have ingested and accumulated in their hepatopancreas toxin produced by algal blooms (see BLUE-GREEN ALGAE). The toxin responsible is known as Brevetoxin. Symptoms are milder than those of PARALYTIC SHELLFISH POISONING and include gastroenteritis, myalgia, vertigo and a burning sensation around the rectum.

See also: AMNESIAC SHELLFISH POISONING; DIARRHETIC SHELLFISH POISONING

neurotoxin A poisonous substance that is capable of adversely affecting the nervous system.

See also: toxin.

neurotransmitter A substance that facilitates the transmission of the signal from a nerve to the cell. Depletion of the substance reduces the capacity for transmission. Parkinson's disease is a condition arising from a depletion of the neurotransmitter known as dopamine.

neutron An elementary particle in the NUCLEUS of an ATOM. It has a unit mass of approximately 1 and no electric charge.

new chemical Any chemical that is not listed by the European Inventory of Existing Chemicals (EINECS) between January 1971 and September 1981. A chemical that does appear on the list is defined as an 'EXISTING CHEMICAL'.

newton The SI unit measure of force. One newton is that which produces an acceleration of 1 metre per second on a mass of 1 kilogram.

See also: joule.

Newtonian fluid One whose viscosity does not change with flow rate.

NHS number In the UK, a unique 10-digit number allocated for the purposes of identification to an individual upon registering with a general practitioner.

NICE The acronym for the NATIONAL INSTITUTE FOR CLINICAL EXCELLENCE.

nicotine The brown fluid active principle in tobacco. In small one-shot doses it can act as a stimulant, but in larger doses this effect is modified to that of a narcotic depressant. Habitual exposure, such as through continued tobacco usage, translates the nicotine effect to that of a sedative, contributing to its addictive effect and enhancing dependency by the habitual smoker.

See also: ALKALOID

NIMBY syndrome A (generally) derogative term derived from the acronym for Not In My Back Yard. It is characterised as applying to residents in a given geographical area who object to the personal detriment they perceive as arising from the (proposed) introduction of a new development into their neighbourhood, albeit that the development might actually be of benefit to the wider population. There is an implicit suggestion that the objection is generated solely from self-interest and that the objectors would consider the same development elsewhere as being satisfactory.

There is a danger in accepting NIMBY claims too readily. Developers can, as a smokescreen, suggest a NIMBY perspective is being created simply because there are local objections to a proposal. This should not be allowed to mask genuine concerns or objections that a particular development really would be better sited elsewhere.

nitrate Together with NITRITE, is part of the nitrogen cycle. Nitrate is used predominantly as an inorganic fertiliser. Nitrate concentrations in groundwater supplies can be elevated due to water run-off from agricultural land. Nitrate can be toxic to humans, principally due to its reduction to nitrite. At 2002 there is no evidence of an association between exposure to nitrate and risk of cancer in humans.

nitrification The process whereby nitrogen compounds are converted by the oxidation of ammonia into NITRATE in the soil by the action of nitrifying bacteria such as *Nitrosomonas*.

nitrite Together with NITRATE, is part of the nitrogen cycle. Sodium nitrite is used as a food preservative, especially for cured meats. Chloramination can lead to the formation of nitrites in water distribution systems. Nitrite is toxic to humans because it is part of the process in the oxidation of HAEMOGLOBIN to methaemoglobin that cannot carry oxygen. If concentrations of methaemoglobin in the blood reach around 10 per cent of normal haemoglobin this can cause the condition known as METHAEMOGLOBINAEMIA, in which clinical symptoms such as CYANOSIS can be manifest. Higher concentrations can lead to ASPHYXIA.

The haemoglobin in young infants is more susceptible to the formation of methaemoglobin, and as a consequence children will form more methaemoglobin from a given intake of nitrate (which is reduced in the body to nitrite). Bottle-fed infants living in areas of high nitrate concentration in drinking water are therefore at greater risk of developing methaemoglobinaemia.

nitrogen fixation The process whereby bacteria living in the soil or in the roots of certain plants (such as clover) convert the free nitrogen in the atmosphere into ammonia (NH_3) and subsequently into more complex nitrogen compounds. These latter can then be used by the plants as nitrogenous fertilisers.

See also: NITRIFICATION

nitrogen oxides The term applied collectively to any of the seven combinations of nitrogen and oxygen. These are as follows:

- Nitric oxide (NO).
- Nitrogen dioxide (NO_2).
- Nitrogen trioxide (NO_3).
- Nitrous oxide (N_2O) (also known as laughing gas).
- Dinitrogen trioxide (N_2O_3).
- Dinitrogen tetroxide (N_2O_4).
- Dinitrogen pentoxide (N_2O_5).

Of these only nitric oxide and nitrogen dioxide are thought to have any health significance to humans.

No Observed (*or* Observable) Adverse Effect Level *see* NOAEL

NOAEL The acronym for No Observed (*or* Observable) Adverse Effect Level. It refers to the maximum level at which a substance can be administered to a living organism at which no toxic effect is discernible in relation to that organism.

noble gas *see* INERT GASES

nocturnal A term that is applied to animals and plants meaning that they are active during the hours of darkness.

See also: DIURNAL

nogging A short horizontal timber used to brace the vertical timbers in a stud partition. The term also refers to brickwork used to infill the vertical studs in timber frame construction.

noise Most commonly defined as unwanted SOUND. It is also rendered as any unwanted electric or electromagnetic energy. This latter broadens the definition to include that unwanted energy which has the capacity to distort or downgrade the quality of signals or data within a system.

noise and number index (NNI) A technique developed at Heathrow Airport, London, for assessing the disturbance to humans living near air terminals due to the noise generated by aircraft. The index is derived by combining perceived noise levels with the number of aircraft movements per day.

noise weighting The term applied to techniques in which a SOUND is attenuated by means of filtering out or reducing the energy component of specified frequencies in order to mimic the perceived sound level to a given recipient. One of the most commonly used is the A weighting, which allows a noise monitor to produce a response approximating that of the human ear.

noisette In culinary practice refers to small round individual portions or slices of meat.

nominal scale One that apportions data into unordered qualitative categories, e.g. Tom and Sue are both adults.

See also: INTERVAL SCALE; ORDINAL SCALE

nominal size (of timber) The size of a piece of timber before it is dressed (see DRESSED TIMBER). Most timber sizes from a timber merchant are quoted as nominal sizes.

non-ionising radiation Any form of RADIATION that does not cause ionisation. They are often commonly regarded as 'safe' because they are insufficiently powerful directly to damage DNA. However intense non-ionising radiation can cause physical damage to other parts of the body, such as the eyes. LIGHT, INFRARED RADIATION, ULTRAVIOLET RADIATION and RADIOFREQUENCY RADIATION are all examples of non-ionising radiation.

non-sporing Relates to the inability of a micro-organism to form a SPORE.

non-tariff trade barrier Any economic, political, administrative or legal barrier to free market trade that does not specifically involve the imposition of monetary charges such as importation taxes or duties or the use of quotas in an effort to prevent the free flow of goods between one country and another.

normal distribution An alternative name for GAUSSIAN DISTRIBUTION.

normative needs Those health needs of a patient as defined or determined by a health professional.

Norwalk agent *see* SRSV

Norwalk-like virus The term originally given to the micro-organisms now more commonly known as Small Round Structured Viruses (SRSV). The name derives from the town in Ohio, USA, where they were first properly described. For general purposes the terms Norwalk-like virus and SRSV are synonymous.

nosing The overhanging portion of the TREAD on a staircase. The total distance from the front to the back of the tread comprises the nosing and the GOING.

nosocomial Refers to that of, or relating to, a hospital. The term is most frequently applied to a disease that arises as a result of a stay in hospital or is aggravated by such a stay. The term 'nosocomial infection' is often used as a synonym for a 'hospital-acquired' infection.

notifiable disease In the UK, a named disease that a registered medical practitioner is required by STATUTE to report formally to a designated authority (usually the PROPER OFFICER of the local authority) whenever such a disease is identified or suspected.

NRPB The acronym for the NATIONAL RADIOLOGICAL PROTECTION BOARD.

NSF The acronym for NATIONAL SERVICE FRAMEWORK.

nuclear fission The splitting of the nucleus of an atom with accompanying release of energy. It is the principle under which NUCLEAR REACTORS generate power.

See also: NUCLEAR FUSION; RADIATION

nuclear fusion The term applied to the process whereby two smaller nuclei of an ATOM are combined to create a new nucleus of a larger atom with the concomitant release of energy. The process has been demonstrated in the use of nuclear fusion weapons. The process has also been promoted for many years as an environmentally friendly method of energy production as it would not generate the radioactive by-products of NUCLEAR FISSION processes. Unfortunately, controlling the process sufficiently to enable energy generation has never been demonstrated as feasible in practice and there is considerable scepticism as to whether the process could ever be commercially viable.

nuclear reactor A device for the production of nuclear energy under controlled conditions. It uses a self-sustained chain reaction NUCLEAR FISSION to generate heat, which is then used to power machines (usually turbines) for the generation of electricity.

nuclear winter The term applied to describe the theoretical outcome of a major nuclear war. The theory postulates that the detonation of many nuclear warheads would create widespread firestorms. These, together with the explosions, would throw sufficient particulate material up into the atmosphere to reflect enough solar radiation to reduce global temperatures and produce winter conditions across the globe.

nucleation The process whereby matter coalesces around a focal point or object.

nucleic acid One of the four major classes of compounds that constitute the living body (the others are CARBOHYDRATES, LIPIDS and PROTEINS). They occur as one of two macromolecules, DEOXYRIBONUCLEIC ACID (DNA) or RIBONUCLEIC ACID (RNA), and contain the genetic information necessary for reproduction.

nucleus In atomic physics, the central structure of an ATOM comprising the PROTONS and the NEUTRONS. In cell biology the nucleus is the controlling centre of a cell and contains the DNA. The cells of bacteria notably do not contain a nucleus.

nuclide The nucleus of an atom, comprising protons and neutrons.

numerator That part of a vulgar fraction which appears above the line. Used extensively in STATISTICS the numerator is essentially that part of the whole (the DENOMINATOR) which defines the fraction.

nutraceutical The term used to describe a food that has (or is claimed to have) medicinal or health-giving properties. Although the term strictly has a different meaning from the term NUTRICEUTICAL the two are sometimes used interchangeably.

Nutrasweet *see* ASPARTAME

nutriceutical A hybrid word derived from 'nutrient' and 'pharmaceutical'. The term is used to describe those foods that contain some added active ingredient which has (or is claimed to have) some medicinal use.

See also: NUTRACEUTICAL

nutritional claim A term that is used in relation to food to describe the presence or absence of a specified level of nutrient within a food. It is usually expressed as a numerical value in the food as a whole, or per unit weight, or as a comparative value with other foods.

See also: FUNCTIONAL CLAIM

nux vomica The term usually applied to the seed of an Indian tree, *Strychnos nux-vomica*, from whence the bitter tasting ALKALOIDS brucine and STRYCHNINE are derived. Brucine has similar properties to strychnine, but is much weaker.

O

oakum A collection of fibres (often originally of hemp) obtained by unpicking strands of rope. It was used in conjunction with pitch to seal the joints in wooden sailing boats in an operation known as CAULKING.

obesity A clinical term used to describe a condition of excess weight of an individual due to the deposition of fatty tissue. The generally accepted medical definition of obesity is an excess of weight of at least 20 per cent over the normal weight for a person of that particular sex and height.

objective assessment One in which that being judged is assessed dispassionately (usually but not necessarily) using a standardised set of criteria against which the assessment is conducted. The judgement should therefore remove any influence of prejudice by the person conducting the assessment but also thereby excludes the dimension of preference.

See also: SUBJECTIVE ASSESSMENT

obligatory/obligate A term indicating a compulsory requirement. It is usually applied to a living organism's respiratory requirement in respect of specified gas, e.g. an obligatory AEROBE requires the presence of free oxygen to respire.

See also: ANAEROBE; FACULTATIVE

obscenity A legal term usually defined in STATUTE. In the UK the term is defined under the Obscene Publications Act 1959 (S1(i)) as follows – 'an article shall be deemed to be obscene if its effect... is, if taken as a whole, such as to tend to deprave and corrupt persons who are likely... to read, see or hear the matter contained or embodied in it'.

observational study A type of epidemiological study in which the raw data relating to that being examined are gathered without any intervention other than that necessary to obtain the information.

See also: EXPERIMENTAL STUDY

occlusion A meteorological condition in which an area of warm air is trapped between two converging cold fronts and forced to rise as a consequence. As the warm air rises it begins to cool causing the water vapour within it to condense and fall, dependent on the temperature, as rain, sleet or snow.

Occupational Exposure Limit (OEL) The maximum advisory concentration of a chemical substance to which a worker could repeatedly be exposed or could inhale that would not, within current scientific knowledge, produce a risk to health within the lifetime of the worker. Occupational Exposure Limits are specified for many chemicals usually on the recommendations of expert advisory committees – in the UK this is primarily the Advisory Committee on Toxic Substances and in the EC it is the Scientific Committee on Occupational Exposure Limits. If a substance in the workplace is controlled to an Occupational Exposure Limit value then it is deemed that adequate control has been achieved. The EC produces Indicative Occupational Exposure Limit Values (IOELVs) to which Member States are obliged to have regard in setting legal limits within their own countries.

octane number An indication of the antiknock properties of a particular grade of PETROL; the higher the number the greater the antiknock property of that particular fuel. The number is generally determined variously in the laboratory, although there are also methods of calculating this on the road. The octane rating is generally taken as an equivalent to that of varying blends of iso-octane, which in pure form has a rating of 100. An octane rating of 90 therefore rates a fuel as having the equivalent antiknock properties as a blend of fuel containing 90 per cent iso-octane.

See also: ANTIKNOCK COMPOUND; KNOCK

odds A statistical calculation derived by comparing the number of events divided by the number of non-events. As a hypothetical example, in tossing a coin a hundred times there were fifty heads and fifty tails, the odds would be one; in other words odds of one mean that there is an even chance of an event happening or not. Odds of greater than one indicate that an event is more likely to happen than not, while odds of less than one mean it is less likely to happen at all.

See also: ODDS RATIO

odds ratio A statistical technique used in epidemiological CASE CONTROL STUDIES derived by dividing the ODDS of an event happening in an exposed group by the odds of the same event happening in the control group. Although the odds ratio has been quoted for many years in scientific studies, it is now increasingly common for researchers to quote the RELATIVE RISK.

odorant A chemical added to another substance to give it a distinct and recognisable odour. Odorants are often used to provide a warning mechanism in case of spillage of leakage.

See also: MERCAPTAN

oedema An excessive collection of fluid in a tissue of the body.

oestrogen A HORMONE responsible for the development and maintenance of female characteristics. The US spelling is estrogen.

See also: PHYTOESTROGEN

ogee An 'S'-shaped curve, usually elongated.

ogive A diagonal structural rib of a vault. It is also the term used to describe a window or arch that is pointed at its highest part.

ohm A unit measure of electrical resistance. One ohm is the resistance in which a potential difference of one VOLT produces a current of one AMPERE.

ohmic heating A process that sterilises foods using heat that has been generated by the passage of an electronic current. It was first suggested as a method of food preservation in the latter part of the 19th century but it required the improvements in electrode design of the 20th century to be able to introduce it into commercial applications. The final product is aseptic and can be stored for months or even years at ambient temperatures. The process is expensive to introduce and run, and is therefore economically best suited to high added-value type products such as cook-in-sauces, pie fillings and liquefied fruit. The process does not appear adversely to affect the taste of the product. The process involves food being pumped between a series of specially designed electrodes with an alternating electric current being passed through the food. The food is held normally for not more than 90 seconds before passing on to cooking and packaging stages.

The potential advantages of ohmic heating include greater control of the process, greater speed and more even heat distribution and improved heat penetration. The absence of a heat transfer surface reduces deposit formation within the equipment that in turn leads to easier cleaning and disinfecting. Ohmic heating is sometimes known as Direct Resistance Heating.

oil A neutral liquid that contains one or more TRIGLYCERIDES. Oils are distinguished from FATS in that the latter are solid at room temperature, whereas oils are liquid at room temperature. There are three main classifications of oils. Fixed (or fatty) oils are ESTERS of fatty acids derived from animal or vegetable material. Mineral oils are hydrocarbon-based compounds derived from petroleum, coal and shale. Essential oils are hydrocarbon-based compounds that are derived from vegetable material. They (usually) possess distinctive odours.

oil acne An occupational skin disorder also known medically as oil folliculitis. The condition is usually initiated by petroleum oils. These irritate the hair follicles on exposed parts of the body, such as on the hands, arms, legs or face. Irritation leads to the blockage of the follicles, which in turn leads to the formation of blackheads. If untreated, secondary pustules or inflamed spots may develop.

See also: CHLORACNE

oil dispersants Chemical agents used to disperse oil in water, especially following accidental spillage from oil tankers and the like. They work by reducing the interfacial tension between the water and the oil, thereby allowing the creation of a large number of droplets that are small enough to be dispersed in the water. In the sea the droplets have a tendency to sink and the sea's mechanical action assists in the natural dispersion and dilution process. Dispersants are not usually able to treat all of the oil in a heavy spillage under field conditions (rates vary from 30 per cent to 90 per cent dependent on weather conditions and application) and it is not appropriate to use dispersants for all types of oil. Very light oils such as diesels disperse naturally and evaporate quite readily. Heavy crude oils and fuel oils are not particularly amenable to treatment. Light and medium-weight oils may however be treated. It is important to protect coastline and vulnerable bird life from the potential effects of the dispersants themselves during the spraying process.

oleic acid Oleic acid ($C_{17}H_{33}COOH$) is an organic acid found in animal and vegetable oils. It is used in the manufacture of SOAP.

oligotrophic Refers to a poor level of nutrients (generally) in bodies of fresh water such as to reduce the capacity of this environment to support plant life.

See also: EUTROPHIC; EUTROPHICATION

-ology The suffix to a word denoting the study of something. The prefix denotes the particular subject. A list giving examples of these appears below.

Scientific term	*Is the study of*
Aetiology	Causes of disease
Agrostology	Grasses – classification and use
Anthropology	Man and mankind

Astrology	Prediction using the stars
Astronomy	Heavenly bodies
Autecology	Organism/environment relationship
Biology	Life
Cardiology	The heart and its diseases
Cytology	Cells
Dendrology	Trees and shrubs
Dermatology	The skin and its diseases
Ecology	Life/environment relationships
Entomology	Insects
Etymology	The origin of words
Epidemiology	Disease in defined populations
Geology	The earth's crust
Geomorphology	The form of the earth's surface
Gerontology	Ageing and the problems of age
Gynaecology	Diseases specifically of women
Haematology	Blood and its diseases
Histology	Minute structures of the body
Hydrology	Environmental water dynamics
Immunology	The body's immune system
Limnology	Freshwater environments
Meteorology	The earth's atmosphere and weather
Morphology	Structure of micro-organisms
Mycology	Fungi
Neurology	The nervous system
Neuropathology	Diseases of the nervous system
Nosology	Scientific classification of disease
Oncology	Malignant disease, e.g. cancer
Ontology	Metaphysics and the nature of being
Ophthalmology	The eye and its diseases
Pedology	Soil morphology and distribution
Pharmacology	Drugs – their character and use
Phenology	Climate and life form seasonality
Philology	Language
Phrenology	Health via head shape (*discredited*)
Physiology	Healthy organs of the body
Radiology	X-rays and similar radiation
Rheology	The flow of materials
Rheumatology	Rheumatic diseases
Scientology	Religion through science (*belief*)
Serology	Antibody/antigen reactions
Sociology	Human social behaviour
Toxicology	Poisons
Tribology	Moving surface interactions
Urology	Urogenital tract and its diseases
Virology	Viruses
Zoology	The natural history of animals

omasum The third stomach of a RUMINANT. The others are the RUMEN (first stomach), RETICULUM (second stomach) and ABOMASUM (fourth stomach). The purpose of the omasum is to grind the food down using the rough internal structure of the stomach.

See also: RUMINANT DIGESTION

omnivorous Means deriving sustenance from a diet consisting of both meat and plant or vegetable material.

See also: CARNIVOROUS

open-cast mining A technique of extraction from the environment of those minerals or deposits that lie relatively close to the surface. The process is also known as strip mining in recognition of the characteristic in which the overlying layers of soil and rock (known as the 'overburden') are stripped away from the deposit in question, which is then dug out directly from the ground. Although the process is generally less expensive than underground mining (and is therefore more economically viable in relation to the extraction of poorer-grade deposits) it is highly visually disruptive to the environment with sites often covering many hectares of land.

Open-cast mining can also have a significant and deleterious effect on the environment, both of the site in question and of surrounding land. The destruction of the integrity of the overburden makes it susceptible to erosion by wind and rain, and its exposure to the air can result in the oxidation of minerals such as iron sulphide, potentially creating acid run-off. The activities of the removal of the overburden and the intensive nature of the extraction process usually involve large-scale vehicle movements, and considerable disruption to the lives of surrounding residents due to noise, dust and vehicular movements is possible. Remediation of the site on completion of the extraction process is expected, but even the best remediation schemes are likely to require many years to mature before the site can be considered to have been returned to a 'natural' state. Open-cast mining is also sometimes known as open-pit mining.

See also: FINGERPRINTING

open tender A TENDER in which any contractor is invited to bid for a specified contract. Such exercises are usually conducted by advertising publicly for bids.

See also: SELECTIVE TENDER

operational Generally taken as that which refers to those activities or actions necessary to ensure the direct implementation of a strategy (see STRATEGIC).

See also: TACTIC

operator Generally considered to be a person who has managerial or overall control of an activity or operation.

opportunistic Refers to that which occurs or happens as a result of conditions being favourable at the time. An opportunistic infection is one that occurs due to a reduced capacity in the immune system to resist infection. An organism that causes an opportunistic infection is one, therefore, that would not be expected, under normal conditions, to cause a similar infection in a healthy person with a fully reactive immune system. Opportunistic screening of patients is that which occurs because the patient is in a position to be screened as a result of some other activity (e.g. they were visiting their medical practitioner for some other reason).

opportunity cost The non-monetary cost of undertaking an action or performing a function. Such a cost comprises the use of resources, locations and time that would otherwise be available for other purposes. Generally for something to be described as an opportunity cost is an indication that, if it were not being used for that to which it is ascribed as an opportunity cost, it would otherwise be deployed elsewhere. Some sources suggest that resources or locations that would otherwise be held in storage or unused can be held to be opportunity costs if they are used.

order of magnitude An expression used to describe a quantity to its nearest power of ten. Reducing a quantity by one order of magnitude decreases it to a tenth; increasing it by one order of magnitude raises it tenfold.

ordinal scale One that apportions data into ordered qualitative categories, e.g. Tom is older than Sue.

See also: INTERVAL SCALE

ore Generally considered to be any MINERAL that contains metallic compounds of economic value.

orf The name of a viral infection that principally affects sheep, cattle and goats. Symptoms include the production of vesicles on the lips; these may develop into ulcers and infection may spread to other parts of the body. Zoonotic infection in humans is possible but the disease is generally benign and self-limiting. The disease in humans primarily affects shepherds, veterinarians, slaughtermen and butchers. Transmission is by direct contact of abraded skin with animal lesions, infected wool or hides, or contaminated pastures.

organic Means pertaining to, or deriving from, nature (i.e. they arise or are derived from plants or animals). Organic chemicals are broadly defined as those containing carbon. They can be highly complex, consisting of long molecular chains or ring structures. When classified by the arrangement of their carbon atoms organic compounds are broadly grouped into four classifications as follows:

- Aliphatic compounds have open carbon chains that may contain single, double or triple bonds.
- Alicyclic compounds have closed rings of carbon atoms that may have single or multiple bonds.
- Aromatic compounds have at least one BENZENE RING.
- Heterocyclic compounds have closed rings containing atoms of carbon together with one or more other elements (typically nitrogen, oxygen or sulphur). The carbon atoms may have single or multiple bonds with other atoms in the ring.

organic chemistry That branch of science which deals with the compounds of carbon. Much of the study and practice of organic chemistry deals with the structures of the various ORGANIC molecules and of techniques employed to synthesise these.

organic food A term used to describe one generally that is produced without the use of synthetic (i.e. non-organic) feed ingredients or fertiliser, growth stimulants, antibiotics or pesticides. The aim is to try, as far as is possible, to secure food that is produced in accordance with 'natural' principles and in a system that coexists with the natural environment, without harm being sustained by either. Internationally, the organic movement looks towards the International Federation of Organic Agriculture Movements (IFOAM) for direction. In Europe, the European Community has regulated (since 1993) the use of the term 'organic' as a marketing device and it has achieved this by requiring that all producers and processors are registered and receive regular checks to ensure compliance with the standards. In the UK, the scheme is administered by the UK Register of Organic Food Standards (UKROFS) who, in turn, approve a number of bodies to undertake registration and inspection.

See also: SOIL ASSOCIATION

organism Any individual or separately identifiable life form. Micro-organisms are those that can only be seen through a microscope.

organochlorine insecticides Used extensively in the UK in the period following the Second World War. They comprise a group of chemicals, including DDT, γHCH and Dieldrin, whose molecules contain a carbon-chlorine bond. They were found to have a long-term capacity to accumulate in the environment and to produce residues in animal tissue. Dieldrin was banned in the UK in 1965 and other organochlorine insecticides were gradually withdrawn from use and replaced by ORGANOPHOSPHOROUS INSECTICIDES.

See also: SYNTHETIC PYRETHROIDS

organoleptic Means capable of being detected by one of the senses, i.e. without specialised analytical techniques. It is often applied to the examination of foodstuffs by means of taste, sight and smell.

organophosphorous insecticides Introduced in the 1960s. They comprise a group of chemicals, including chlorfenvinphos, propetamphos and diazinon, whose molecules contain a carbon-phosphorus bond. They are less persistent in the environment than ORGANOCHLORINE INSECTICIDES but require more frequent application to achieve the same effect. Paradoxically, although they are less persistent in the environment than are organochlorine insecticides, they are potentially more toxic to wildlife and need careful handling and disposal to minimise their environmental impact.

See also: SYNTHETIC PYRETHROIDS

orogenesis A process whereby mountain ranges are created due to the movement of TECTONIC PLATES. The collision of two tectonic plates can cause the land at the junction to push upwards creating a range of mountains such as in respect of the Alps and the Himalayas. Alternatively the edge of one plate is pushed below the edge of the other in a process known as subduction. In such a phenomenon a trench between the two plates is created, this gradually accumulating abraded material as the plates rub against each other. If the plates continue to move orogenesis will eventually occur.

osmosis The process whereby a SOLVENT diffuses from a lower concentration solution, through a SEMI-PERMEABLE MEMBRANE to a higher concentration, thereby balancing the concentrations on either side. The key component of osmosis is the semi-permeable membrane, so called because it will permit the passage of the solvent but not the passage of the substances dissolved in it. Over time the phenomenon of osmosis will (unless influenced by other factors) tend to ensure that the solutions on either side of the membrane are of equal concentrations.

See also: REVERSE OSMOSIS; TRANSPIRATION

OSPAR Convention The Convention for the Protection of the Marine Environment of the North-East Atlantic. The UK is a party to the Convention. The Convention has the stated aim to:

> Prevent pollution of the maritime area by continuously reducing discharges, emissions and losses of hazardous substances with the ultimate aim of achieving concentrations in the marine environment near background values for naturally occurring substances and close to zero for man-made synthetic substances.

osteoporosis A condition in which the mineral content of bones is depleted to the extent that they become especially fragile and may even become deformed.

See also: PEAK BONE MASS

otoxic Refers to a capacity to be harmful or have a detrimental effect on any of the constituent parts of the auditory system.

See also: PARACUSIS

Otto cycle Another name for the complete cycle of a four-stroke internal COMBUSTION engine.

outbreak Two or more diseases of common AETIOLOGY linked (or thought to be linked) to a common cause.

outpatient A person in receipt of treatment at a medical facility but is not required to become hospitalised for purposes such as rest, recuperation, observation or further treatment.

overburden The collective term used to describe the totality of the overlying layers of soil and rock that need to be stripped away to gain access to a deposit in the extraction technique known as OPEN-CAST MINING.

oxidation In chemistry can be defined in three ways as either:

1 The addition of oxygen to a substance, or
2 The removal of hydrogen from a substance or
3 The loss of electrons from a substance.

See also: REDUCTION

ozone A colourless gas with an odour said to be reminiscent of the seaside. It is an ISOTOPE of oxygen and is produced naturally as a result of UV radiation on chemicals in the atmosphere or by the electric discharge of lightning on atmospheric oxygen. It is used as a disinfectant in certain applications, such as in larger leisure pools where the absence of a residual allows people to stay for longer periods than might be acceptable were chlorination to be used. As a gas it has an adverse effect on human respiration.

See also: OZONE LAYER

ozone layer A part essentially of the STRATOSPHERE that is rich in OZONE. The ozone acts as a barrier to ultraviolet (UV) radiation from the sun. Ozone depletion in the stratosphere was first noticed in the 1970s and has subsequently been recorded, notably in polar regions, as an anthropogenic effect particularly of the discharge to atmosphere of CHLOROFLUOROCARBONS. Depletion therefore increases exposure to UV radiation and could result in increased risk of certain types of skin cancer.

P

p value *see* PROBABILITY (P) VALUE

PAH compounds The anagram for POLYCYCLIC AROMATIC HYDROCARBON compounds.

paired t-test A test of statistical significance used to compare the MEAN of two measurements in respect of an individual such as in a CROSSOVER TRIAL.

pandemic A worldwide EPIDEMIC.

papule Another name for a pimple.

paracasein The US name for CASEIN.

paracetamol An ANALGESIC and fever-reducing drug, similar in effect to ASPIRIN. It is estimated that around 6,000 tonnes of paracetamol is used annually in the UK. Paracetamol is toxic in larger doses, the critical health effects of paracetamol overdose being congestion and NECROSIS in the liver, and OEDEMA and necrosis in the proximal tubes of the kidney.

paracusis Any abnormality or dysfunction in the auditory system.

See also: OTOXIC

paraffin A generic name given to ORGANIC aliphatic hydrocarbons with a double carbon bond. The name paraffin derives from the Latin words *parum* and *affinis* meaning together 'lack of affinity' indicating that paraffins are relatively stable in the presence of other chemicals. Paraffins are also known as alkenes. Paraffins are used as solvents and fuels.

paraffin wax The name given to a wax derived from PETROLEUM. The name distinguishes such products from other waxes of natural origin such as beeswax and those derived from plants.

paralytic shellfish poisoning (PSP) An illness caused by the consumption of shellfish, primarily bivalve molluscs, which have ingested and accumulated, in their hepatopancreas, toxin produced by ALGAL BLOOMS. The toxin responsible is known as Saxitoxin. Symptoms include tingling of the tongue, lips, arms and neck followed by loss of muscular co-ordination and possibly by respiratory distress. In extreme cases death may follow.

See also: AMNESIAC SHELLFISH POISONING; DIARRHETIC SHELLFISH POISONING; NEUROLOGICAL SHELLFISH POISONING

paraphasia A condition in which a patient or sufferer is unable properly to place words in a sentence or in which the wrong words are use. The term refers especially to a condition that arises as a result of damage or a lesion affecting the speech region of the brain.

paraplegia A condition in which the lower limbs of humans (the hind limbs of animals) are paralysed. The condition is usually accompanied by paralysis of the bladder and the rectum.

See also: QUADRIPLEGIA

parasite An organism that derives nourishment (energy) directly from another organism (the host). The term can be taken very widely and there are many different types of parasitic relationship. In general (although not exclusively) the term is commonly applied to organisms that live directly off the host's body, tissues, fluids or gut contents without killing the host. That some parasites do kill the host is true, but such is likely to be 'accidental', as there is usually no long-term benefit in doing so since the parasite loses its living food supplier.

There are many ways of classifying parasites. ECTOPARASITES live on the external body of the host, whilst ENDOPARASITES live within the body of the host. Parasites may be facultative (i.e. are not compelled to survive solely as parasites) or obligatory (i.e. they cannot live otherwise). Some parasites are permanently so, while others may be temporary or intermittent, using parasitism only as a transitory part of their life history or as an occasional means of obtaining nutrients.

The parasite's relationship with the host may not always noticeably affect the latter in any way. However, many parasites can cause disease (i.e. they are PATHOGENIC) directly in the host (e.g. *Plasmodium* spp. cause malaria) and other parasites may act as transmitters of pathogenic organisms from host to host (e.g. the tsetse fly may transmit the organism responsible for trypanosomiasis).

parboiling The culinary term for the part boiling of a foodstuff for around half its normal cooking time. The aim of part cooking the item is so that it can be finished by cooking with another method.

parching The browning of an item of food (such as peas) in a dry heat.

parenteral Refers to that which is part of the body but not of the alimentary tract. It is especially used to denote medicines administered to parts of the body other than the alimentary tract such as those administered as subcutaneous, intravenous, intramuscular or intrasternal.

Parliament In the UNITED KINGDOM, the totality of the legislature comprising the HOUSE OF COMMONS and the House of Lords. The Parliament together is responsible for the passage of new legislation and debate on issues of the day.

See also: GOVERNMENT

paring The culinary term to describe the process of peeling or trimming an item of food.

Parkinson's disease *see* NEUROTRANSMITTER

paroxism Sudden, time-limited but aggressive attack such as a convulsion or spasm that can affect a person as a result of suffering from a disease or other condition of ill health.

parr A stage in the life history of the SALMON.

partial pressure The pressure exerted by any single gas in a mixture of gases. The total pressure is the sum of the partial pressures of all the gases in the mixture.

particulate matter The term applied to small portions of solid or liquid material capable of being carried in the atmosphere. It is generally classified according to size based on the individual particle's ability to pass through size selective barriers used to measure the various aerodynamic diameters of the particle in question. The size is reported as particulate matter (PM) of *X* diameter. Thus, for example, PM10 relates to particles with a 10 microns diameter or less, and PM2.5 to particles with a diameter of 2.5 microns or less.

parturition The natural process of giving birth by the expulsion of the foetus and its membranes from the uterus following contraction of the surrounding muscles.

pascal The SI unit of pressure. One pascal is the pressure equivalent to a force of one NEWTON applied to an area of one square metre.

passive smoking The unavoidable inhalation of tobacco smoke by a person who is not smoking but who is sharing an atmosphere polluted by tobacco smoke. Tobacco smoke contains a range of toxic pollutants including nicotine, tar, carbon monoxide and hydrogen cyanide. Passive smoking is known to be capable of causing lung cancer in non-smokers.

pasteurisation A process whereby foodstuffs (notably milk) are exposed to temperatures, usually of less than 100°C, for sufficient time so as to reduce the presence of pathogenic or spoilage micro-organisms to acceptable levels. Satisfactory pasteurisation depends on adequate heat penetration to achieve a specified temperature for a specified time. The time/temperature balance is crucial to the process as equivalent levels of protection can be afforded by alternative higher temperature/shorter time or lower temperature/longer time combinations – these latter are specified in legislation for some foodstuffs, including milk. The process is not one of STERILISATION since it cannot guarantee the destruction of all microflora.

patent An official document granted by a government that confers to an inventor a specific right or title to an article, formulation or other invention such as to prevent anyone else making, using or selling that invention without the permission of the inventor. Patents are time-limited and can be granted by a government only in respect of the geographical territory under its jurisdiction. Patents cannot be given in respect of discoveries, theories, aesthetic creations, artistic performances, information systems or computer programmes.

See also: INTELLECTUAL PROPERTY

pathogen An organism that is capable of causing or inducing a disease or illness in another living being.

See also: BACTERIUM; FUNGUS; HELMINTHS; PATHOGENIC; PROTOZOAN; VIRUS; YEAST

pathogenesis The summation of the processes whereby a PATHOGEN is capable of producing symptoms of a disease.

pathogenic Means disease causing, or capable of causing disease.

See also: PATHOGEN

pathogenicity A measure of the capacity of a micro-organism to produce disease.

See also: VIRULENCE

patina The surface coating caused by deterioration of the constituent materials brought about by ageing and exposure to the atmosphere, referring and occurring principally on structural or artistic artefacts.

See also: VERDIGRIS

patulin A natural toxin that is produced by certain mould species, including *Penicillium expansum*. It causes a number of adverse effects, including producing changes to genetic information within cells, which in turn can lead to adverse effects on the developing foetus, the immune system and the nervous system.

paunching The culinary term used to describe the process of evisceration or drawing (see DRAW) of rabbits or hares.

PC SUM *see* PRIME COST SUM

PCB The acronym for polychlorinated biphenyl. PCBs are a group of chlorinated hydrocarbons, some of which are similar in chemical structure to some chemicals in the DIOXIN and furan group. They are generally stable and have excellent electrical insulating properties, good fire resistance and low volatility. They were widely used in electrical systems and wiring between the 1930s and 1970s, especially before 1976. Any electrical equipment that was manufactured before that date is likely to contain PCBs to some degree.

PCBs are soluble in fats and oils, and accumulate in fatty tissues of living organisms and are strongly absorbed in soil. In humans health effects following contact have included the formation of cysts, pustules and blackheads on the skin. They can enter the body through a number of different pathways, including inhalation, skin contact or ingestion. It is estimated that over 90 per cent of human exposure to PCBs however is by ingestion through dietary intake. The use of new PCBs has been prohibited in the UK since 1986 and was required, by international treaty, to be completely phased out by the year 2000. There are currently 209 known separate PCBs with a variety of uses; the most common brand names under which they have been sold are Aroclor, Askarel, Clophen and Kaneclor.

PCR *see* POLYMERASE CHAIN REACTION

peak bone mass (PBM) The condition when the mass of an individual's bone tissue is at its greatest. Bones develop throughout childhood and adolescence so that, for both males and females, by the age of 18 years, up to around 99 per cent of peak bone mass has been achieved. The bone mass then stabilises until the age of around 35–40 years of age, when bone mass begins to deteriorate as bone tissue is lost, principally through an imbalance created between the natural rate of bone generation and resorption. Loss of bone mass due to age makes the individual more susceptible to fractures. In severe cases OSTEOPOROSIS may develop. Correct diet during the formative years is considered to be the crucial factor in achieving correct peak bone mass, with the contribution delivered by micronutrients such as calcium, magnesium, phosphorous, zinc and vitamins D and K being particularly important.

peak sound pressure The greatest sound pressure level that occurs at a particular location and within a specified frequency range within a given time.

pelagic A term applied to describe those fish such as tuna and mackerel that live in the upper waters of the open sea.

See also: DEMERSAL

pendulum arbitration *see* ARBITRATION

penetrating damp(ness) That moisture which crosses a building material to affect the fabric or internal surfaces. Most penetrating dampness arises as a result of deterioration of the building structure with age (resulting in higher porosity and therefore increased transmission potential) or as a result of incorrect installation of materials or poor construction technique. Occasionally interstitial CONDENSATION is mistaken for penetrating damp.

peptide A compound formed by the combination of two or more amino acids.

perigee The point in the orbit of a satellite when it is nearest to the earth. The term used to describe the point at which the satellite is furthest away from the earth is the APOGEE.

See also: PERIHELION

perihelion The point in the orbit of a satellite when it is nearest to the sun; the point at which it is furthest from the sun is known as the APHELION.

See also: PERIGEE

perinatal Refers to the period in humans from the 28th week following conception to the 28th day after birth.

periodic table A classification of all the elements arranged with respect to their atomic numbers and grouped according to the similarity of their properties.

peritoneal Refers to that which is of, encompassed or enclosed by the PERITONEUM.

peritoneal dialysis The process of DIALYSIS performed by circulating a dialysing fluid through the abdominal cavity, using the PERITONEUM of a patient as a SEMI-PERMEABLE MEMBRANE, to remove nitrogenous waste material from the BLOOD. Such dialysis is not performed directly in the bloodstream.

See also: HAEMODIALYSIS

peritoneum The membrane lining the abdominal cavity and the organs contained therein.

See also: PLEURA

permafrost An abbreviation of the phrase 'permanently frozen ground' and refers essentially to the sub-surface of the ground, the temperature of which never rises above 0°C. The term does not necessarily imply that the ground holds ice.

Permanent Secretary The head of a civil service department.

permanent threshold shift A permanent condition of damage occasioned to an individual's hearing capability. The name derives from the characteristic whereby the sufferer usually develops an inability to hear higher frequency sounds (especially between 4,000–6,000 Hz (see HERTZ)) at the sound level that could be heard previously. Permanent threshold shift can also result in a distortion of the sound such as to make certain received messages, such as speech, indecipherable.

See also: PRESBYCUSIS; TEMPORARY THRESHOLD SHIFT

permethrin Permethrin (chemical formula $C_{21}H_{20}Cl_2O_3$) is a synthetic pyrethroid pesticide with widespread uses, including in fly sprays, in cleaning water mains and in mothproofing. Although originally considered to have low toxicity to both humans and animals, more recent research suggests that it is an endocrine disruptor in that it can be oestrogenic and is suspected of having effects on male hormone binding sites.

pernicious anaemia Primarily a disease of middle age in which the body fails adequately to generate red blood cells. The disease can be treated by injections with vitamin B_{12}, but this is not a cure and treatment will be required for life.

See also: ANAEMIA; APLASTIC ANAEMIA

persistence A measure of the ability of a substance to remain unchanged in the environment. The term is usually applied to those substances that are or are potentially toxic. Persistence is a significant consideration when assessing the HAZARD PROFILE of a substance. Long-term persistence of a mildly toxic substance can lead to potentially more cumulative damage than a more toxic but less persistent substance. The long-term persistence is especially important when considering those chemicals that have profound effects on the environment, such as GREENHOUSE GASES.

personal protective equipment (PPE) Any piece of equipment or clothing (including that specifically designed to offer protection against weather conditions) and which is intended to be worn by a person at work in order to protect them from a risk to that person's own health and safety.

PEST analysis A term derived as an acronym and relates to a systematic evaluation (usually of a business). It is conducted through the consideration of the respective aspects of Political, Economic, Social and Technological parameters. It focuses predominantly on those external factors that can affect that which is under consideration.

See also: SWOT ANALYSIS

pesticide Any chemical substance that can be used to kill pests. The term includes rodenticides, insecticides, herbicides, fungicides and bactericides. The choice of pesticide is important. Whilst it may be impossible to identify an ideal, the preference should be to use one that is as specific as possible to the target organism, which has as short a life as possible, is not metabolised into secondary harmful chemicals and does not exhibit bioaccumulation or does not accumulate in the environment. The pesticide should also be used in the lowest practicable quantity.

petechiae Small rash-like red spots on the skin that do not disappear when pressed. They are caused by haemorrhage of the small blood vessels and are one of the classical symptoms of MENINGITIS, although they are also produced by a number of other diseases.

petrochemical Any chemical found in or derived from CRUDE OIL or NATURAL GAS. Petrochemicals are composed principally of varying combinations of carbon and hydrogen but may also include other elements in varying quantities.

petrol A fuel used primarily for powering internal COMBUSTION engines and comprising a mixture of highly VOLATILE hydrocarbons derived from the FRACTIONAL DISTILLATION of PETROLEUM. In the USA petrol is known as gasoline.

See also: DIESEL

petroleum Another name for CRUDE OIL although the term strictly includes both crude oil and natural gases containing hydrocarbons. The name petroleum literally means 'oil from rock'. It consists primarily of hydrocarbons although around 10 per cent of the total consists of aromatic hydrocarbons and other substances such as oxygen and nitrogen. Sulphur is a common contaminant ranging in content up to around 5 per cent. There is no standard composition of petroleum, and natural deposits contain varying degrees of solid, liquid and gaseous elements dependent on source. There are three broad classifications. Asphaltic type petroleum contains naphthenes, paraffin type petroleum contains predominantly paraffin hydrocarbons and the mixed-base type both paraffin hydrocarbons and naphthenes.

Although commercial exploitation began in earnest in the mid 1850s petroleum has been exploited on a major scale for only a little over a century and its impact on industrial development has been remarkable. The economics and politics of oil production have had a major influence in shaping modern society, which is largely dependent on petroleum and its products. Although petroleum is used as a major primary fuel source, petroleum products are used in generating electricity as well as in the production and manufacture of medicines, fertilisers, foodstuffs, plastics and a wide range of industrial and domestic chemicals.

See also: NAPHTHENE; PETROL; SUSTAINABLE DEVELOPMENT

pH Used to denote the degree of acidity or alkalinity of a substance on a logarithmic scale of 0 to 14, where 0 is highly acid, 7 is neutral and 14 is highly alkaline. Distilled water has a pH of 7.0. The term pH is an abbreviation of 'Potential Hydrogen' and the scale reflects the hydrogen ION concentration of the substance, i.e. the scale is numerically equivalent to the negative logarithm to base 10 of the hydrogen ION concentration.

phage type A variety of bacteria within a single species determined by its differential response to a range of parasitic bacterial viruses known as bacteriophages. An individual strain of bacteria is frequently susceptible to attack by several bacteriophages, the pattern of attack determining the phage type. Bacteria that can be typed in this way include certain types of *Salmonella*, *Listeria monocytogenes* and *Staphylococcus aureus*.

phagocyte A cell that forms part of the body's immune system and which acts by enveloping and digesting foreign organisms or particles.

See also: LEUCOCYTE; MACROPHAGE

phagocytosis The process in the body's defence mechanism whereby invading bacteria (see BACTERIUM) are destroyed by LEUCOCYTES. The process starts by a slowing of blood flow around the site of invasion and the accumulation of leucocytes at the site, in response to the detection of the bacteria. In a successful defensive response the leucocytes then attack and destroy or digest the bacteria.

See also: ABSCESS; PUS

pharynx The region between the back of the nose and mouth lying above the larynx and oesophagus.

phenol Phenol (chemical symbol C_6H_5OH) is a toxic, white crystalline compound. It is derived from BENZENE and is used in the manufacture of synthetic resin, plastics, pesticides and disinfectants. It is also used as an industrial solvent.

phenotype An organism differentiated or described by its physical appearance and/or physiological properties. The process of so describing organisms is known as phenotyping.

phlegm Thick mucus, especially that which affects the respiratory passageways.

photochemical reaction One that is brought about by the action of sunlight, especially one brought about by the action of ultraviolet light.

photochemical smog That caused by the action of sunlight on certain pollutants in the atmosphere (typically nitrogen dioxide and volatile organic compounds) leading to the production of ground-level OZONE, a gas that has the potential to cause respiratory problems.

See also: SMOG

photochromic glass Glass that incorporates a reactive filter which responds variably to the amount of light falling on it to progressively darken the glass as the level of light increases. It is used widely in reactive lenses in protective and optical spectacles, and in glazing for buildings.

photometer A scientific instrument used to measure light by comparing the measured amount of illumination with that provided by a standard source.

photon An indivisible quantum of ELECTROMAGNETIC RADIATION. Light is composed of streams of photons. The energy carried by the photon is related to its frequency. Higher-frequency radiation, such as ULTRAVIOLET RADIATION, carries more energy than does lower-frequency radiation, such as infrared radiation.

photosphere The layer from which all visible light emitted by the sun is generated. The photosphere consists of PLASMA at a temperature of around 6,000°C.

photosynthesis A complex biochemical process whereby green plants produce chemical energy. CHLOROPHYLL in the plant converts solar energy (sunlight) into adenosine triphosphate (ATP), which is then used in a series of reactions using CARBON DIOXIDE and water to produce CARBOHYDRATES (in the form of SUGARS such as GLUCOSE) and oxygen.

phototropic Means attracted to light. It is usually applied to certain plants and vegetative matter to indicate that they grow towards, or whose flowers move to face, the light.

phylloxera Phylloxera (*Phylloxera vastatrix*) is an insect pest of vines that feeds on the roots. It was introduced into Europe in the 1870s when it virtually destroyed the entire European vineyard. Nowadays every European vine is grafted onto US vine rootstock that is immune to its native pest.

phytoestrogen A chemical produced naturally by some plant species that can act as an ENDOCRINE DISRUPTOR if eaten by an animal or human. Phytoestrogens are similar to the female hormone OESTROGEN but they are much less potent.

phytoplankton Microscopic plants found in salt and freshwater environments. They are mostly forms of unicellular algae such as DIATOMS.

See also: BLUE-GREEN ALGAE; ZOOPLANKTON

phyto-sanitary A hybrid term in increasingly common usage to describe food safety issues in relation to the consumption of food of plant origin, or measures designed to protect human health consequent upon such consumption.

piecework A system whereby payment at an agreed rate is made to an employee for each unit produced, irrespective of the length of time it has taken.

pilaster A column built into a wall so that only a portion is seen projecting from the wall's surface.

pile A columnar structural component of a building driven into or constructed in the ground for the purpose of strengthening or acting as a foundation. Piles may be of concrete, timber, steel or other durable material. A 'driven' or 'hammered' pile is one that is forced into the ground by blows; a 'vibrated' pile is driven in by vibration; a 'jacked' pile is pushed in by force; and a 'cast' or 'bored' pile is one constructed IN SITU within a framework or bore hole.

pile driver A device used to hammer or drive a PILE into the ground by force.

pinion The smaller of a pair of intermeshing toothed wheels that together make up a GEAR.

Pink's disease An alternative name for the condition known as ACRODYNIA.

pinna The external, fleshy part of the ear.

piscivorous Means feeding on, or deriving sustenance from, fish.

pithing A stage in one type of process of the slaughter of animals for food in which a (usually flexible, metal) rod is inserted through the hole in the skull made by a CAPTIVE BOLT PISTOL to destroy a portion of the central nervous system. The operation is performed in order to prevent the animal from kicking as a reflex action following STUNNING.

See also: BLEEDING

placebo A Latin term that means 'I will please'. It is applied historically to those substances for which no pharmaceutical effect was known or could be demonstrated but which proved to be beneficial in making a patient feel that something had been administered to improve their condition, thereby possibly inducing improvement through auto-suggestion. Placebos are now used predominantly in clinical trials to test the efficacy of new drugs or therapies.

See also: DOUBLE BLIND TRIAL

plan A type of drawing used in architecture in which the details of a building or similar construction are reproduced as if viewed from above. Frequently plan drawings are used to show the relative positions and sizes of internal arrangements and features. The drawing is done with mathematical accuracy but does not attempt to create a three-dimensional effect such as through incorporating shading or perspective.

See also: ELEVATION

plankton Any of a group of small, often microscopic, drifting plants or animals that live on or near the surface of bodies of both salt and freshwater.

See also: ALGAE; DIATOM; DINOFLAGELLATE

plasma In biology, the colourless fluid of the blood in which the blood corpuscles are suspended. In astronomical terms plasma is a gas that contains free ions and electrons, and is thereby capable of conducting electricity.

plasmid A circular piece of DEOXYRIBONUCLEIC ACID (DNA) located outside the chromosome and found in bacteria. Plasmids are used in genetic engineering as a vehicle to insert new genetic material into other micro-organisms or plants.

plaster of Paris A white powder that sets hard when mixed with water. It comprises principally a hemihydrate form of calcium sulphate ($CaSO_4.2H_2O$). It is used to produce casts from moulds or in medical procedures for producing the casts used for setting broken bones.

See also: GYPSUM

plasticiser An additive to paint, varnish, cement or similar to increase its workability. Some plasticisers can adversely affect the longevity, strength or finish of the final product and should be chosen with care.

See also: RETARDER

plat du jour In a restaurant refers to the main dish of the day.

plate glass Glass of superior quality to SHEET GLASS, usually being thicker and with a smoother surface finish, originally making it the material of choice for larger, especially display, windows. Nowadays it has largely been superseded by FLOAT GLASS, although the term plate glass is sometimes incorrectly applied to this commodity.

platelet A specialist type of blood cell involved in providing an effective clotting effect in the event of injury or similar.

plenary Describes something as being complete or full. A plenary session of an assembly is one in which all members are entitled to be present.

plenum chamber A compartment attached to a distribution system, which is kept at a pressure higher than the distribution system itself in order to supply that which is to be distributed (e.g. air) to the supply pipes or outlets.

pleura (or pleural membrane) The lining of the cavity of the chest that covers one of the lungs. Since mammals have two lungs each mammal has two pleurae.

See also: PERITONEUM

pleural Refers to that which is of, encompassed or enclosed by the PLEURA.

pleurisy An inflammation of the PLEURA. The condition is also known as pleuritis.

plumb A designation in construction indicating that a surface or fixture is vertical. The position is determined using a 'plumb line', which is simply a length of line or similar with its top held against that which is to be measured. It terminates in a 'plumb bob', being a weight upon which gravity acts to stretch the line out vertically.

plumbago *see* BLACK LEAD

plumber's solder A specialist SOLDER used especially for forming joints in pipework. There are two main types. Coarse solder is a 75 per cent lead and 25 per cent tin mixture with a melting point around 250°C. Fine solder is a 50:50 mixture of lead and tin with a melting point of around 190°C.

plutonic The term used to describe IGNEOUS ROCK derived from MAGMA that has cooled to solidification below the surface of the earth.

See also: GRANITE

plywood A material constructed by bonding thin layers of wood (plies – singular ply) together to form a sheet with the grain of alternative layers usually running at 90° to each other. There may be any number of layers, but these are usually described in the name, such as in 3-ply (3 layers), 5-ply (5 layers), etc. The alternation of the wood grain produces a material that has no inherent weakness in any particular direction. Specialist plywood includes that made for decorative purposes such as by incorporating a final top layer of veneer. Marine ply (or marine plywood) is a specialist plywood designed to be especially durable in MARITIME use or on exposed surfaces.

PM *see* PARTICULATE MATTER

pneumoconiosis A disease of the lung caused by the inhalation of dust.

pneumonia An inflammation of the tissues of the lung. It is generally classified dependent on the cause. Pneumonia caused by allergic response to pathogenic micro-organisms such as bacteria or viruses is generally known as alveolitis. Pneumonia produced in response to chemical or physical agents is generally known as pneumonitis.

pneumonitis *see* PNEUMONIA

poaching The term used to describe the culinary practice of the gentle cooking of a food by immersing it in a hot liquid such as water, milk or similar.

podzol A soil from which much of the life-sustaining material such as organic material and metal oxides and hydroxides has been leached by the action of rainfall. Such soils are generally located in cooler temperate climates such as Canada, Russia and Scandinavia where bacterial activity is reduced because of low temperature and the potential for replenishment of leached material is thereby limited. Much of the podzol has little agricultural potential and is currently covered by coniferous forest. Because it is already by nature acidic the podzol is highly vulnerable to a decrease in PH and considerable damage has occurred in some areas as a result of the effects of ACID RAIN.

pointing The finishing of the bonding material in a brick or masonry structure on the external or exposed surface. Historically pointing has been done using decorative or coloured mortar – nowadays it is usually undertaken using the same mortar as is used for the bonding process. Pointing can be undertaken for either decorative or weatherproofing purposes; in the latter case the joint is deemed to have been 'weather struck' or simply 'struck'.

polar rain A stream of electrons impelled by the solar wind and attracted to the earth at its poles by its magnetic fields.

polarised light Light whose waves have been restricted to a single plane. This is usually achieved intentionally by passing normal light through a polarising filter.

polio *see* POLIOMYELITIS

poliomyelitis Also known as polio or infantile paralysis, an infectious viral disease affecting the brain and spinal cord. It can cause partial paralysis and convulsions. Death might occur dependent on the site of paralysis, especially if the muscles used in respiration are affected. Polio has been virtually eliminated in many parts of the world following the successful introduction of poliomyelitis vaccination programmes, especially following the development and introduction of an orally administered vaccine.

Polkinghorne Committee The name commonly applied to the Committee on the Ethics of Genetic Modification and Food Use set up in the UK, under the chairmanship of the Reverend Polkinghorne, to study the ethical issues arising from the application of genetic modification in food supplies. The report was published in September 1993.

pollutant Any substance present in some other substance or medium that is likely to cause harm or to exceed a statutory quality standard. A pollutant is essentially a CONTAMINANT that is likely to cause harm; the contaminant may, but does not necessarily, possess this characteristic.

'Polluter Pays' Principle The ethos whereby a person or persons whose business from whence pollution originates is held to be responsible for the costs of measures to prevent, control or reduce the said pollution. The principle encompasses both the producer and the consumer. It is not inherently required that polluters necessarily undertake protective measures on their own behalf. Indeed, some such actions may be performed or commissioned by enforcement agencies. The principle would however hold that the costs of such actions are borne by the polluter; this is likely to be reflected in the costs of goods or services.

See also: 'USER PAYS' PRINCIPLE

Pollution Prevention and Control (PPC) A system of integrated pollution prevention and control that is being introduced in the UK, with a target date for completion by 2007, to replace both INTEGRATED POLLUTION CONTROL and LOCAL AIR POLLUTION CONTROL systems.

polychlorinated biphenyl *see* PCB

polycyclic aromatic hydrocarbons (PAHs) Organic chemicals, the carbon atoms of which are arranged in a series or rings. They have a high molecular weight and mostly derive from the incomplete combustion of organic fuels such as coal and oil. Coal accounts for around 68 per cent of emissions and motor vehicles for around 25 per cent. The polycyclic aromatic carbon group includes complex mixtures of several related compounds. They are important from the human health viewpoint as they are toxic in high concentrations and a number are considered to be carcinogenic.

polymer A long molecular chain combination of an unspecified number of chemical building blocks known as MONOMERS.

polymerase chain reaction (PCR) Is primarily laboratory-based technique in which multiple copies of a fragment of DEOXYRIBONUCLEIC ACID (DNA) are produced by a process known as amplification.

In essence double-stranded DNA is heated to break down the hydrogen bonding to produce single DNA strands. Primers are then introduced to isolate the specific DNA sequence required. The isolated portion of DNA (the sequence) is then synthesised to produce as many copies as needed.

polypeptide A naturally occurring POLYMER created by the combination of amino acids.

See also: PROTEIN

polysaccharide A long-chain POLYMER formed from MONOSACCHARIDE units (SUGARS). Polysaccharides are members of the CARBOHYDRATE group and include storage carbohydrates such as GLYCOGEN and STARCH, and structural compounds such as CELLULOSE and CHITIN as well as many gums.

ponding The accumulation of rainwater or drainage water on a surface that is supposed to be free draining. The phenomenon is indicative of either faulty workmanship or of deterioration of the surface.

Pontiac fever A mild form of LEGIONNAIRES' DISEASE characterised by malaise, headache, chills and (occasionally) fever. The disease does not present as pneumonia. It was so named after the first outbreak of the disease, which occurred in Pontiac, Michigan, USA, in 1968.

pop rivet *see* BLIND RIVET

porcelain Glazed pottery ware, often of fine quality, manufactured from CHINA CLAY.

portfolio A term used to describe the allocated responsibilities of a Minister of the Crown or of a Cabinet member of local government. The term is also applied to the complete investments of a financial or business organisation or of an individual or sometimes to the total assets of a business or individual.

Portland Cement *see* CEMENT

portico A roofed entranceway to a building. It may or may not be partly enclosed.

positron A positively charged BETA PARTICLE.

posterior Means located at or near to the back or rear.

See also: ANTERIOR

postulate Also known as an axiom, something taken as being self-evident and for which proof is not needed and (perhaps) might not even be capable of being provided or obtained.

potable Refers to water (or sometimes to steam) and means of quality suitable for drinking. The term is usually applied to drinking water and the standard for assessing quality may be guidance or STATUTORY. The parameters measured usually include pathogens, dissolved solids, colour, taste and smell.

power In the statistical sense refers to the ability of a clinical trial or similar to detect a true difference between the effect of an INTERVENTION on a study group when compared to a control group.

See also: TYPE I ERROR; TYPE II ERROR

pozzolanic Means having properties like pozzolana, a type of porous volcanic ash used in the production of HYDRAULIC CEMENT.

See also: FLY ASH

PPE *see* PERSONAL PROTECTIVE EQUIPMENT

precautionary principle The ethos whereby preventative measures are introduced in circumstances where there are reasonable grounds for concern that the introduction of a substance, energy, procedure or practice may cause damage to human health, the environment, ecosystems, amenities or legitimate usage of facilities. The principle applies even in circumstances where there is no conclusive evidence to link possible adverse effects to a given input. The principle was originally adopted by world governments at the 1992 Earth Summit (see UNITED NATIONS CONFERENCE ON ENVIRONMENT AND DEVELOPMENT) in the Rio Declaration on Environment and Development, the original wording being:

> In order to protect the environment, the precautionary approach shall be widely applied by states according to their capabilities. Where there are threats of serious or irreversible damage, lack of full scientific certainty shall not be used as a reason for postponing cost-effective measures to prevent environmental degradation.

The principle is usually applied in practice to limit the emission of potentially damaging pollutants or the operation of potentially damaging technologies even where knowledge is inconclusive, but only provided the balance of likely costs and benefits means that the preventative action is justified.

precedent *see* JUDICIAL PRECEDENT

precision A characteristic to describe the likelihood that the performance of an operation or procedure will lead to subsequent results of the same or similar quality. It is sometimes taken as a synonym for reproducibility, especially in terms of clinical trials or similar.

pre-clinical A stage following infection by a PATHOGENIC organism in which the clinical symptoms have not yet become manifest.

presbycusis The normal progressive loss in hearing capability occasioned by advancing age.

prescriptive Means that which has been specified or laid down. In the UK many agencies (including local authorities) are empowered by Government to act in accordance with prescriptive legislation, i.e. they can only act or take such action as is specified in STATUTE.

See also: PROSCRIPTIVE

preservative Any substance added or incorporated into another with the purpose of increasing the latter's useful life or staving off deterioration.

pressurised water reactor (PWR) A NUCLEAR REACTOR powered by ENRICHED URANIUM and using water as the MODERATOR.

presumptive In relation to a case of disease refers to one that has been determined (usually) on physical examination as fulfilling the criteria sufficient to be described as a particular condition. A confirmed case will only usually be declared on receipt of positive laboratory identification.

prevalence That proportion of people in a defined population which has a specified disease or condition at any one time. The prevalence rate is usually expressed in terms of chronic illness, and it is not particularly useful in considering acute infections.

See also: INCIDENCE

prevention of disease One of the primary objectives of PUBLIC HEALTH. There are classically four ways of disease prevention. These are as follows –

Primordial, i.e. stopping infection coming into a community.
Primary, i.e. avoiding the spread of disease:

Secondary, i.e. providing early and effective treatment for sufferers.
Tertiary, i.e. rehabilitation of long-term effects or SEQUELAE.

prima facie A Latin term meaning literally 'on first appearance'. It is used in legal terminology to describe evidence that is sufficient to establish a case unless or until such is disproved. *Prima-facie* evidence therefore is that which is sufficient to prove a contention in court and to justify a decision in its favour, unless subsequent evidence can be provided to rebut it.

primary colours Those that can be mixed in varying proportions to create any SECONDARY COLOUR. Primary colours can only exist as themselves; they cannot be created by mixing any combination of other colours. In light the primary colours are red, green and blue. In pigment or paint the primary colours are red, yellow and blue.

See also: SECONDARY COLOURS

Primary Legislation A piece of main statute law passed by a legislature. Primary Legislation is passed in two forms. Public General Acts have general applicability, while Local and Personal Acts have specific and limited application.

See also: ACT; BILL; REGULATION; SECONDARY LEGISLATION

prime cost sum (or PC sum) Originally a specific cost in a BILL OF QUANTITIES allocated to the purchase of an item or a specified quality of work upon which the contractor had no discretion. The term has also come to include items that have not been determined at the time of pricing.

principal In legal terms, the party in a contract with whom a contractor holds the contract.

prion The name derived from 'proteinaceous infectious particle', the generally accepted infectious particle thought to be responsible for the transmission of BOVINE SPONGIFORM ENCEPHALOPATHY (BSE). A prion is a fragment of a protein chain and as such is not a living entity but a chemical molecule, a characteristic that means that it cannot be 'killed' but has to be denatured in order to be inactivated. The term 'proteinaceous infectious particle' is often shortened to 'PrP'. Cells have been found naturally to contain various types of PrP but most can be destroyed by ENZYME activity. These latter are called 'normal PrP'. They are distinguished from 'abnormal PrP', which cannot be destroyed by enzymes. It is these latter that are thought to be responsible for BSE.

probability (p) value A measure of the likelihood that a described statistical association is true. Starting at 1.0, the smaller a p-value becomes, the greater the probability that the thing being described did not arise by chance. Conventionally statistical significance is usually accepted if there is a less than 1 in 20 likelihood that the event arose by chance, i.e. if the p-value is less than 0.05. It must be remembered that this is a statistical test only and, although they are highly reliable tools, such tests do not prove CAUSALITY. Probability is also closely associated with sample size; a p-value above 0.05 might, for example, be indicative that an insufficient number of samples have been available for analysis for statistical significance to be proved.

prodromal Refers to the period following the incubation of a disease when the symptoms are beginning to be manifest but these are not as yet sufficiently developed to enable a differential diagnosis to be made.

prokaryote Any unicellular organism that does not possess a cell nucleus. The classification includes bacteria, BLUE-

GREEN ALGAE, actinomycetes and mycoplasma.

prolamine A PROTEIN that is insoluble in water but soluble in a 70–80 per cent concentration of aqueous ethanol.

propanone An alternative (and chemically more correct) name for ACETONE.

propellant A volatile gas or liquid used as a transportation vehicle to carry another chemical or mixture in aerosol or droplet form from a pressurised container. Many propellants such as butane and propane are derived from PETROLEUM and are inflammable. Care should therefore be taken when using pressurised aerosol dispensers containing unknown propellants near potential sources of ignition.

Proper Officer The title given to an individual to whom a local authority is required to designate the power to undertake defined statutory duties on its behalf. It is useful to recognise that in undertaking these duties the officer is deemed to be acting as if the whole council was taking the action. The term is frequently applied to the appointment of (usually) a medically qualified practitioner to undertake certain of the local authority's functions in respect of communicable disease control, but it must be remembered that this is not the only 'Proper Officer' role.

prophylactic A medicine or treatment administered to ward off an infection.

See also: PROPHYLAXIS

prophylaxis The administration of medication or a procedure to an individual in advance of clinical symptoms or infection to prevent the occurrence of a disease.

proprietary product One that has a unique formulation or characteristic and is therefore clearly distinguishable from other products. A proprietary product is usually one covered by a PATENT and is either manufactured by a single company or by other companies under licence. The term proprietary is often applied to medicines to denote the product of an individual company. Proprietary products are often sold under a branded (i.e. company or trade) name.

proscriptive In the UK in legal terminology is that which specifies activity that may not be undertaken. An agency (such as a health authority) operating under proscriptive rules may therefore undertake any activity, not otherwise illegal, in pursuit of its stated objectives unless it is specifically prohibited from doing so. This is in contrast to the method of operation necessarily adopted by agencies operating under PRESCRIPTIVE requirements.

protein A naturally occurring POLYPEPTIDE comprising AMINO ACIDS with or without other groups. Proteins are one of the essential building blocks of animal life. They have a large range in molecular size, some proteins consisting of several million atoms. Proteins may be classified in a number of different ways. By their varying solubility they are known as ALBUMINS, GLOBULINS, GLUTELINS or PROLAMINES. Fibrous proteins are known variously as COLLAGEN, ELASTIN, FIBRIN and KERATIN.

See also: CARBOHYDRATE; FAT; LIPID

proteomics The science devoted to the identification and characterisation of cell proteins.

protocol Any plan, specified procedure or systematic method for undertaking or carrying out an allotted task, the aim being to produce consistency.

proton An elementary particle with a unit mass of approximately 1 and a positive electric charge. Together with the NEUTRON, the proton forms the NUCLEUS of an ATOM.

See also: ELECTRON

protoplasm The totality of the content of a living cell including the nucleus (if present) and the CYTOPLASM.

See also: EUCARYOTE; PROKARYOTE

protozoan (plural protozoa) Simple single-celled organisms that, unlike bacteria (see BACTERIUM), contain a nucleus. Many are capable of movement; indeed, some are highly mobile within their own environment, possessing short threads (cilia) or long threads (flagella) extending from their bodies; these are agitated to propel the organism around. Some protozoa are parasitic (see PARASITE) and/or PATHOGENIC. MALARIA is a disease caused by protozoan parasites (Plasmodium spp.).

provisional sum An ESTIMATE included and identified as such in a QUOTATION inserted due to inability to give a more exact figure at the time of producing a price. Provisional sums are included for various reasons, the most common being due to uncertainty as to what might be exposed or shown to be necessary as the work proceeds or due to matters that cannot be determined accurately at the time of quotation. The final price will be adjusted once the provisional sum can accurately be decided upon.

proximate analysis The description of an analysed sample in terms of its major constituents such as water, fat, protein and ash contents rather than in terms of its chemical composition.

proximity principle A philosophy that requires the disposal of waste generally to be undertaken as near as possible to its place of production or generation. The principle was suggested as possibly improving the way in which waste disposal be undertaken by necessarily involving those who generate the waste in considering disposal options. The principle also ensures that those who generate waste cannot simply transport it elsewhere to protect their own local environment at the expense of someone else's.

prussic acid A weakly acidic, but highly poisonous, aqueous solution of hydrogen cyanide.

psittacosis Also known as chlamydiosis or ornithosis, a primarily avian disease caused by infection with *Chlamidia psittaci*. It is one of the most common causes of atypical pneumonia in humans although detection of the source of the infection is unlikely in the majority of cases unless larger-scale outbreaks help epidemiological investigation to suggest a common source. *Chlamidia psittaci* has also been noted as an important pathogen of domestic animals. The name psittacosis was originally applied to the disease in birds because it was thought to affect predominantly the parrot (or psittacine) family. It is now known that over 130 different bird species worldwide (including pigeons, doves, turkeys, finches and game birds) have been proven to be capable of carrying the infection, and investigation of potential sources should not overlook this possibility. Transmission to humans is likely to be as a result of aerosols derived from infected birds. There is evidence to suggest that direct person-to-person spread may be more significant than has previously been acknowledged. The disease in humans usually presents as an infection of mucosal membranes although severe chronic inflammation may also occur.

psocids Small (1–2 mm) wingless insects, most commonly encountered as pests of stored foods, especially cereals. The term is a generic one as there are several different species encompassed by the name. Psocids range in colour from pale yellow, sometimes almost translucent, to dark brown or black. They naturally feed on moulds, mildew and yeasts that can be found growing on paper or cardboard products, hence their common appellation, 'booklice'. They thrive in damp conditions and elimination of damp is a good control measure. They do not present any direct threat to man, but they are potential carriers of PATHOGENS, and foods contaminated should be regarded as unsuitable for human consumption.

psychosomatic In reference to disease relates to an AETIOLOGY that has both physical and mental components. The term is especially applied popularly to those adverse health outcomes that can be or have been influenced by the beliefs, perspectives or mental state of the patient.

psychrometer A scientific instrument for measuring RELATIVE HUMIDITY. The term is another name for a HYGROMETER.

pthalates Should more properly be called pthalic acid diesters. They comprise a group of organic chemicals with extensive use as plasticisers, household goods, lubricating oils, and some plastics. In the cosmetics industry they are used as vehicles for carrying perfumes. Their use in food packaging is now very limited, generally being confined to the manufacture of some adhesives and printing inks. They are no longer used in the manufacture of cling film or most other food contact plastic materials.

They are widely distributed in the environment and found at low levels in many foods. They are carcinogenic and can exhibit weak oestrogenic activity. Their presence in the environment is normally due to their release during other production processes or derived from the disposal of the products in which they are used. Some pthalates occur naturally in coal, crude oil and shales, but the environmental contribution from these sources is thought to be negligible.

See also: ENDOCRINE DISRUPTOR

public health Generally accepted as 'the science and art of preventing disease, prolonging life and promoting health throughout the organised efforts of society' (defined by Winslow, 1920, and accepted by the Acheson Committee, 1988). It is important to recognise that in practice the term covers a wide definition of health and that it is the health of the population rather than the individual that is the crucial element.

puddle A stiff mixture of clay and water, sometimes with the addition of sand, used to produce an impervious layer. Many artificially created ponds have puddle installed over the open surface prior to filling to prevent water leakage in use. The process of installing the puddle is known as 'puddling'.

pulmonary oedema An abnormal collection of body fliud within the air spaces of the lung.

pulverised fuel ash *see* FLY ASH

pumice (or pumice stone) A light, porous acidic rock formed predominantly by the gaseous lava emitted by a volcano. The stone is used as an abrasive, as a polishing medium and as a lightweight AGGREGATE in specialist concrete.

punkah A large swinging fan originating in Asia and used predominantly indoors to generate air movement for cooling purposes. The punkah is usually constructed as a rectangular or shaped panel with one end attached to the ceiling and the bottom end operated backwards and forwards (traditionally manually) by a cord.

pupa The transitional stage in the life cycle of an insect between LARVA and IMAGO. During the pupa stage the insect does not feed and is usually cocooned in a protective casing, which gives the appearance that the insect is dormant. This is not so as, internally, the insect is undergoing a dramatic change in cell rearrangement in the transitional process known as

METAMORPHOSIS. The pupa is sometimes known as the chrysalis and the stage as the chrysalid stage.

See also: IMAGO; LARVA

purée In cooking, any ingredient (including vegetable meat or fish) that has been chopped very finely and sieved to produce a smooth, homogenous pulp.

purlin A horizontal structural timber in a pitched roof running across the line of RAFTERS along one pitch and joining them at some point along their length.

See also: JOIST

purse seine net A fishing net in which an area of sea is enclosed by a net run out in a circular formation. The bottom of the net is then drawn together to form an enclosed netting sac (the 'purse'), which is then drawn back to the fishing vessel along with the catch.

pus A thick fluid, usually white, yellow or green in colour, which forms to create an ABSCESS. It may also be present at various sites of injury or infection such as ULCERS or other damaged sites on the body. The fluid is mainly comprised of dead cells of both bacteria and the fatty degeneration and death of LEUCOCYTES that form the body's main defence against bacterial invasion.

See also: PHAGOCYTOSIS

putty A stiff paste used to fasten and/or waterproof the edges of the glazing in a window or to seal holes and KNOTS in woodwork. The most common contain chalk or whiting mixed with linseed oil, and harden after application.

putty powder A powder, usually comprising tin oxide or tin with lead oxide, used as a polish on glassware or metal ware.

pyogenic Means capable of promoting the formation of PUS.

pyrethroid and pyrethrin Widely used as insecticides. They are of low acute toxicity to humans although they may act as sensitising agents in some instances. They can be irritants to the upper respiratory tract, to mucous membranes, eyes and to the skin. Ingestion can lead to nausea, vomiting and diarrhoea. The term pyrethrin (or pyrethrum) is derived from the *Pyrethrum* chrysanthemum, from which six naturally occurring insecticidal compounds (pyrethrin I, pyrethrin II, cinerin I, cinerin II, jasmolin I and jasmolin II) can be obtained. Pyrethrin I and pyrethrin II are the most potent insecticides of the six pyrethrins. The term pyrethroid is applied to synthetically manufactured substances that are similar to the naturally occurring pyrethrins.

pyrethrum *see* PYRETHROID AND PYRETHRIN

pyrexia The medical term for fever.

pyrolysis A system in which combustible gases are generated from (usually, but not exclusively, waste) organic material by heating it in a vessel in the absence of air or oxygen. Dependent on the composition of the source organic material, the process can also generate combustible char and a variety of liquid wastes including combustible oils.

See also: GASIFICATION; INCINERATION

pyrometer A scientific instrument used for measuring temperature. The term is particularly applied to an instrument used for measuring very high temperatures.

Q

QALY An acronym for Quality Adjusted Life Year. It is an attempt to place a value on the quality of life by ascribing a numerical indicator to different health states, e.g. death = 0, perfect health = 1. QALYs are used to provide an indication of the benefits or otherwise of (primarily medical) interventions, especially in situations where there is a choice, and they are also used as a tool in resource allocation. The QALY is a subjective measure, and, for example, some people consider that there are worse health states than death (i.e. on the scale of 0–1 above, they should have a negative value). An individual QALY is calculated in respect of the changes in health status both with and without the proposed intervention and includes negative effects. For example, if intervention *X* is likely to provide an additional 3 years of life with an improved quality, but ultimately is likely to provide an additional 1 year of life with reduced quality – calculating the QALY might aid deciding whether the intervention can be considered to be worthwhile.

See also: DALY; EUROQOL/EQ5D

quadrant A SECTOR of a circle in which one quarter of the circumference of the circle forms the ARC. The angle formed by the two radii will be 90°.

quadriplegia A condition in which the four limbs of the body are paralysed. The condition is also sometimes known as tetraplegia.

qualitative Refers to that which can be described by an adjective (i.e. it is descriptive of a certain character or quality).

See also: QUANTITATIVE

quality Applied to a description of goods and services, and is usually defined in formal quality systems as 'fitness for purpose'. Quality is essentially for ensuring that the expectations or demands of the customer are met. Quality is not strictly about ensuring that the product or service is necessarily 'better' or of higher quality than alternative products.

Quality Adjusted Life Year *see* QALY

quality assurance A system for ensuring the attainment of an acceptable standard of goods or services by the assessment of QUALITY at all stages in the production process. In a formal system most adhere to the principles laid down by the International Standards Organization under ISO 9000. Many systems of quality assurance pay particular attention to those stages in the production process that are critical to the quality of the final product

– these stages are known as CRITICAL CONTROL POINTS.

See also: HAZARD ANALYSIS CRITICAL CONTROL POINT; QUALITY CONTROL

quality control A system for ensuring the attainment of an acceptable standard of goods or services, usually achieved by undertaking an evaluation of a statistically representative sample of the finished product. Its main disadvantages are that it is not until the end of the production process that an assessment is made (by which time many inferior products could have been made or even dispatched). Second, assessment at the end of the production line is often more expensive. It has now largely been superseded by QUALITY ASSURANCE.

Quality of Well Being (QWB) An index for measuring the health-related quality of life.

See also: QALY

quantitative Refers to that which can be described by a number – i.e. it is a measure of size, extent or quantity.

See also: QUALITATIVE

quantity surveyor The professional responsible for overseeing and managing the use of materials on a building site. The individual is also responsible for determining how much material is needed for a job and for pricing the work prior to the submission of a QUOTATION or a TENDER.

See also: BILL OF QUANTITIES

quarry tile A hard, durable and unglazed tile made from fired clay.

quartz *see* SILICA

queen closer A brick cut in half lengthways and incorporated into the bond of a wall with its end outwards in order to ensure that the wall achieves its proper design dimensions allowing for the dimensions of the bricks.

See also: HALF BAT; KING CLOSER

Queen's Bench Division In the UNITED KINGDOM, one of the Divisions of the HIGH COURT.

quenching The process of rapidly cooling a heated material by immersing it in a liquid, usually water or oil. The process is often used in association with TEMPERING in order to alter the surface characteristics of a construction material. Marquenching is a slower cooling process, used to reduce the possibility of distortion or cracking during cooling.

quicklime A common name for calcium oxide (chemical symbol CaO), the material from which the HYDRATED LIME (i.e. calcium hydroxide, or slaked lime) used in lime mortar is made.

quicksand is that in which water moves upward with sufficient force so as to hold the grains of sand (or similar) in suspension. It is popularly portrayed as a treacherous stretch of land that can trap the unwary, sucking them down to certain doom. In practice it is simply an area of essentially negligible load-bearing capacity that will allow more dense objects to sink. It is theoretically (and reportedly) possible to 'swim' in quicksand, and recorded loss of life to such a phenomenon possibly had a considerable component of blame attributable to panic rather than to any enhanced intrinsic danger attributable to the quicksand itself.

quinine A bitter alkaloid extracted from the bark of the cinchona tree (a native of South America). Its salts were used in the preparation of tonics and medicines. It is perhaps most celebrated for its use in conjunction with chloroquine and other drugs in the treatment of malaria. Quinine is often cited as an example of a naturally occurring therapeutic medicine, many others of which could be lost to science unless the destruction of the world's rain forests is halted.

quinsy An inflammation of the tonsils and surrounding tissues accompanied by the presence of abscesses. In extreme cases the swelling and the presence of abscesses can cause blockage of the air passages.

quintile An astronomical expression describing an aspect of 72° between two heavenly bodies. The term is used in EPIDEMIOLOGY to describe a fifth portion of the population (e.g. the lowest or highest quintile would be that fifth portion of the population with the attributable lowest or highest of a specified factor).

quitclaim A formal renunciation of rights over a piece of land or of a legal claim of liability against a person.

quod erat demonstrandum (or QED) A declaration appearing at the end of given or stated evidence (originally especially in geometry). It means that which was to be proven or demonstrated, i.e. the evidence preceding the declaration has proved the contention.

quoin The external corner of a wall, or a stone forming a part thereof (in which case it is also known as a 'cornerstone').

quorum The minimum number of an assemblage, according to its rules or regulations, which needs to be present for business to be transacted. Decisions taken in the absence of a quorum can be challenged as *ultra vires*.

quorate Means having a QUORUM.

quota A proportionate share of the whole as allocated to an individual or group. It is often used to describe that proportion or quantity which an individual or group may manufacture or procure in accordance with national or international agreements or laws.

quotation A precise figure for the amount that a builder or other tradesman will accept to undertake a job. The quotation can be varied by agreement with the customer and can include provision for uncertainty and unforeseen work that might be necessary to complete a contract. In some circumstances the quotation can be upheld in the courts as the final price to be paid by the customer.

See also: ESTIMATE; FIXED-PRICE CONTRACT; QUANTITY SURVEYOR

QWB The acronym for QUALITY OF WELL BEING.

R

r-value A measure of the THERMAL RESISTANCE of a substance or body.

rabbet (or rabbit) *see* REBATE

rabies A zoonotic (see ZOONOSIS) viral disease spread predominantly through the bite of an infected animal. All warm-blooded animals are susceptible. It is endemic in many parts of the world although the UK has been rabies-free for many years. In mainland Europe the disease in the wild appears especially adapted to carriage in the fox. Man is not highly susceptible to the disease although infection can prove fatal. Vaccines are largely effective in preventing the development of the infection provided they are administered from the day of infection.

Rachmanism The unethical exploitation of tenants by a landlord, characterised by slum conditions, lack of maintenance, extortion and the heightening of racial fear. The term derives from the notorious operational techniques used by Perec Rachman (1920–62), a British property owner, born in Poland.

rack rent The rental that the owner of a property or land might reasonably be expected to receive were it to be available for such purpose on the open market.

radar A system for the detection of the position and speed of distant objects such as aircraft through the detection of reflected very high-frequency radio pulses emitted specifically for the purpose. The name derives from Radio Detection And Ranging.

See also: MICROWAVE

radiation The transmission of radiant energy in the form of electromagnetic waves, streams of particles, sound or heat. The term is particularly applied to those particles or energy emitted during the process of nuclear decay.

See also: NUCLEAR FISSION; NUCLEAR FUSION

radiation sickness The totality of the sub-lethal adverse health events arising as a result of exposure to IONISING RADIATION. A radiation dose of between 1–3 SIEVERTS to the average adult human is likely to produce severe radiation sickness. Symptoms include nausea and vomiting followed by diarrhoea, hair loss and haemorrhaging. Pregnancy may be terminated or damage to the foetus is possible. In the longer term, damage to tissues such as the bone marrow can result in cancers such as leukaemia and other cancers such as lymphoma might also occur.

radical An atom or group of atoms that has one or more unpaired electron.

radioactivity The phenomenon exhibited on the spontaneous disintegration of the atoms of certain elements whereby they emit IONISING RADIATION in the process. The radiation is emitted variously in three ways, being as alpha particles (comprising two protons plus two neutrons), beta particles (an electron) or as gamma radiation. The disintegration of larger atoms can lead to the creation of other elements, some of which may themselves be radioactive. For example, the emission of an alpha particle by uranium-238 leads to the creation of radioactive thorium-234 and so on; this is known as the decay sequence.

radio frequency radiation (RF) ELECTROMAGNETIC RADIATION used for the purposes of telecommunication. The wavelength is longer than that of infrared radiation.

radioisotope An elemental ISOTOPE that emits radiation. Tritium (^3H) is a radioisotope of hydrogen and Carbon-14 (^{14}C) is the radioisotope of carbon.

radiological protection The science and art of protecting or limiting the harm to human beings and other living creatures from the harmful effects of RADIATION.

radionuclide A type of nucleus of an atom (nuclide) that is unstable and disintegrates spontaneously, emitting IONISING RADIATION in the process. The nuclei of radium and uranium are radionuclides.

radiotherapy The application of controlled RADIATION for treating disease. It is used especially in the treatment of certain types of CANCER.

radon A naturally occurring inert radioactive elemental gas formed on the disintegration of uranium. It was discovered by the Polish-born physicist Marie Curie (born Manya Sklodowska) in 1898. It is of environmental health importance predominantly because of its natural occurrence in certain geological strata and subsequent ingress into buildings whereupon it has carcinogenic potential for the lung. Apart from the potential to cause lung cancer there is no known link with other types of cancer. There are twenty-nine known types of radon although only three of these occur naturally, and these latter are radon-219 (commonly known as actinon), radon-220 (commonly known as thoron) and radon-222 (commonly known as 'radon). Only radon-222 and radon-220 are of radiological consequence as radon-219 has an extremely short HALF-LIFE (3.96 seconds).

The UK government has adopted an Action Level for radon in dwellings based on the annual average for the entire dwelling. This has been set at 200 Becquerels per cubic metre of air (Bq/m^{-3}). If a dwelling is found to be above the Action Level the occupier is advised to reduce the radon level. The average radon level for the UK is 20 Bq/m^{-3} with maximum values in the UK (principally in Cornwall and Devon) of around 10,000 Bq/m^{-3}. Where the Action Level is exceeded in 1 per cent or more of houses in a given area, that area is designated as a Radon Affected Area. There are designated Radon Affected Areas in parts of England, Scotland, Wales and Northern Ireland. The Building Regulations have also been modified to require precautions to minimise radon entry in new homes being built in designated Radon Affected Areas. The risk of contracting radon-induced lung cancer is greatly increased for smokers; the risk is generally assessed as being about ten times higher than that for non-smokers.

raft foundation The slab-like base consisting of equal thickness throughout of (usually) reinforced concrete upon which the total load of a building is borne. The design is an attempt to distribute the weight of the building equally across the entire foundation and is used particularly where the bearing capacity of the soil or sub-soil is poor.

rafter Generally, a sloping structural roof timber of a pitched roof extending from the RIDGE to the WALL PLATE. In some forms of roof construction a horizontal timber brace is incorporated across the rafters running across the roof void from part way along the rafter on one side of the roof pitch to the same position on the corresponding rafter on the other side – this is known as a 'collar rafter'.

See also: JOIST; PURLIN

raking shore *see* SHORE

ramekin (or ramequin) Both a savoury dish comprising a baked cheese mixture and also the container within which such a dish is baked.

random error An artefact in an epidemiological study or clinical trial in which observed values differ from true values as a result of chance. A study in which there is a large component of random error indicates a potential fault in the study design and that the results are unlikely to be capable of duplication in other studies.

random rubble masonry *see* RUBBLE MASONRY

randomised controlled trial A type of epidemiological investigation or experiment in which the efficacy of an INTERVENTION is assessed by comparing the effects on those receiving the intervention with the effects on those not receiving the intervention or receiving a PLACEBO. The recipients and controls are selected from a COHORT who are ascribed as such randomly, usually using random numbers generated by computer or special tables to avoid selection BIAS.

rate A key epidemiological tool for monitoring disease in populations. The rate is calculated as the number of specified events within a defined population over a given time. By comparing the rates over different time periods it is possible to provide an estimation as to the track of infections or diseases seasonally and year-on-year. The comparison of rates allows early detection of potential outbreaks or epidemics as well as facilitating monitoring of the success or otherwise of disease prevention or health promotion strategies.

rated output Usually the maximum steady output of a device (such as a generator) achievable whilst it is in normal operation.

RCD *see* RESIDUAL CURRENT DEVICES

reagent Any substance or combination of substances used in a chemical reaction. The term especially applies to those substances that are used to test qualitatively or quantitatively for another substance.

real estate A term used chiefly in Northern America to describe land, either with or without buildings or other structures upon it. In the UK the term is sometimes rendered as 'real property'.

realtor A term used chiefly in Northern America to describe someone who deals in the sale or letting of land or buildings, especially someone with a recognised accreditation. A realtor is roughly equivalent to the term 'estate agent' used in the UK.

reamer A woodworking tool used to smooth to size the walls of a roughly bored hole in a piece of timber.

rebate (or rabbet) A (usually) rectangular groove or step cut or built into the surface or edge of a piece of timber to receive a door, window or similar.

receptor In molecular biology, a specific area on the surface of a cell that allows connection with a specific molecule in order to initiate a reaction in the workings of the cell.

reciprocating engine An internal COMBUSTION engine in which the pistons move back and forth (i.e. they 'reciprocate') (impelled by an explosive mixture) inside a cylinder to produce a rotating motion in a crankshaft.

See also: ROTARY ENGINE

recombinant Refers to a large molecule, usually but not exclusively of DNA, which has been created by combining fragments from other molecules. Recombinant DNA therefore is a DNA molecule that comprises fragments of DNA from more than one source. The term is also used to describe a fragment that will itself be incorporated into the larger molecule.

recreational water Any body of water that is used for the purposes of bathing, swimming, water sports or similar. Recreational waters include both natural resources such as coastal and fresh waters whilst man-made recreational waters include swimming baths and spa pools.

recycling The process whereby waste or discarded material, which would otherwise be sent for disposal, can be reprocessed into usable raw materials or saleable items. Recycling is seen as a key component in reducing the reliance on LANDFILL OR INCINERATION as a means of waste disposal and of furthering the aims of SUSTAINABLE DEVELOPMENT.

See also: REUSE

red blood cell *see* ERYTHROCYTE

Red Data Books Those comprising a series within which all known endangered species of plants and animals are listed. The books are sponsored by the Swiss-based World Conservation Union.

See also: GREEN DATA BOOK

red list A list of chemicals that are particularly hazardous to aquatic environments and whose discharge should be minimised. The list is based on European Directives and contains pesticides such as DDT, Dichlorvos, Dieldrin, Endosulfan and Malathion.

reducing A culinary term used to describe the process of boiling or simmering a mixture in order to remove surplus liquid to produce a more concentrated product.

reduction In chemistry can be defined in three ways as either:

1 The removal of oxygen from a substance or;
2 The addition of hydrogen to a substance or;
3 The gain of electrons by a substance.

See also: OXIDATION

reference nutrient intake (RNI) The amount of nutrient that is sufficient to sustain a large majority (usually taken as 97 per cent) of a given population. Provided the average intake for a group is maintained at the reference nutrient intake level it is unlikely that individuals within the group will exhibit deficiency.

refining The process of deriving or obtaining a substance by the appropriate processing of a raw material. Refining might involve a range of activities, either in isolation or in combination, such as distillation, chemical or electrical separation, centrifuging and filtration.

refraction A change in direction of a propagating wave such as light or sound. Refraction usually occurs when the propagating wave moves from one medium to another, especially where the velocity of the wave is necessarily changed. The ratio of the sine of the ANGLE OF INCIDENCE to that of the ANGLE OF REFRACTION is known as the 'refraction index' of the material.

refractive index The ratio of the velocity of light of a specified wavelength in air to its velocity in another substance. The measurement of the refractive index can be used to assess the purity or composition of the substance under examination.

refractory Applied to describe material that does not fuse or deteriorate upon exposure to high temperature, such as that used in the refractory lining of kilns, BLAST FURNACES or retorts.

refrigerant The fluid in the circulatory system of a refrigerator that carries the heat energy away from the storage compartment and acts as the transfer medium within the cooler. Many refrigerants are gases at normal temperature and pressure, and many in historic use have been found to cause damage to the OZONE LAYER.

regolith A loose, non-coherent covering comprising various materials such as rock, sand, soil and alluvium that lies on top of the bedrock. Dependent on location it can be of various depths, ranging from little more than a few millimetres to many tens of metres. The regolith essentially includes the totality of the fertile mantle above the earth's crust upon which terrestrial life exists and depends. Because it is non-cohesive and is not an integral part of the bedrock it can be particularly susceptible to EROSION, especially if it is disturbed such as by mining or DEFORESTATION.

See also: COASTAL EROSION

regression analysis A statistical technique that uses mathematical assumptions to test a relationship between an exposure to a certain factor and a biological response. The technique is very powerful and requires the use of a specifically designed computer programme. Regression is used to adjust for CONFOUNDING FACTORS but should be used with caution as the technique is not capable of testing whether the assumptions, under which the analyses are undertaken, are or are not correct.

Regulation A piece of legislation made under the powers of an ACT. In practice more than one Regulation is usually contained in a single document and collectively they are known as 'Regulations' with each discrete section or paragraph being denoted as a Regulation. Regulations are subservient to the Act under which they are made. Regulations are also known as SECONDARY LEGISLATION.

See also: ACT; BILL

relative density A synonym for SPECIFIC GRAVITY.

relative humidity The ratio of the amount of water vapour contained within a given volume of air to the maximum amount of water vapour the same volume of air could actually hold at the same temperature, assuming constant pressure.

See also: ABSOLUTE HUMIDITY; AIR CONDITIONING; HUMIDITY

relative risk (RR) An expression of the occurrence (such as either as a percentage or ratio) of a particular phenomenon in an exposed population compared to the occurrence of the same phenomenon in a population that has not been exposed. A relative risk of 1 means that the probability or risk of an event occurring in either population when compared to the other is even. If the risk is doubled the relative risk is 2; if it is halved it is 0.5. Relative risk is usually a more understandable way of describing and comparing risk than is the ODDS RATIO.

relay In electronics, differentially a device that automatically controls the setting of another component of the electrical system or alternatively one in which a small change in current or voltage switches other systems or devices on or off.

reliability When applied to epidemiological studies refers to the degree to which they manage to produce consistent results.

See also: SENSITIVITY; SPECIFICITY

renal Means of or pertaining to the kidneys.

rendering The process of extracting fat from animal tissue by the application of heat and separation of the fats so melted. It is also a process of applying a covering of mortar or plaster over the facing of a wall or similar structure for the purposes of weatherproofing or of improving the aesthetic appearance.

rennet Described variously as, first, the internal lining of the fourth stomach (ABOMASUM) of a calf; second, the secretion (containing RENNIN) that derives from the lining of the stomach of a calf; or, third, an artificially manufactured substance that is used commercially to curdle milk in the manufacture of cheese or JUNKET.

rennin The active enzyme ingredient of RENNET that curdles milk. It is sometimes known as chymosin.

repeatability An expression of the ability to obtain a result from one study that is similar to the result achieved by a different study or studies of the same thing. Repeatability is important in research for determining or indicating the likelihood that a finding is a true one or not. The more studies that confirm the finding, the more likely it is to be true.

See also: MACRO-ANALYSIS

repetitive strain injury (RSI) A chronic condition characterised by stiffness, swelling and reduced movement capacity, usually of a limb or digit, arising as a result of the inflammation of a tendon (known as tendinitis). It is caused by frequent and repeated movement (usually of limited extent) of part of the body and is most usually encountered in those whose employment essentially involves the use of keyboards, such as for typewriters and computers. For many years the existence of the condition was controversial, with many cases being ascribed to idleness or as a bid for compensation, but most countries now recognise it as a legitimate industrial disease. Repetitive strain injury is also known as (work-related) upper-limb disorder (ULD).

Research Council A public body in the UK set up to fund research or to administer the funding of research.

residual current devices (RCD) A piece of equipment that can detect certain faults in an electrical system when it is in operation and which responds to these by switching the system off. Residual current devices can be incorporated into the main switchboard of the system or incorporated as a separate addition. This is usually plugged into a socket between that and the plug of a piece of operational equipment or incorporated into the plug or the (usually portable) equipment. Those residual current devices that are designed for protecting people have a sensitivity of around or below 30 milliamps (mA).

resin An imprecise term used predominantly to describe a range of organic solid or semi-solid exudates obtained from plants.

See also: ROSIN

resonance A sympathetic vibration (including those capable of acoustic measurement) induced in a body or system in response to an external vibration.

Resonance is often variable with frequency and that at which the greatest response with the minimum loss of energy is elicited is known as the 'resonance frequency'.

respirable particles Those portions of PARTICULATE MATTER that are able, by virtue of their small size, to penetrate to the gas exchange region of the lungs.

retarder A chemical agent that is used to reduce the rate of progress of a chemical reaction. They are used for example INTER ALIA to delay the hardening of cement or mortar to extend the period during which it can be worked.

See also: ACCELERATOR

reticulum The second stomach of a RUMINANT. The others are the RUMEN (first stomach), the OMASUM (third stomach) and ABOMASUM (fourth stomach).

See also: RUMINANT DIGESTION

retina A multi-layer structure covering the interior surface of the back of the eye. It contains the light sensitive cells known as 'rods' and 'cones' that together produce the image transmitted via the optic nerve to the brain where it is registered as sight. The rods provide differentiation between black and white, and the cones provide information to allow the differentiation for colour vision. The human eye contains around 100 million rods and around 6 million cones.

See also: CORNEA; CONJUNCTIVA; VITREOUS HUMOUR

return A change in direction of a wall, moulding or similar, especially one at right angles to the original.

reuse A system of reducing waste by ensuring that items or materials can be used more than once. Such a quality needs to be planned into the item as part of the production of manufacturing process and is one way of furthering the aims of SUSTAINABLE DEVELOPMENT by reducing reliance on raw materials, energy and transport costs. Items that can be designed as reuse include reusable packaging and refillable containers.

See also: RECYCLING

reveal The inner surface, exposed to view, of a JAMB.

revealed preference An actual indication of people's willingness to pay for a non-marketed commodity as determined from their actual spending preferences rather than by what they say they prefer. Since it is a measure of what actually happened it is clearly more accurate than a prediction, but it can only be conducted after the event (i.e. EX POSTE) and therefore has limitations in use.

See also: CONTINGENT VALUATION

revenue That which is received as income.

See also: CAPITAL

reverberation time The length of time it takes, after the source has ceased, for the time-mean-square sound pressure in an enclosure to reduce by 60 dB (see DECIBEL).

reverse osmosis The process whereby a more highly concentrated solution is pressurised to force the SOLVENT through a SEMI-PERMEABLE MEMBRANE to a less concentrated or pure solvent on the other side. This process is employed in the DESALINATION of seawater.

See also: DIALYSIS; OSMOSIS

rheostat A component in an electrical circuit used to vary the amount of total resistance. A dimmer switch used to alter the intensity of electric lighting is essentially a rheostat.

rhinitis The inflammation of the mucous membrane of the nose.

ribonucleic acid (RNA) A complex organic polymer. Its function in the body is to receive genetic information from DEOXYRIBONUCLEIC ACID (DNA) in a process known as transcription and then to synthesise PROTEINS. There are several different forms of RNA classified according to function. Examples are messenger RNA (mRNA), which carries genetic information to the protein synthesis mechanism, ribosomal RNA (rRNA), which forms part of the structure of the protein-making ribosomes, and transfer RNA (tRNA), which carries AMINO ACIDS for protein synthesis.

Richmond Committee The name given to the Committee on the Microbiological Safety of Food, set up in the UK by announcement to the House of Commons on 21 February 1989 in relation to the microbiological safety of food. The Committee produced two reports – 'The Microbiological Safety of Food, Part I' was published in 1990 and 'The Microbiological Safety of Food, Part II' was published in 1991.

Richter scale One named after C.D. Richter, who devised it to provide a measure of earthquakes based on the comparative quantity of energy released. The scale ranges from 0 to over 8, the largest recorded earthquake measuring 8.6 on the Richter scale.

See also: EPICENTRE; MERCALLI SCALE

rickettsia (plural rickettsiae) A type of rod-shaped bacteria. They are principally obligate (see OBLIGATORY) intracellular parasites found in arthropods. They are responsible for a range of infections including typhus and Q-fever.

ridge The horizontal junction where two portions of a sloping roof meet.

See also: EAVES; HIP ROOF; GABLE

right-handed A shortened descriptor of those constructional items considered to be more convenient to people who are right-handed. The term applies to staircases whose handrail is on the right-hand side when ascending. In the UK it applies to a door whose hinges are on the right-hand side of one that swings towards you (in the USA it applies to a door whose hinges are on the right-hand side of a door opening away from you). A screw with a right-hand thread is one that enters the hole when turned clockwise as viewed from the head end.

rim lock A device for securing a door or similar, the locking portion of which is positioned on the door surface, causing a barring device to slide behind the JAMB to prevent it being opened.

See also: MORTISE LOCK

ringworm An infection of the skin that, in spite of its name, is generally caused by the fungus *Trichophyton verruscosum*. It is usually associated with contact with cattle but the infection can affect horses, pigs and dogs. Fungi of the *Microsporum* spp. may also cause the condition. Direct contact with infected animals is the usual mode of spread but the spores can remain viable for several months so it is possible for the fungi to be passed on by contact with contaminated objects such as fence posts, walls and from equipment such as cattle crushes. Ringworm is not particular serious although the round crusty skin lesions and associated hair loss may be disfiguring and aesthetically displeasing. Ringworm is regarded as a chronic benign infection and is relatively common among farmers, veterinarians, slaughtermen and those working with horses.

Rio Declaration The shorthand used to describe the declaration of twenty-seven principles made following the United Nations Conference on Environment and Development (UNCED) in Rio de Janeiro in 1992.

See also: AGENDA 21; PRECAUTIONARY PRINCIPLE

riparian Refers to that which belongs to or inhabits a river bank. The term is also applied to those who own land adjoining a river. A riparian authority is one whose jurisdiction includes the area of the river bank.

riser The vertical portion of the step in a staircase.

See also: TREAD

rising damp(ness) Moisture that is drawn up a porous building material by capillary attraction from moist soil or from other moisture at its base. Traditional wisdom suggests that dampness in even highly porous or absorbent materials cannot rise much above a metre from the moisture supply. Other sources claim that much dampness is frequently incorrectly attributed to rising damp and that CONDENSATION or PENETRATING DAMPNESS are really the true cause of much dampness in buildings.

rising main A vertical pipe within a building used to supply gas or water, or to remove sewage.

risk Essentially the chance of any given HAZARD being realised. For environmental health purposes it is most frequently used in relation to health and safety or to food. In terms of the latter, in 1998, the Codex Alimentarius Commission defined risk as 'a function or the probability of an adverse health effect and the severity of that effect, consequential to a hazard(s) in food'.

risk analysis A complete and structured approach to dealing with and minimising risk. Risk analysis embraces all aspects of risk and includes RISK ASSESSMENT, RISK COMMUNICATION, RISK MANAGEMENT and RISK MONITORING.

risk assessment The process of determining the actual risk in respect of a given situation or set of circumstances in relation to a defined HAZARD.

risk communication The process of imparting understanding in relation to risk to those who are, or might be, affected by it in order that they can make judgements about or form options for its control.

risk factor Any phenomenon, behaviour, exposure or act that is known to be or has been shown to be associated with risk.

risk management The process whereby those with responsibility for the evaluation and control of risk decide upon, implement and/or review controls for managing risk using the available evidence or their best judgement to inform the process to minimise risk.

risk monitoring The process of assessing the effectiveness of control measures that have been developed to control risk.

rivet A metal pin, usually with a preformed head at one end, used to fasten two items together by creating a head at the unformed end as it projects through the materials in order to hold the pieces together in compression. Hot riveting was a technique used to fasten steel or iron plates together using red-hot pins and hammering the unformed end until the head and joint was made. Aluminium pins can be used in cold riveting for smaller jobs.

RNA *see* RIBONUCLEIC ACID

RNI *see* REFERENCE NUTRIENT INTAKE

rodenticide A chemical substance that kills rodents (usually applied to substances that kill rats and mice).

See also: ANTICOAGULANT

rods *see* RETINA

Roman tile Also known as a mission tile, originally a roofing tile of curved or semi-circular appearance laid longitudinally along the downward sloping face of the roof, which alternatively curves upwards and then downwards. More modern tiles sometimes referred to as 'Roman' incorporate both convex and concave surfaces in a single tile such as to present an 'S'-shape in cross section.

Röntgen rays A synonym for X-RAYS.

rosacea An abbreviation for the condition known more fully as acne rosacea. It is a chronic condition in which the skin of the face and forehead becomes flushed with blood and the blood capillaries become engorged or rupture. In more chronic forms this flushing can be accompanied by the formation of pimples and enlargement of the SEBACEOUS GLANDS. This latter can cause gross swelling, known as rhinophyma, of the nose. The condition has historically been associated with the frequent excessive consumption of alcohol but can also be exacerbated by excessive food consumption and exposure to sunlight.

rosin Most correctly a translucent brittle substance, amber in colour, obtained as a residue in the distillation of turpentine oleoresin. Also known as colophony it is used as a constituent of varnish, paints, certain inks and sealing waxes, and as a lubricant for the bows of stringed instruments. The term is also sometimes used as a synonym for RESIN.

rotary engine An internal COMBUSTION engine in which the piston (usually comprising three curved sides) rotates eccentrically around a central shaft, the combustion chamber being created in the space between the curved sides of the piston and the internal walls of the cylinder within which it is housed.

See also: RECIPROCATING ENGINE

rotavirus First identified as human pathogens in the early 1970s, they are now recognised as the major cause of viral gastroenteritis of children in all countries. All healthy children have protective circulating and mucosal antibodies by the age of 5 years. Most infections occur in children under 5, although adults are occasionally diagnosed. It is estimated that every child in the world on average experiences around two attacks per year and that there are around 5 million deaths annually due to the infection. In the USA annually around 3 per cent of all hospitalisations of children under the age of 5 years are due to disease caused by rotavirus. In temperate zones the disease tends to be predominant during the winter months. The virus acts by attacking the villi of the intestine, degrading and destroying them, resulting in the inability to absorb food and causing diarrhoea that, in turn, can lead to dehydration.

The virus is detected by electron microscopy under which it has the appearance of a wheel, which is where the name derives. Transmission is generally by the faecal/oral route although there is evidence for supposing that respiratory spread is possible. There is little evidence of food-borne spread. Personal contact must, therefore, remain as the most likely method of source and transmission. The infective dose is low and a single virus may be capable of causing infection. The incubation period is around 2–3 days.

roulade In culinary practice, a meat roll or gelatine, or a foodstuff that is rolled or comes with added gelatine.

Roundup The commercial name of one of the most commonly used herbicides. Its active ingredient is known as GLYPHOSPHATE.

router A woodworking tool used to produce grooves or mouldings in timber.

Royal Environmental Health Institute of Scotland (REHIS) An equivalent professional body in Scotland to the CHARTERED INSTITUTE OF ENVIRONMENTAL HEALTH in the rest of the UK. It was formed in 1983 following the amalgamation of the Scottish Institute of Environmental Health and the Royal Sanitary Association of Scotland. Its membership includes Environmental Health Officers, Red Meat Inspectors, White Meat Inspectors and Food Safety Officers.

RSI *see* REPETITIVE STRAIN INJURY

rubber Also known as India rubber, the term originally applied to an organic LATEX obtained from TAPPING the sap of certain plants, especially *Hevea brasiliensis*. Nowadays much rubber is produced synthetically.

rubber mounting A device for isolating an engine or machine from its housing or support in order to reduce vibration.

rubber process dust That which arises during the processing stages (such as handling and mixing) in the manufacture of rubber. The term does not apply to dust arising from the abrasion or wear of the end product.

rubble An assemblage of irregular shaped solid objects comprising pieces of broken stones, bricks or similar. It is usually used as a packaging material in hard construction such as base packing for roadways or occasionally as AGGREGATE.

rubble masonry (sometimes called random rubble masonry) A technique whereby walls and similar are constructed from irregular pieces of stone cemented together to form a single structure. Such walls were often covered in RENDERING to disguise their surfaces and produce a more aesthetically pleasing finish. Although using random rubble introduced a cost saving by cutting out the time needed for dressing the stone, the irregular pieces used can create void spaces and produce a weaker structure.

See also: ASHLAR

rubella Also known as German measles, a mild viral disease of worldwide distribution with seasonal prevalence in winter and spring. The disease is characterised by fever and a diffuse rash, which can sometimes resemble that of measles or scarlet fever. Although the disease in children is generally mild, in adults the disease may present stronger symptoms that can include, for example, a combination of fever, headache, conjunctivitis and malaise for a period of around 1–5 days. Clinical diagnosis of rubella can be difficult or inconclusive so laboratory confirmation may be necessary. Occasional complications can arise from rubella infection; these include arthralgia (pain in a joint, in the absence of arthritis or other swelling) and (rarely) encephalitis. From the public health point of view the most important aspect of rubella is the potential to produce abnormalities in the developing foetus if the mother contracts the infection, particularly during the early part of pregnancy. Defects are rare when maternal infection occurs after the 20th week of gestation.

The reservoir of infection is man. Transmission is through droplets or direct contact with patients. Infants with congenital infection produce a large number of viruses in their pharangeal secretions and in their urine. The incubation period is between 14–23 days, typically being between 16–18 days. The disease is highly communicable from about 1 week before the rash to around 4 days afterwards. Exclusion from school or work should be imposed for 7 days after the onset of the rash. Immunisation by live attenuated vaccine is frequently conducted in association with live virus vaccines for measles and mumps.

See also: MMR VACCINATION

rudaceous Generally refers to SEDIMENTARY ROCKS and means composed of coarse-grained material.

See also: ARENACEOUS; ARGILLACEOUS

rumen The first stomach of a RUMINANT. The others are the RETICULUM (second stomach), the OMASUM (third stomach) and ABOMASUM (fourth stomach).

See also: RUMINANT DIGESTION

ruminant An animal (e.g. cattle and sheep) that has a RUMEN (the first stomach in a compound stomach).

See also: RUMINANT DIGESTION

ruminant digestion The process whereby an animal that possess compound stomachs (i.e. a RUMINANT) digests its food. The process involves the intake of quantities of vegetable material (e.g. grass), which passes to the RUMEN where it is stored. When convenient the animal regurgitates the food for a more complete mastication (i.e. the food is re-chewed) before swallowing it again to pass into other stomachs in the digestive system. Bacteria in the rumen help break down the CELLULOSE and STARCH components of the food, and convert them to fatty acids. The fatty acids and the bacteria are subsequently digested by protozoa (see PROTOZOAN), which in turn are digested further along the digestive tract to provide nutrients to the animal.

See also: RETICULUM; OMASUM; ABOMASUM

ruminant feed ban A measure introduced in the UK in 1988 in order to prohibit the use of meat and bone meal (MBM), unless derived from non-ruminants, in the feed for any RUMINANT.

running repairs (or maintenance) Those that are undertaken whilst that which is being repaired or maintained remains in use.

run-off That rainwater displaced by the action of gravity from a sloping surface or inclined body.

rust A reddish-brown oxide of iron (chemical symbol $Fe_2O_3.3H_2O$) formed by the action of oxygen on iron in the presence of moisture. The name is also applied to a range of fungal infections of vegetative matter that have the reddish-brown appearance of rust.

S

SAE The acronym for the Society of Automotive Engineers. The Society sets many standards in the automotive and aviation industries. One of the most frequently seen is the standard viscosity classification system for engine oils (e.g. SAE 20W50).

Saint George's Respiratory Questionnaire (SGRQ) An index used to measure the relative quality of life of various patients suffering from respiratory diseases or disorders.

See also: QALY

saline Means salty or containing SALT.

See also: DISTILLATION; SALINITY

salinity The measure or degree of the amount of SALT contained in a liquid (usually water). The measurement is usually expressed quantitatively (see QUANTITATIVE) as parts per million (ppm). The measurement of 'total dissolved solids' is an expression of the totality of salts that the solution contains rather than a quantification of the contribution made by any particular salt or other constituent.

The largest salt component of seawater is contributed by sodium chloride (NaCl), although salts of potassium (K) or magnesium (Mg) are also usually present. Salinity varies, with most natural waters (even those considered as 'freshwater') usually containing some dissolved solids. Water from the sea contains the most total dissolved solids of natural waters, although this varies dependent on source and sea temperature. 'Typical' seawater contains around 30,000 ppm dissolved solids whilst that from the Dead Sea has the highest concentration at around 240,000 ppm.

See also: DESALINATION

salmon A fish of the family *Salmonidae* of which there are a number of species. The salmon has a complex life cycle, being hatched from eggs laid in a freshwater river. The young fish hatches, remaining in the river to mature; after about 2 years it will be about 6 inches long. At this stage the fish will have distinct blue bands on its body and is known as a 'parr'. The marks gradually fade as the fish grows and it is then known as a 'smolt'. It is at this stage that the fish migrates to the sea. If the fish returns from the sea within a year it is known as a 'grilse'. Mature fish return from the sea to breed, usually to the same river in which they were born.

salmonella A bacterium of the intestinal tract of numerous species of both animals and birds, and is a major contributor in many countries to the incidence of food poisoning. There are over 2,000 serotypes of salmonella and there is considerable variation in the VIRULENCE and pathogenicity between the different types. There is even variation in these respects within the same types across different countries. Serotypes are generally (although not exclusively) named after the location where they are first found. Both typhoid and paratyphoid are due to infection with the organism (*Salmonella typhi* and *Salmonella paratyphi* respectively).

Salmonella has a worldwide distribution and is generally one of the human pathogens routinely screened for in faecal specimens submitted to laboratories in the UK. Although predominantly a food poisoning organism, person-to-person spread is not uncommon and infection arising from environmental contact or transmission via intermediary articles is not unknown. A high proportion of reptiles carries salmonella organisms asymptomatically. Faecal carriage rates in some species of reptile can be more than 90 per cent.

salt A compound (not water) formed by the chemical reaction of an ACID and a BASE. In common parlance the term is usually applied to common or table salt also known as sodium chloride (chemical symbol NaCl).

salt glaze A hard glaze incorporated into the surface of pottery or stoneware achieved by adding salt to the kiln when firing.

salting An alternative name for a salt marsh, being an area of land regularly washed by saltwater tides.

sandblasting The method of cleaning the surface of a material by blasting it with abrasive sand.

saponification The process of making SOAP.

saprophyte A micro-organism, plant or fungus that derives its sustenance from dead or decaying organic material.

sarcoidosis A chronic, progressive disease of unknown AETIOLOGY. It is characterised by an abnormal GRANULOMA-like increase in cell growth of many organs, but especially the skin, lungs, lymph nodes, heart, salivary glands, eyes and the bones of the feet or hands. The disease is frequently self-limiting but may require medical treatment in more serious cases. More recently it has been postulated that genetic and environmental factors might be implicated in its generation, but the evidence, especially for the latter, is not conclusive.

sarking A waterproof membrane or layer incorporated into a roof structure immediately below the slates, tiles or other roof covering in order to exclude wind or wind-driven rain or snow.

sash bar A GLAZING BAR that divides the viewing portion and to which the individual panes of glass are fastened in a SASH WINDOW.

sash window An opening in a wall comprising two glazed panels (the 'sashes') that slide parallel to each other to provide the opening portion, the whole being housed in a frame fixed in the building structure. Since the sashes always, in normal operation, stay within the frame, the opening portion is never more than 50 per cent of the total window area.

There are two types of sash window. The first is the 'vertical' or 'balanced' type in which the sashes are supported by counterweights attached to pulleys by cords and move up and down within the frame, each sash resting as positioned by being balanced against the counterweight. This type is also known as the 'double-hung' sash window. The second and less common type is the 'sliding sash' in which the sashes move horizontally across each other. These do not require counterweights as their position is maintained by gravity and friction.

saturated hydrocarbon A hydrocarbon in which the molecules have utilised all of the bonding electrons to make single bonds with other atoms – it does not contain any double or triple bonds. As a consequence the molecule possesses a full complement of hydrogen atoms and is therefore unable to combine with the atoms of other elements without giving up at least one hydrogen atom. A saturated hydrocarbon molecule is therefore more chemically stable than an UNSATURATED HYDROCARBON atom.

saturated solution One in which a liquid is incapable of dissolving any more of a solid material than it already contains at that temperature. A supersaturated solution can sometimes be created by dissolving (particularly a crystalline) solid in liquid at a higher temperature and then cooling the resultant solution to below its maximum saturated solution temperature. Provided this is done in clean conditions and in the absence of any of the solid itself, the solid will remain in solution even beyond the saturated solution level. However, if a piece of the solid is introduced into the supersaturated solution this will act as a focus around which the solid can reform and the excess of the solid will precipitate from the solution.

saturation temperature The temperature at which that air cannot take in any more water vapour (i.e. the temperature at which it has achieved 100 per cent RELATIVE HUMIDITY).

sauté A culinary term used to describe the process of cooking something in fat without causing it to become browned.

sawn timber That which has been roughly shaped or sized from a larger piece of wood or from the original trunk, but has not been planed or finished for decorative or aesthetic purposes. Much constructional timber is sawn timber. In the USA such timber is known as 'lumber'.

See also: DRESSED TIMBER

SBM *see* SPECIFIED RISK MATERIAL

SBO *see* SPECIFIED RISK MATERIAL

scabies An allergy that develops in response to the excreta and saliva of the parasitic mite *Sarcoptes scabiei* that burrows into the skin of infected persons. The mites are white and transparent with four pairs of extremely short legs. The mites are totally adapted to their life in the warm and humid tunnels in the skin and soon die of cold and dehydration if they leave them. Transmission of mites is therefore normally by direct skin-to-skin contact. Merely shaking hands is not sufficient to transmit the mite; a skin-to-skin contact of several minutes' duration is usually necessary transmission to occur.

See also: ALLERGY; PARASITE

scampi The name prescribed in law in the UK for the crustacean species *Nephros norvegicus*, commonly called the Dublin Bay Prawn or Norway Lobster. It is generally sold as a breaded product and a large proportion of sales consists of reformed scampi recovered from minced meat extruded to form scampi-like shapes. There can be significant adulteration of the breaded product with fish other than scampi. Adulteration has been reported by the inclusion of a different, closely related species of shellfish, *Metanephrops andamanicus*, commonly called Korean scampi, and other substitutes have included mixtures of scampi, warm-water prawn species and squid.

scarf joint One in which the ends of two members are shaped (otherwise than flat, end-on) to match each other on contact when the joint is made. A common version is where both the ends slope or taper and are fastened together by adhesive or by bolts. A more rigid joint can be achieved by notching steps into the ends to achieve a 'stepped' or 'hooked' scarf joint.

See also: BUTT JOINT

scarification An alternative term for the process of scratching, abrading or making rough the surface of something.

scarlet fever An infectious disease caused by the ERYTHROGENIC toxin produced by certain types of streptococcus. Symptoms include a typical symmetrical rash that does not itch and is usually associated with fever, headache and vomiting; in younger children convulsions or delirium may also occur in the early stages. The INCUBATION PERIOD is variable, being usually 2–3 days but this can be up to 1 week. Historically in Western society scarlet fever was a life-threatening illness, but it is now regarded as generally mild. Complications such as infection of the kidneys can arise and it is always advisable for patients with such conditions to be attended by a medical practitioner.

scavenger A chemical that readily combines with an unwanted contaminant in order to secure its removal.

scavenging precipitation The removal of atmospheric contaminants by the action of rainwater.

sclera The white, opaque covering of the majority of the eye, most of which is hidden from view within the eye socket. The visible portion is usually referred to as the 'white of the eye'.

See also: CORNEA; RETINA

scombrotoxin poisoning A form of FOOD POISONING caused by the consumption of fish of the scombroid group (e.g. tuna, mackerel and pilchards) that have been adversely affected by the production of toxin following decomposition or spoilage. Symptoms include a burning sensation in the mouth followed by nausea, vomiting and respiratory distress. Symptoms typically last for between 8–12 hours.

scoria Another name for SLAG. The term is also applied to lightweight lava infused with steam holes.

scoring In cooking, the process of making a number of shallow cuts in the surface of a foodstuff, such as meat or fish, in order to reduce cooking time and to improve the flavour, especially in order to allow marinade to penetrate more deeply into the food.

Scottish Environment Protection Agency The public body responsible in Scotland for taking the lead in implementing the Government's environmental policy. It is broadly analogous to the ENVIRONMENT AGENCY of England and Wales.

scrapie A progressive brain disease of sheep and goats. The name derives from the intense itching that the condition appears to produce, causing affected animals to scrape off their wool in an effort to seek relief. Scrapie is a transmissible spongiform encephalopathy and was first identified in 1730. Although scrapie has never been known to lead to spongiform encephalopathy in humans it has been postulated that BOVINE SPONGIFORM ENCEPHALOPATHY (BSE) was derived originally from the agent responsible for scrapie. It is thought that the epidemic of BSE in the UK originated as a result of the inclusion in cattle feed of protein derived from scrapie-infected sheep.

screed Can be variously described as:

1 The finishing coat of either cement, mortar or plaster on a wall or floor.

2 A bed of mortar into or upon which ceramic tiles are laid.
3 A layer of concrete laid on a flat surface to produce a FALL for drainage.
4 A strip of material affixed to a wall or floor to act as a guide to the correct depth of cement, plaster or mortar required, or the board of similar used to remove excess filling material from between such formers.

screening The systematic application of a test or examination to assess the health status of individuals in respect of specified morbidity. This is in a defined population that has been determined as being especially at risk of developing that condition. Screening is undertaken particularly to identify those cases of disease in an early stage of development (and that would benefit from early treatment) such as would not normally cause the individual to seek medical advice.

scrubber A device for removing airborne contaminants from a stream of gas by passing it through a specially designed series of water sprays. They can be expensive to operate and, dependent on the nature of contaminant(s) being removed, the water is often (to some degree) recycled to conserve resources and reduce waste.

scupper A protective covering installed to prevent clogging over the opening of a drain. The term is also applied to describe a drainage opening in a parapet through which rainwater is conveyed.

Sea Empress An oil tanker that ran aground just outside the mouth of Milford Haven, UK, on 15 February 1996 spilling around 70,000 tonnes of North Sea light crude oil. Although an environmental disaster, the potential for damage was greater than that associated with the TORREY CANYON due to the proximity of the spill to land. Lessons learned from previous oil spills, however, resulted in far less overall environmental damage than had been feared and the clean-up operation is generally considered to have been highly successful.

searing A process used in cookery in which the surface of a foodstuff, such as meat, fish or vegetables, is browned in fat prior to grilling or roasting. The purpose is to seal the surface to retain both moisture and flavour.

seasoning (of timber) The process of removing or reducing the moisture content of unseasoned or 'green' wood prior to its use in order to prevent cracking, distortion or warping arising from drying when in use. Seasoning can be accomplished naturally (and some claim most effectively) by keeping the timber dry and allowing evaporation to complete the process, but for many timbers this can take years and it is both inconvenient and expensive. In practice most timbers in modern use are seasoned artificially in specialist kilns.

sebaceous glands Small glands located in the skin next to hair follicles. They secrete an oily substance known as SEBUM. Blockage of these glands by sebum causes BLACKHEADS and can lead to ACNE or ECZEMA.

sebum A natural lubricant of the skin and hair secreted by the SEBACEOUS GLANDS.

secondary colours Those colours of either light or pigment that can be made by mixing various proportions of PRIMARY COLOURS.

Secondary Legislation Legislation made under the powers contained in PRIMARY LEGISLATION. It is generally subservient to the legislation under which it is made.

See also: ACT; BILL; PRIMARY LEGISLATION; REGULATION; STATUTORY INSTRUMENT

secret nailing A technique of fastening timbers by driving nails through non-exposed parts of the structure to ensure that the nail heads cannot be seen on the external surface on completion. The technique is used particularly on varnished or polished decorative timber cladding. The technique is also known as blind nailing.

sector (of a circle) Any portion of a circle comprising an ARC, each end of which is attached to a radius of the circle from whence the arc derives.

See also: SEGMENT (OF A CIRCLE)

sedimentary rock One that is formed from accumulated deposits at the bottom of lakes, seas and oceans or on land sufficient to coalesce into a cohesive and recognisable whole.

See also: IGNEOUS ROCK; METAMORPHIC ROCK

seed bank A resource in which the seeds of threatened plants are housed specifically to safeguard against potential extinction of the species. In the UK the Millennium Seed Bank Project was launched by the Royal Botanic Gardens, Kew, as one of the largest of its type, aiming to safeguard the future of over 2,400 species.

segment (of a circle) Any portion of a circle comprising an ARC whose ends are joined by a straight line, this latter being known as a 'secant'.

See also: SECTOR (OF A CIRCLE)

seismic Means of or relating to an earthquake or (alternatively) in relation to vibrations in the earth's crust generated thereby or to those arising from man-made explosions (the latter are usually generated deliberately in order to explore the structure of rock strata etc.).

seismograph A scientific instrument used to detect, measure and record vibrations within the earth's crust such as those generated by earthquakes.

selection bias A form of SYSTEMATIC ERROR known as BIAS in an observational epidemiological study or similar in which the study design is flawed and thereby allocates 'exposed/cases' and 'non-exposed/controls' incorrectly. This creates a fundamental error in the study and, although it will provide consistent and repeatable results, these are nevertheless wrong.

selective medium A variety of culture medium used in the laboratory and has been formulated specifically to promote the growth of a target organism or group and/or to inhibit the growth of other micro-organisms or groups.

selective tender One in which quotations are sought solely from a pre-specified group of contractors. This group is usually pre-selected because they have been judged to satisfy certain specified preconditions. The advantage of this type of TENDER is that the client will be satisfied that each contractor is capable of meeting basic requirements. The main disadvantage is that such procedures can be perceived by those not on the list as restricting work availability to an artificially limited CARTEL.

See also: OPEN TENDER

semiconductor A material that does not always conduct electricity but can be induced to do so. Some such materials (e.g. silicon) are used in applications in which they variously act as both conductor and insulator.

semi-permeable membrane A solid (usually thin) barrier that will allow the passage of smaller molecules from one side to the other but will prevent the passage of larger molecules. Semi-permeable membranes are usually used to separate dissolved solids from solution.

See also: DESALINATION; DIALYSIS; OSMOSIS

sensible heat That portion of energy that is absorbed or released when a temperature changes without a change from gaseous, liquid or solid state.

See also: LATENT HEAT

sensitivity In relation to an epidemiological study, a measure of the proportion of people in a screened population who are identified by the test as being ill compared to the total number of people in that population who are actually ill. It is a measure of the false negative (see TYPE II ERROR) component of the study – i.e. a measure of those who are falsely determined as not suffering from the disease under consideration.

See also: SPECIFICITY

sensitivity analysis An assessment of a particular outcome through the analysis of the most important variables in order to determine the degree to which the outcome could have been affected by marginal changes in those variables or by the inclusion/exclusion of same.

sentinel practice One of a number of designated health facilities in the UK operated by a general (medical) practitioner (GP) in which particular recordings of a wide range of clinical diagnoses are made. Comparing the diagnoses to the populations served enables an estimate of the RATE of illnesses in the population to be calculated for overall monitoring purposes.

separate drainage system A wastewater disposal system in which foul water and surface water are conveyed by different systems to different points of disposal. Surface water is usually conveyed directly to natural watercourses; foul water is conveyed to a treatment facility.

separation distance The space that is required to be maintained between a genetically modified crop and conventional crops. The separation distance is intended to reduce to a minimum cross-pollination between the two crops. It should be noted that the maintenance of a specified separation distance cannot altogether rule out the possibility of cross-pollination. For example, a study in Mexico published in 2001 suggested that wild maize had been contaminated with genetically modified DNA from crops at least 100 kilometres (62 miles) away.

See also: FARM SCALE EVALUATION; GENETICALLY MODIFIED ORGANISM

septic tank A self-contained reception and treatment facility for SEWAGE with the resultant effluent flowing out for disposal – usually to a soakaway system. The degree to which the tank acts as a treatment facility is controversial, with some claiming that they do little more than allow solid material to break down and others claiming that they promote or facilitate bacterial digestion of the contents.

sequelae Those health consequences (such as long-term effects, becoming IMMUNOCOMPROMISED or recurrent infection) that may be manifest following the occurrence of a disease.

sequencing Any technique whereby the order of constituent building blocks comprising smaller molecules is determined in respect of a larger molecule within which they are found. The term is usually used in reference to the application of the technique to the location of amino acids or nucleotides, respectively, to proteins or to DEOXYRIBONUCLEIC ACID (DNA)/RIBONUCLEIC ACID (RNA) molecules.

serodiagnosis The identification of a micro-organism by the use of SEROLOGY.

serogroup A broad band of micro-organisms differentially identified or characterised by their response to an ANTIBODY.

See also: SEROTYPE

serology A means of establishing infection by the detection of ANTIBODIES. The term also applies to the IN VITRO examination of micro-organisms using the differential response to specific antibodies as a technique for characterisation.

serotype A differentiated species of bacteria determined by their reaction to antisera (i.e. ANTIBODY). Bacteria that can be typed in this way include *Salmonella*, *E. coli*, *Listeria monocytogenes*, *Clostridium perfringens* and *Bacillus cereus*. The process of determining a serotype is known as serotyping.

serous Relates or pertains to that which resembles or contains SERUM.

serum The clear portion of any biological fluid.

serum antibodies ANTIBODIES that can be found in the SERUM portion of the blood.

set aside The controlled withdrawal from use of productive agricultural land as part of the procedure of regulating production. The system was introduced following concern that fit crops were being ploughed into land, otherwise destroyed or stockpiled simply to maintain price structures within controlled agricultural systems. The system relies on the payment of subsidy to compensate for the economic loss to the farmer. The procedure has now largely lost favour as the land set aside was often simply left untended and the (by some) 'hoped for' increase in wildlife habitat largely failed to materialise as there was insufficient time allowed for a natural ecosystem to develop. It is likely that the system will in future give way to one in which subsidies are paid on the basis of countryside and wildlife management, with the aim of improving the habitat rather than one of essentially promoting neglect.

settlement tank A container in which solid material is allowed to separate out from suspension in a liquid. The term is synonymous with the term 'detritus tank'.

sewage The foul or waste material that is carried away in a system of pipes to a disposal or treatment facility.

See also: SEWER

sewage sludge The solid and semi-solid material produced during the treatment of SEWAGE. Disposal of sewage sludge is problematic. It can be used to produce gas (such as METHANE) for use as a fuel, but such practices can be expensive to install. Much sewage sludge is disposed of to land as a fertiliser, but this practice requires careful regulation to avoid nuisance from smell, the pollution of watercourses or AQUIFERS and the potential for disease transmission by PATHOGENIC organisms.

See also: CRYPTOSPORIDIUM

sewer A drain or pipe through which SEWAGE and/or surface water is conveyed for disposal. An interconnected system of sewers is known collectively as a 'sewerage system'.

sex chromosomes Those CHROMOSOMES that determine the sex of the offspring of sexual reproduction.

SGRQ The acronym for SAINT GEORGE'S RESPIRATORY QUESTIONNAIRE.

shallow well One in which the water supply is derived from ground higher than the first impermeable stratum. In theory such water would be more likely to be contaminated by microflora deriving from surface contamination than would water obtained from lower strata.

See also: ARTESIAN WELL; DEEP WELL

shard A broken piece of the whole, especially of pottery.

sharps A collective term to describe waste such as injection needles that are potentially contaminated with human (or, in the case of veterinary waste, animal) tissue or fluid, such as to pose a risk of injury to waste disposal operatives due to stabbing (known as a 'stick' injury). Such waste should be kept separately from other waste in specially designed and distinctive containers, and, in the UK, is usually removed and disposed of as a special collection.

shear force That which is applied at 90° across a structural component or object tending ultimately to cause it to part across the line of stress. The shear force is usually designated by the symbol '*V*'.

See also: COMPRESSION; TORSION

sheep scab Also known as psoroptic mange, a contagious disease of sheep caused by the mite *Psoroptes ovis* that lives on and in the skin. It is a year-round infection but is more prevalent during the winter months. The mites set up an intense irritation that the animal tries to relieve by scratching and biting, causing damage to the hide, loss of wool and potentially the development of open sores that can subsequently become infected, which in turn can lead to death.

sheet glass The generic term used for glass of normal domestic glazing quality.

See also: FLOAT GLASS; PLATE GLASS

shelf life The period during which a particular food product is expected to retain its quality and safety if conditions since manufacture are maintained to specification. Many manufacturers allow a margin of safety between the expiry of the given shelf life and the time when, normally, commencement of detectable signs of deterioration could be expected. Consequently a product that has passed its shelf life is not automatically unfit for human consumption, but prudently should not in any event be eaten.

shellac Originally a yellowish-coloured resin secreted by a tropical insect known as the lac. When dissolved in ethanol or similar solvent it was used in polishes, varnishes and as a dressing for leather. In modern times shellac has been produced commercially but many preparations incorporating this artificially manufactured material still refer to it using the same term.

Shiga-like toxin (SLT) Any toxin with similar biological profile to the toxin produced by the bacteria *Shigella dysenteriae* type 1 (the 'Shiga bacillus'). In practice this toxin is indistinguishable from VEROCYTOTOXIN and the terms are used synonymously.

shim A washer or thin plate used to pack a joint and adjust or position the various components correctly in relationship to one another.

shore Temporary support or BRACE used to strengthen an existing structure, formwork or excavation. A 'dead shore' is one that braces vertically, a 'flying shore' is one that braces horizontally and a 'raking shore' is one that braces as a sloping support. The last is the most common.

See also: NEEDLE

short circuit An inadvertent or accidental breaching of an electrical circuit causing two points to make contact and to carry current across a shorter electrical pathway of lower resistance than the original circuit.

short-wave radiation Electromagnetic radiation with a wavelength below 0.5 microns. In terms of light the term usually applies to those in the ultraviolet spectrum, i.e. those with a wavelength below 0.4 microns. Short-wave radiation contains more energy than does LONG-WAVE RADIATION.

shotblasting The method of cleaning the surface of a material by blasting it with metallic pellets (usually of steel), fired under pressure or impelled by CENTRIFUGE.

shuttering *see* FORMWORK

SI The acronym for Système International, an international standard of units adopted in 1960 by the 11th General Conference on Weights and Measures. The system uses *inter alia* metres, kilograms and seconds – it is also known as the metric system.

See also: AVOIRDUPOIS

sick building syndrome The term applied to adverse health effects associated with buildings in which the occupants exhibit or complain of a range of disparate symptoms, allergenic in nature, whose cause is attributed to the occupation of the building itself. The symptoms are usually relatively mild in manifestation but some are claimed to be seriously debilitating. Although there was scepticism that the syndrome actually existed when it was originally proposed in the early 1980s, it was officially recognised by the World Health Organization in 1982. Identification of remedial action can sometimes be more of an art than a science; most authorities now accept that certain types of building are predisposed to affect the health of their occupants adversely. In the UK the Health and Safety Executive (HSE) has published advice. Affected buildings are usually those used for work and most of these were erected during and since the 1960s. It has been suggested that some psychological factors associated with boredom, stress or repetitive work may contribute to the likelihood of complaints of sickness.

sievert (Sv) The SI unit of measurement used to categorise received dose of radiation. For practical purposes the sievert is perhaps too large a unit to use and dose is often expressed in terms of microsieverts (mSv). The average annual whole-body dose from natural radiation in the UK is around 2.2 millisieverts.

See also: EQUIVALENT DOSE

significance level In statistical terms, a measure that an event or phenomenon could have arisen by chance. Statistical significance is achieved when a predetermined significance level is achieved or exceeded. For many clinical trials or epidemiological investigations this is when the PROBABILITY (P) VALUE is less than 0.05. This point is achieved when the significance is at the 95 per cent level – this can alternatively be expressed as the probability that the event occurred by chance is less than 1 in 20.

significance test Any one of a number of statistical tests used to determine whether the outcome of a variety of investigations is true or could have (or the degree to which it could have) arisen by chance.

See also: ANALYSIS OF VARIANCE; CHI-SQUARED TEST; T-TEST; WILCOXON TEST

silage A crop harvested for animal fodder. It comprises grass that is cut whilst still green and then stored in a silo or other airtight facility. This mass then undergoes partial fermentation of the sugars in the forage in the absence of air (i.e. under anaerobic conditions), which produces the organic acids necessary to keep the silage sweet until used. The liquor that derives from making silage has a high Biological Oxygen Demand

and accidental spillages can cause severe nuisance and heavily pollute affected areas, usually killing large quantities of aquatic life in the process.

silencer A device, usually constructed as a duct, used to reduce a sound level by the absorption or reaction of the sound energy.

Silent Spring The title of a book first published in 1962 and written by Rachel Carson. The book was one of the first of its kind to deal with issues of chemical pollution of the environment by pesticides, herbicides and fertilisers, and was a landmark in the progress of the environmental movement. The book was greeted with considerable scepticism and opposition at the time it was published, but it was instrumental in identifying and eventually leading to controls on the use of pesticides such as DDT. The title of the book refers to the scenario in which chemicals had destroyed bird life to the extent that spring arrived with no attendant bird song.

silica A common name for silicon dioxide (SiO_2). In crystalline form it is known as quartz. Silica makes up around 60 per cent of the earth's crust and is a major constituent of sand and clay.

silly season The term applied to the summer months in the UK during parliamentary recess when journalists are reputed to report particularly frivolous or nonsensical events in order to fulfil work output expectations.

silt In popular parlance, any fine particulate deposit in a watercourse deriving from mineral degradation. More specifically the term is applied to particles of between 2–50 micrometres in diameter, i.e. of intermediate size between clay and sand.

silver solder A hard SOLDER with a melting point at a higher temperature than normal solder. It has a higher than average strength and is used predominantly in plumbing. The name derives from its silver content, which is usually between 1–5 per cent by weight.

silviculture The branch of forestry concerned with the growing and tending of trees.

simmering The process in cooking in which a liquid is maintained at a temperature of just below boiling point. The liquid is first brought to the boil and then the heat is adjusted to maintain the correct temperature. Simmering is used to cook those foods that would otherwise lose their flavour or texture by being cooked at a higher temperature.

simulation The method whereby a real-life activity is undertaken in the form of a MODEL in order to test a hypothesis or to research processes or outcomes. Simulations can be undertaken physically or mathematically, although increasingly they are progressed using computer technology.

sine die Literally means 'without a day (appointed)'. It is sometimes used in legal proceedings to denote the postponement of a hearing, in which case it usually refers to an indefinite postponement.

sink A natural depository within which various chemical materials or solids are stored such as to remove them from transmission or use elsewhere for the duration of their time in the sink. The oceans are the largest of the sinks although soil, natural waters, glaciers and living tissue (especially trees) make additional significant contributions to the storage capacity. If, for example, large areas of ice are melted, the carbon dioxide trapped within it is released and, if this is done in large enough quantities, can (and indeed does) impact on the earth's climate by contributing to global climate change.

Similar effects can occur if large areas of forest are cleared and burnt, thereby releasing large quantities of stored carbon back into the wider environment.

Sintra statement The name applied to the statement by the Ministers of the European Commission made at their meeting in Sintra, Portugal, on 23 July 1998 supporting the OSPAR Commission for the protection of the marine environment of the North-East Atlantic. The statement committed the European Community to preventing and eliminating pollution, protecting human health and ensuring sound and healthy marine ecosystems. The statement includes a list of actions to which the Member States are committed in pursuit of these aims and of producing a sustainable approach to the marine environment.

sisal A coarse fibre derived from the leaves of (originally) the tropical plant known as 'sisal'. It was used extensively as a material for manufacturing rope and for reinforcing a variety of building materials such as plaster. It has now largely been superseded by modern materials.

size A thin solution used to seal or act as a binder on a wall or similar prior usually to the application of further coats of protective or decorative finishes. The term also applies to solutions used to provide a glaze to paper or to stiffen textiles.

skillion roof Any roof with a single sloping surface – sometimes known as a monopitch or single-pitch roof.

skin The largest single organ of the body. It serves several functions including physically protecting the inner organs and body structures, regulating the body's temperature and providing a barrier to infectious agents. It comprises two distinct portions, the EPIDERMIS (or EPITHELIUM), which is the outer protective layer, and the corium. The latter is also known as the 'cutis vera', 'true skin' or 'dermis' and contains many nerve fibres and blood vessels.

skin cancer Generally accepted as being predominantly caused by or associated with exposure to sunlight. Prevention strategies therefore centre on reducing exposure or by covering the skin with a physical or chemical protective barrier. There are four main types of skin cancer:

- Basal cell cancer (affects mainly the elderly and accounts for around 90 per cent of skin cancer patients).
- Squamous cell cancer (affects around 5 per cent of skin cancer patients; most of those affected are elderly).
- Malignant and non-malignant melanoma (affects predominantly younger adults, is highly aggressive and can be fatal – incidence has increased 30 per cent in the last 30 years).
- Others (these are rare).

The component of the sun's rays that contributes to skin cancer is ultraviolet radiation. Although popular opinion suggests that UVA is the radiation that is associated with skin cancer, all varieties are thought to be potentially carcinogenic.

See also: ULTRAVIOLET RADIATION

skirting The name given to the moulding positioned to provide a decorative finish at the base of a wall at its junction with the floor. The skirting also provides protection for the wall against impact damage. The skirting is sometimes referred to as the baseboard.

slag A term that refers generally to a by-product of blast furnaces during steel manufacture and comprising primarily calcium silicates and alumino-silicates. It is used for a variety of purposes including as Portland Cement (see CEMENT) substitute, as an AGGREGATE or spun into MINERAL WOOL.

slaked lime *see* HYDRATED LIME

slash and burn A technique of land clearance or DEFORESTATION in which the vegetation is cut down and burnt IN SITU, thereby returning nutrients such as potash to the soil to enable farming to take place. This fertility in the soil does not usually last beyond a few years and farmers find it necessary to clear fresh virgin land to sustain their food production levels. The land once cleared is abandoned and develops into grassland and scrub.

slate A naturally occurring rock formed in the earth's crust by the physical compression of clay, silt, shale or similar. It is very fine grained and durable. It is used in block form for constructing walls and can be split laterally to produce roofing and cladding material (i.e. 'slates').

See also: METAMORPHIC ROCK

slide rule A device comprising tabulated numerical scales (usually expressed as logarithms) that can be moved against each other to perform calculations primarily of multiplication and division. They have largely been superseded by the advent of electronic pocket calculators.

sliding sash The horizontally moving glazed portion of a sliding SASH WINDOW.

slope A reference angle of an inclined surface, usually to the horizontal.

SLT *see* **Shiga-like toxin**

slurry A viscous fluid created by holding a finely ground solid in suspension in a liquid. The term is especially applied to describe liquefied manure.

smallpox Once a severe, life-threatening, viral disease that had plagued humanity for centuries. The causal organism was a member of the variola virus group. In the late 1950s the World Health Organization (WHO) commenced a campaign to eradicate the disease completely. The campaign proved to be highly successful, the last known case of naturally acquired smallpox being recorded in October 1977. Two years later, in 1979, the WHO certified that global eradication had been completed This was the first time that a disease organism had been entirely eliminated by organised human endeavour from any live natural reservoir of infection.

The live virus had not, however, been entirely destroyed as stocks were still being held under strict security at two known sites in the USA and Russia. It is also possible that other countries have retained stocks for use in biochemical weaponry research programmes. It is argued that such stocks could lead to accidental or terrorist release.

Small Round Structured Viruses *see* SRSV

smelt (or smelting) The process of melting a metallic ore in order to obtain the metal. A smelt is also a fish (*Osmerus eperlanus*) related to the salmon. It is sometimes known as a sparling.

smog An atmospheric phenomenon whereby PARTICULATE MATTER or gaseous chemicals derived from pollutants becomes incorporated into mist or fog in sufficient quantity to be capable of producing clinical effects. There are three basic types of smog, characterised by the pollutants found. Summer smog is caused by the accumulation of ozone or nitrogen dioxide, vehicle smog is caused by nitrogen oxides and winter smog is caused by sulphur dioxide or nitrogen oxides.

smoke *see* AEROSOL; FUME; SMOG

smoke test A method of testing the integrity of drainage systems. There are two basic methods of smoke testing. The first is rather rudimentary and requires the use of a cartridge (known as a smoke rocket) to generate smoke, which then enters the system, without significant pressure being used. Leakage signifies the presence of a (usually significant) breach in the system. The second method requires the use of a specialist machine that introduces smoke, under pressure, into the drainage system. Pressure testing, especially of existing systems, can be counterproductive as the test can cause damage that would otherwise not have occurred.

See also: AIR TEST; COLOUR TEST; WATER TEST

smolt A stage in the life cycle of the SALMON.

smørbrød (or smørrebrød) A name of Norwegian or Danish origin respectively to describe a HORS D'ŒUVRE served on bread and butter.

smörgåsbord A table laid out with HORS D'ŒUVRES and other dishes from which one helps oneself. The name is of Swedish origin.

Snow, John Famous in the annals of epidemiology as being the man who identified the source of an outbreak of CHOLERA in London in 1894 as arising from contaminated water delivered from a public water supply pump in Broad Street. He is reported as having removed the handle of the pump, thereby making it inoperable, preventing consumption of the water and ending the outbreak. His identification of the source of the outbreak was achieved by plotting the cases of cholera on a map and noting that the area around this particular pump was the most affected.

soap An emulsifying agent that facilitates the dispersal of an insoluble substance in water. The process of manufacturing soap is known as saponification and involves breaking down fat using a BASE to produce glycerol and soap – which is the salt of a fatty acid. The fatty acid salt is usually of sodium or potassium. When mixed with water, soap forms spherical groups of molecules known as micelles. The soap molecule itself comprises a sphere with a tail. The sphere is hydrophilic and attracts water; the tail is hydrophobic and is attracted to (for example) the hands. When soap is rubbed on the hands during washing the soap molecules surround the dirt, preventing it sticking to the hands, and this is then washed away. Soap can only act in this way when it is rubbed; the more it is rubbed the more effective the cleaning effect.

socket The female union portion of a mechanism for joining together two units of equipment or a system. The human hip is a ball and socket joint in which the ball of the femur fits into the socket of the pelvis. The socket of an electrical system is the female union into which the 'male' plug is inserted to obtain power.

socio-demographic factors Those VARIABLES that relate to the basic characteristics of a defined population such as age, sex and ethnicity.

socio-economic factors Those VARIABLES that relate to the basic characteristics of a defined population bearing principally on financial and economic issues such as employment, income and assets. A socio-economic burden is that which is calculated by totalling the financial and OPPORTUNITY COSTS that would otherwise not have occurred if the burden had not been present.

Sod's Law A humorous expression denoting the perception of perversity usually summarised as 'anything that can go wrong, will go wrong'. It is alternatively known as 'Murphy's Law'.

soffit The exposed underside surface of a structural timber, archway, EAVES or ARCHITRAVE.

soft water Water that is low in its content of calcium or magnesium bicarbonates and sulphates. Hardness of whichever cause is generally expressed as the equivalent of milligrams per litre of calcium carbonate (mg/l $CaCO_3$). Water with less than 100 mg/l is regarded as soft.

See also: HARD WATER

software The generic term applied to the operating programmes of a computer.

See also: HARDWARE

softwood Any wood derived from a CONIFEROUS tree. The term does not inevitably describe accurately the wood's physical characteristics of softness or hardness, but as a generality softwood can be considered as softer and less durable than HARDWOOD.

Soil Association An organisation, formed in 1946 in the UK, which campaigns, promotes and certifies the production of ORGANIC food and organic methods of farming.

soil drain A system of pipes or similar used to convey SEWAGE or contaminated commercial or industrial wastewater (including trade effluent) to a point of discharge or treatment. Such a system of conveyance should be regarded as a soil drain if it carries this type of waste exclusively or in addition to rainwater.

See also: STORM DRAIN

solarium A room or area (especially in a sanatorium or treatment facility) specifically constructed or organised to enable the occupants or attendees to benefit from the supposed therapeutic benefits of exposure to the sun's rays. The term also applies to such facilities supplied with artificial sunlight. Health concerns emerging over recent years regarding the potential carcinogenic (see CARCINOGENICITY) potential of ultraviolet rays has considerably reduced the appeal and use of such facilities.

solar flare A rapid expulsion of energy from the sun, often associated with areas surrounding SUNSPOTS. A solar flare can result in the earth receiving higher than normal levels of X-RAYS and high-energy ions as a consequence.

solar noon *see* MERIDIAN

solder An alloy used to join two metals together (especially in wiring or electrical systems). Importantly the solder has a lower melting point that the metals being joined. It is essential that the surfaces to be joined are exposing bright metal and are clean and free from contamination. Cleaning is usually accomplished by means of a 'flux', which melts at a lower temperature than the solder and frees the surface of oxides and similar contaminants.

See also: PLUMBER'S SOLDER; SILVER SOLDER

soldier course A horizontal line of brick or masonry in a wall in which the units of construction are laid with their longest sides lying vertically on the exposed surface(s) of the wall.

See also: STRETCHER

solenoid A means of converting electrical force into mechanical force. It consists of a wire coil, usually surrounding a cylindrical metal former, through which an electric current is passed to create a magnetic force. This latter is used to operate switches in electrical circuitry.

Solicitor General In the UK is the officer of the Crown ranking immediately below the ATTORNEY GENERAL (in Scotland the LORD ADVOCATE).

solstice That time of year when the sun is at its greatest angle from the vertical at the equator and is consequently directly overhead at the Tropic of Cancer (in the north) or Capricorn (in the south). The solstice therefore produces the longest and shortest days of the year in northern and southern hemispheres – this occurs on or around 21 June or 22 December.

soluble wheat protein (SWP) Used as an emulsifier in a range of food products such as breakfast cereals, processed meat products and sauces and dressings. Although not a food in its own right, it has been used extensively as a product ingredient for a number of years. There are reports that some forms contain high proportions of GLUTEN and could promote allergic reactions in sensitive people.

solute That which is dissolved by a SOLVENT.

solvent Any liquid that has the ability to dissolve a solid and hold it in solution. No solvent will dissolve everything and the different capacities of different solvents are used commercially and analytically as a means of separating a mixture of different solids from each other.

See also: SOLUTE

somatic antigen Another name for the 'O' antigen.

See also: LIPOPOLYSACCHARIDE

somatotropin A HORMONE produced by the anterior pituitary gland located at the base of the brain. It is associated with growth. Bovine Somatotropin (BST) has been produced artificially and used, particularly in the USA, as a way of increasing milk production in cattle. It is not licensed for use in the European Union.

sonic boom The name given to the SOUND generated by an aircraft that is travelling faster than the speed of sound. Sound is transmitted by pressure waves. These have a variable velocity dependent on the density of the medium through which they are travelling. In air the density varies with altitude. If an aircraft travels faster than the speed of sound in that environment, it catches up and overtakes the pressure waves as they are generated. This leads to the creation of a shock wave that spreads outwards from the rear of the aircraft in the form of a sonic cone. When the sonic cone passes an observer (such as one standing at the earth's surface) it is detected as a sonic boom. A sonic boom therefore is a continuing event that travels with the aircraft and that will last as long as the aircraft continues to generate the sonic cone, i.e. as long as it is travelling faster than the speed of sound.

sonification The process used in certain types of HUMIDIFIER to generate a water aerosol. The process involves the use of oscillators vibrating at ultrasonic frequencies to generate a particularly fine mist. The process has been used widely in some types of food display units to reduce drying out of the contents. Some manufacturers claim that the process also assists colour and texture retention, as well as improving SHELF LIFE.

soot The product of the incomplete combustion of fossil fuels. It comprises predominantly finely divided particles of carbon, although trace elements or other pollutants may be present dependent on the nature of the source fuel and the combustion process.

sorbitol MacConkey agar A type of SELECTIVE MEDIUM used in the laboratory growth and detection of the bacterium ESCHERICHIA COLI O157:H7.

sound A form of energy, transmitted by pressure waves in air or other material, such as potentially to produce the sensation of hearing. The pressure waves are generated by vibration. In humans the range of detection of frequencies is generally taken as being between 20–20,000 Hz (see HERTZ). The speed (i.e. velocity) of sound varies with the density of the atmosphere and consequently with altitude. At sea level the speed of sound is around 1,220 km/hr (i.e. around 760 miles/hr).

See also: BOOM

sound intensity The rate of the transmission of sound energy per specified unit area in a specified direction.

sound level meter A device used to measure sound in accordance with an accepted national or international standard.

sound power The total sound energy emitted by a source expressed per unit of time.

sound pressure The dynamic variation in atmospheric pressure generated by the passage of sound energy.

sous vide A COOK CHILL food preparation process that involves sealing food in a vacuum package, prior to cooking and subsequent chilling. The system was developed in France in the mid 1970s. The system is intended both to increase the SHELF LIFE of the product and maintain quality in terms of the smell, taste and texture of the food.

soy (or soya) sauce A food additive made from fermented SOYA BEANS. It is very salty and is usually dark brown in colour. It is a staple ingredient of many Chinese dishes.

soya (bean) The fruit of the Asian bean plants *Glycine max* or *Glycine soja*. It is highly nutritious and provides a good food yield per unit area of ground. Soya beans have been hailed as being a health food because of their low fat content. Some sources suggest that soya beans have also been found to contain high levels of ENDOCRINE DISRUPTORS and that any health effects might not necessarily be beneficial.

See also: SOY (OR SOYA) SAUCE

spa A place or resort at which a mineral spring issues. Such places have been popular since Roman times and bathing in or imbibing of the waters has been considered by some as an aid to health.

spa bath A bathing facility in which jets of water and/or air are used to create (sometimes violent) motion in the water content of the pool with the aim of stimulating or soothing the bather. Such facilities are used primarily for leisure or therapeutic purposes. In commercial applications spa pools must be designed for such purposes and operated in accordance with the manufacturer's instructions. Pools designed for domestic use generally are not capable of maintaining water quality for higher demand and poor installation and operation has resulted in LEGIONNAIRES' DISEASE.

span The linear distance of the void space between the two sides of a supporting structure.

special waste In the UK is any waste that is specifically designated as such by statute or more broadly is deemed to be dangerous to wildlife, is potentially explosive or is used as a human medicine.

speciation The term used to describe the phenomenon of changes within a single species such as to produce two new and (in some way) distinctly different species. It occurs as a result of the isolation of two groups of the same species who then go on to develop in different ways. It is thought to be due to random changes in the occurrence of particular genes within a population in a process known as GENETIC DRIFT.

See also: NATURAL SELECTION

species A sub-division of a classification of living organisms within a GENUS, in which there are only minor differences between the members. A species can be sub-divided into SUB-SPECIES or STRAINS.

specific gravity The ratio of a mass of specified volume of a substance to the mass of an equal volume of another substance (usually water), at the same temperature or at a specified reference temperature. Various countries use different reference temperatures, for example 4°C is often cited as a reference temperature in the UK. Unlike density, specific gravity is not described in terms of designated units.

specific heat (capacity) The amount of heat required to raise a unit mass of a substance by a unit temperature under specified conditions (e.g. under constant pressure).

See also: LATENT HEAT

specific humidity The mass of water vapour in a given volume of air divided by the mass of that volume of air.

specific immune response Any immune response initiated by the body in reaction to invasion by a specific ANTIGEN.

specification In construction terminology, an itemised list of materials and the methods to be employed in their use in carrying out a contract. The specification is usually broken down in category by trades – alternatively each trade will be given a separate specification. Even for modest contracts a detailed specification is vital to both customer and contractor as it provides a reference to resolve disputes about workmanship and materials. The specification is also vital in producing a quotation, as it should remove any ambiguities or uncertainties as to what is required, thereby ensuring that each contractor is providing a price for work of the same standard.

specificity In relation to an epidemiological study, a measure of the proportion of healthy people in a screened population that is correctly identified as such. It is a measure of the false positive (see TYPE I ERROR) component of the study – i.e. a measure of those who are determined falsely as suffering from the disease under consideration.

See also: SENSITIVITY

specified risk material That potentially contaminated tissue required by legislation in the UK to be removed from a carcass to promote the control of BOVINE SPONGIFORM ENCEPHALOPATHY. When legislation was first introduced to this effect the material was originally referred to as Specified Bovine Offal (SBO) and later as Specified Bovine Material (SBM).

spectrophotometry The analysis or examination of a substance or material by the quantification of the different colours of the spectrum it possesses when it is in solution.

spectroscopy The examination of the electromagnetic spectra produced when radiant energy is emitted or absorbed by a substance. It is used as an analytic technique.

speed *see* VELOCITY

speed of light The distance covered by the electromagnetic radiation (of which visible light is a part) per unit time. The speed of light was once thought to be a universal constant, but it is now known that this contention is not absolutely true. For all practical purposes however the speed of light is quoted as being 299,796 km/sec. (equivalent to 186,293 miles/sec.).

speed of sound The distance through a defined medium covered by a sound wave per unit time. The speed varies with the medium through which the energy is transmitted and its density. In air at standard temperature and pressure the speed of sound is 344 m/sec. (equivalent to 1,130 ft./sec.).

spigot An end of a length of tube or pipe that is fitted into a socket at the end of another pipe in order to complete a joint.

spirelli Spiral-shaped bacteria (see BACTERIUM).

spirit level A device used in construction for establishing horizontal or vertical planes. The device consists of a flat edge that is placed against the surface to be evaluated. A curved glass partly filled with fluid indicates true horizontal or vertical planes by the correct positioning of a bubble of gas between two markers.

spirits of salt An old-fashioned name for hydrochloric acid.

spirits of vitriol An old-fashioned name for sulphuric acid. Alternative names include spirits of alum and spirits of sulphur.

splice To make a sound joint between two units of the same material, ideally leaving the whole as strong as if the material were a single entity. This is achieved by weaving the ends of fibres together (such as for rope) or by overlapping two ends, attaching them together and strengthening the bond (such as for timber).

spoil Extraneous material excavated from a mine, quarry or similar along with that for which the excavation is primarily undertaken, as such it is a waste material from the process. Spoil can produce environmental contamination as heavy metals and the like become accessible to leaching processes and subsequently can be transmitted to ground or surface water sources.

spoilage organism Any micro-organism capable of acting on or within a foodstuff sufficient to cause a change in its smell or taste such as to render it unacceptable for human consumption. Spoilage organisms are not necessarily PATHOGENIC.

spoiler A device incorporated into an aerodynamic design for the purpose of interrupting the airflow across a surface. Spoilers have variable uses, the principal ones being to increase drag, improve stability, divert airflow or reduce wind noise.

spongiform Means sponge-like or having a structure resembling that of a sponge.

See also: BOVINE SPONGIFORM ENCEPHALOPATHY

spontaneous abortion Also known as 'miscarriage', an ABORTION occurring (in humans) before the completion of 24 weeks of pregnancy and brought about without external intervention. It is estimated that, in the UK, between 15–20 per cent of all pregnancies end in spontaneous abortion, most occurring within the first 12 weeks of pregnancy. The most common cause of spontaneous abortion is considered to be foetal defects such as chromosomal abnormalities.

See also: STILLBIRTH

spontaneous combustion The ignition of a substance or body due to heat generated from within itself by chemical or microbial action. The phenomenon in relation to the human body was a popular storyline in Victorian literary fiction as an explanation for several instances of charred human remains being found indoors in the absence of associated fire damage to the room itself. It is now postulated that such (very rare) instances are actually brought about by sparks or similar from open fires setting alight the clothing of people near to the fire, causing it to act as a wick with natural body fat melting in the heat and acting as tallow. This is thought to result in a slow-burn phenomenon gradually incinerating the body. If such is a true explanation it is highly likely that those affected are already either severely incapacitated, unconscious or already dead when the process starts, otherwise they would clearly be expected to react against the initial ignition.

sporadic case A single case of disease that has apparently no associated causal relationship with any other cases. It is important to recognise that an inability to establish a link is not proof positive that no link exists.

spore A physiological form that certain micro-organisms can assume to protect themselves from adverse environmental conditions. To enter such a stage the micro-organism produces a (usually) highly resistant outer shell and enters a period of dormancy. Spores can survive conditions that would kill the vegetative organism (see VEGETATIVE CELL), such as the action of certain chemical bactericides. Some spores are even capable of surviving boiling for limited periods of time.

sporing Relates to the capacity of a micro-organism to form a SPORE.

SRM *see* SPECIFIED RISK MATERIAL

SRSV An abbreviation of Small Round Structured Viruses, a group of viruses responsible for a range of symptoms of food poisoning, including vomiting and diarrhoea. They were first described in the early 1970s, although the disease they cause was known for many years before that. In 1929, for example, a condition known as *Hyperemesis hiemis* (winter vomiting disease) was described, and it is likely that this was an SRSV infection.

SRSVs have a poorly defined taxonomic (see TAXONOMY) status and so tend to be named after the geographical location of identification. Generally they came to be known as the Norwalk group of viruses, from the town in Ohio where they were first properly identified. The term Norwalk virus (or Norwalk agent) is interchangeable with the term SRSV. SRSVs are second only to ROTAVIRUSES in economic cost. SRSV is also known as NORWALK-LIKE VIRUS, and for general purposes the terms are synonymous.

stack effect A natural air movement up a hollow structure such as a chimney. It arises as a result of differential air pressure at either end of the structure caused by height and air movement.

stained glass A decorative panel comprising units of different coloured glass usually retained in a framework of lead beading (known as 'cames'), the whole presenting a pattern, design or picture. The term is strictly misleading in that the glass used is usually fired with the colour incorporated into its structure rather than added later as a stain. Stained-glass windows comprising coloured glass held in cames are also sometimes known as 'leaded lights'.

stainless steel Specialist STEEL produced by incorporating (predominantly) different percentages of chromium and nickel into the construction material. Stainless steel is used in circumstances where purity of product or durability without deterioration is necessary, such as in cutlery or medical devices and instruments.

stair Also step, the constructional portion of a stairway that allows the user to ascend by one unit rise.

stair lift A powered addition to a STAIRWAY allowing a person to sit on a seat and travel along a railed track from one level to the next. They are usually fitted to existing stairways to facilitate access by those with disability.

stairway The complete structure that allows a person on foot to ascend from one horizontal level to another by a series of ascending stages or steps.

stairwell (or stair well) An open area surrounding and incorporating a stairway.

stakeholder An individual or group that holds an interest in an activity or in the outcome of that activity. In public and environmental health terms the involvement of stakeholders (such as members of the community or interest group) in the decision-making process is increasingly seen as crucial in improving the chances of success of an INTERVENTION.

stanchion A vertical bar or column inserted into or incorporated within a structure to improve its structural strength.

standard deviation The root of the average of the squares of the differences from their MEAN of a number of observations. It is a measure of the spread of a number of observations from their mean.

standard error A measure of the uncertainty of the accuracy of a sample MEAN.

See also: CONFIDENCE INTERVAL

standard plate count (SPC) *see* AEROBIC COLONY COUNT

standardised mortality ratio (SMR) The ratio of deaths that occur in a given population (the 'observed' deaths) to the number of deaths that would, under normal circumstances, be expected to occur in that population. This is usually expressed as a percentage by multiplying by a hundred.

Staphylococcus aureus is a Gram positive (see GRAM'S STAIN) cocci bacterium that is present normally in the nose, throat, pudendum or on the skin of between 20–40 per cent of people. The organism is not usually harmful, but it can cause infection if it gains access to the body through sites where the skin is broken. *Staphylococcus aureus* can, for example, be responsible for enterotoxin-induced food poisoning. The toxin is highly heat stable and is capable of withstanding boiling for 1.5 hours. The symptoms of *Staphylococcus aureus* enterotoxin food poisoning are typically nausea, diarrhoea and vomiting, with a relatively short incubation period of around 2–4 hours (this can actually range from 30 minutes to 7 hours, dependent primarily on the amount of toxin present in the food when it is ingested). The foods most frequently associated with *Staphylococcus aureus* food poisoning are cooked meats, poultry and those types of foods that are handled post-preparation without further cooking.

A particular variant of *Staphylococcus aureus* is defined by its resistance to the antibiotic Methycillin (USA – Methicillin). Methycillin-resistant *Staphylococcus aureus* (MRSA) is potentially the most common cause of surgical-site infections in hospitals. The resistance of some strains to antibiotics is a matter of extreme concern. It does not behave any differently from other strains of *Staphylococcus aureus* and is capable of colonising asymptomatic carriers in the same way as do other strains. Its resistance to certain antibiotics is therefore a cause for concern. MRSA can cause septicaemia, pneumonia and other serious infections. Medical advice is that, whilst MRSA is a problem in hospitals, it poses no particular risk in the community, including to those in residential and most nursing homes. In the event of an epidemic of influenza, however, because of its antibiotic resistance, secondary infection with MRSA could potentially kill more patients than the primary infection with the influenza virus itself.

staple A loop of wire sharpened at each end and driven into another material in order to affix something contained within the loop to its surface.

starch A POLYSACCHARIDE found as a storage CARBOHYDRATE in higher plants. It is a major source of nourishment for humans and other non-ruminant species. Animals do not produce starch. Foods such as maize, potatoes and wheat have a high starch content.

See also: GLYCOGEN

stated preference is an alternative name for CONTINGENT VALUATION.

See also: REVEALED PREFERENCE

static electricity The electric charge that accumulates on a surface as a result of two or more non-conducting solids or liquids rubbing together. The build-up of static electricity can sometimes be sufficient to create a spark onto a surface with a different electrical potential if such is brought into close proximity. This spark can be sufficient to ignite susceptible gaseous materials or finely divided suspended materials in air, potentially leading to fire or explosion. Such a phenomenon is known as electrostatic ignition.

statistics The science of applying mathematical calculation to predict an outcome. The term is also applied to a tabulation or databank of figures used in such calculations.

See also: EPIDEMIOLOGY

status spongiosus A condition in which there has been extensive SPONGIFORM damage to the tissues of the brain.

See also: BOVINE SPONGIFORM ENCEPHALOPATHY

statute An enactment of a legislature (i.e. a law).

See also: ACT; REGULATION

statutory That which is specified in STATUTE (i.e. in law).

statutory instrument A piece of SECONDARY LEGISLATION in the UK made in the form of a Regulation. They generally contain the detailed application required to explain the precise working of a piece of PRIMARY LEGISLATION.

statutory nuisance In the UK, any Act or omission that is specified in STATUTE to be a nuisance. The appellation has nothing principally to do with the nature of the nuisance itself in terms of its severity or effect.

See also: PRESCRIPTIVE

steel An alloy of iron incorporating a carbon content of around 0.1–1.7 per cent. Iron with a higher carbon content is usually produced from a process in which the molten metal is cooled in moulds, the product being known as CAST IRON. Iron with a lower carbon content is termed WROUGHT IRON. Specialist steels with varying properties can be produced by incorporating different metals to produce alloy steel.

See also: MILD STEEL

steeping A culinary process in which either hot or cold water is poured over the surface of a foodstuff and leaving it to stand in order to improve its texture or to extract colour or flavour.

stereoscopic The term applied to a method of sighting in which two images are combined to give an impression of the three-dimensional nature of that being viewed. A stereoscopic camera is one in which two images are generated from vantage points equivalent to the positioning of the human eyes, the whole presenting a three-dimensional image when viewed through a stereoscope.

sterilisation The process whereby heat or chemical means are applied to a surface, equipment or material in order to kill all those micro-organisms and their spores that are present.

See also: AUTOCLAVE; CLEANING; DISINFECTION

steroid A chemical sub-classification of chemical belonging to the LIPID group. In their turn they comprise a number of related substances including STEROLS, sex HORMONES and bile acids.

sterol One of a number of solid alcohols, waxy in appearance, included within the STEROID group. Sterols include the substances CHOLESTEROL and ergosterol.

sterulent A chemical agent that has the power to produce STERILISATION.

See also: DETERGENT; DISINFECTANT

stick injury A traumatic injury caused by the inadvertent penetration of the skin by a sharp object such as that arising from stabbing due to the incorrect handling or disposal of sharp (used or soiled) medical instruments.

See also: SHARPS

stillbirth Accounted as a late foetal death occurring (in humans) after 24 weeks of pregnancy.

See also: ABORTION; SPONTANEOUS ABORTION

stipendiary magistrate A qualified, paid magistrate who arbitrates in a Magistrates' Court but, because of their qualification, usually sits (i.e. arbitrates) alone.

See also: COURTS; MAGISTRATES' COURT

stock size The term applied to a construction item indicating that it should be available from normal stock and therefore would not be expected to need to be specially ordered or commissioned.

stochastic Means random. A stochastic process is therefore one in which the inputs are random in nature, possibly involving elements of chance.

Stockholm Conference The name applied usually to describe the United Nations Conference on the Human Environment that took place in Stockholm, Sweden, in 1972.

stocktake An audit of the quantity and/or quality of goods in store in order that they can be enumerated and compared to existing records to identify whether or not there are any discrepancies. The process is known as stocktaking.

stoichiometric Refers to having the exact proportions for a chemical reaction. A stoichiometric equation is one in which the amounts of chemicals on both sides of the equation balance (also known as a 'balanced equation').

stopping The term applied to the technique of filling cracks or holes in timber with putty or similar before applying decorative of protective finishes.

storm drain A pipework structure or system designed to convey rainwater for disposal separately from a system for foul or soil water.

See also: SOIL DRAIN

strain In biology, a distinguishable grouping of organisms within a SPECIES that can be described by particular characteristics.

strategic The term used to describe that which is produced as a plan or policy to further a particular aim. It is usually taken to describe an overview of a set of defined actions.

See also: OPERATIONAL; TACTIC

strategic waste management assessment (SWMA) A comprehensive set of information produced by the ENVIRONMENT AGENCY in relation to the types of waste produced in a given locality and how these are managed.

stratopause That layer of the earth's atmosphere which forms the junction between the STRATOSPHERE and the MESOSPHERE.

stratosphere That layer of the atmosphere lying between the TROPOSPHERE and the MESOSPHERE. It comprises a lower iso-thermal layer and a higher area in which temperature increases due to the presence of OZONE, which absorbs radiation from the sun.

See also: OZONE LAYER

stretcher A unit construction material such as brick, block or stone, which is incorporated into the structure with its length running parallel to the construction surface.

stretcher bond That in which the construction of a wall is achieved by laying all the unit components as STRETCHERS.

See also: FLEMISH BOND; HEADER

strike Also known as blowfly strike or blowfly myiasis, an infestation of the skin of live sheep with the maggots of the blowfly *Lucilia sericata*, also known as the green bottle fly. Intense infestation can make hides commercially worthless. Control is through good husbandry and the use of insecticidal sheep dips.

strip mining An alternative name used to describe OPEN-CAST MINING.

stroke The name given to a state of unconsciousness or physical incapacity brought about by a diseased condition of the brain. Although there may be some preliminary symptoms such as headache, dizziness or vomiting, the attack itself is of sudden onset. The attack is due to the loss of blood to a portion of the brain caused either by blockage or HAEMORRHAGE of a blood vessel. In survivors, damage to the brain is usually manifest by paralysis of one side of the body or the other (hemiplegia), a condition that can respond varyingly to treatment. Stroke is usually, but not exclusively, a disease of age. Hereditary factors may predispose an individual to attack but cardiovascular (see CARDIOVASCULAR SYSTEM) disease and HYPERTENSION, and the factors that contribute to these conditions, will increase the risk of attack.

See also: CORONARY HEART DISEASE; EMBOLISM; THROMBOSIS

strut A structural component in a state of COMPRESSION, holding two or more components together.

strychnine One of two (the other is brucine) bitter-tasting ALKALOIDS derived from the Indian tree *Strychnos nux-vomica*. Strychnine in pure form is a white crystalline substance. It has been used in both human and veterinary medicine as a muscle relaxant and as a poison, particularly for rats.

See also: *nux vomica*

stud partition A CURTAIN WALL consisting of timber framing (the stud work) covered by sheets of plasterboard and finished with a plaster skim.

stunning That stage in the slaughter of animals for human consumption in which the animal is rendered unconscious. There are several methods whereby this can be achieved, including gas, electricity or physical means.

See also: BLEEDING; CAPTIVE BOLT PISTOL; PITHING

sub judice Refers to that which is still under judicial consideration and therefore cannot legally be commented upon. The phrase derives from the Latin and means literally 'under a judge'.

subcutaneous Refers to that which is below or administered below the skin, such as a subcutaneous injection.

subduction The process whereby one TECTONIC PLATE slides beneath another at the junction where two such plates are moving towards each other.

See also: OROGENESIS

subjective assessment One in which that being judged is assessed from the viewpoint of the person conducting the assessment. The judgement is therefore open to the influence of prejudice but provides the dimension of preference.

See also: OBJECTIVE ASSESSMENT

sublimation The direct conversion of a solid to a vapour or vice versa without it passing through a liquid stage. Examples are the direct conversion of snow into water vapour and the direct deposition of water vapour in the atmosphere onto the surface of ice.

sub-species A closely defined grouping within a SPECIES in which the members have very similar characteristics.

See also: SUB-TYPE

substructure That portion of a construction which is located below that of the main component, usually synonymous with the foundation of a building but can be used to delineate cellars and underground units.

See also: SUPERSTRUCTURE

sub-type A sub-classification of a SPECIES of living organisms. A sub-type defines the membership very closely and is one of the preferred differentiations within microbiology for the identification of (particularly pathogenic) microflorae.

sucrose Sucrose ($C_{12}H_{22}O_{11}$) is a DISACCHARIDE form of SUGAR. It is most commonly found in the kitchen where it is known simply as sugar, common household sugar or table sugar.

sugar One of a number of simple CARBOHYDRATES characterised by their sweet taste. They may be classified as MONOSACCHARIDES (e.g. GLUCOSE) or DISACCHARIDES (e.g. SUCROSE). They are used in cooking as sweeteners. In food commodities such as jam they are used as preservatives because they remove water from the cells of the intended food.

See also: POLYSACCHARIDE

sulphur dioxide Sulphur dioxide (SO_2) is a colourless, water-soluble, acidic gas that is produced from the combustion of fossil fuels (particularly coal and oil that contain sulphur as a contaminant). Major sources of naturally occurring sulphur dioxide are volcanic activity and the oxidation of releases from marine organisms. The largest single source of atmospheric sulphur dioxide in the UK is power generation, but local industrial and domestic sources may be important in urban areas. Vehicles are not a major source of sulphur expressed as sulphur dioxide. Humans tend to be more sensitive to sulphur dioxide than do animals. On inhalation around 95 per cent is absorbed by the upper airways of the lung; however, the remaining 5 per cent or so may be sufficient to cause bronchoconstriction (see BRONCHOCONSTRICTOR) in normal adults. Asthmatics are around ten times more susceptible to the effects than normal adults. Health effects are difficult to describe precisely as exposure usually occurs in combination with other pollutants, but bronchoconstriction and bronchospasm have both been recorded.

sunspot A transient dark area on the face of the sun that appears as thermal activity in the area decreases and temperature drops by as much as 1,000°C compared to surrounding areas. Sunspots are associated with increased electromagnetic activity, sometimes sufficient to disrupt electromagnetic activity (such as communications systems) on the earth. The number of sunspots on the face of the sun is variable but appears to reach a peak around once every eleven years or so. It is postulated that sunspot activity is linked to meteorological conditions such as drought and glaciation on the earth.

superbug Any variant of a pathogenic organism that has developed or acquired resistance to one or more (essentially medicinal) compounds to which it was once susceptible.

superconductor Any material through which an electric current can be passed without loss of energy due to resistance. In practical terms only certain pure metals or alloys at temperatures approaching ABSOLUTE ZERO have been demonstrated as exhibiting no electrical resistance.

supersaturated solution *see* SATURATED SOLUTION

superstructure That portion of a construction which is found above that of the main component.

See also: SUBSTRUCTURE

supertanker An exceptionally large ship used to transport PETROLEUM or similar. There are varying definitions as to what size is sufficient to define a tanker as a supertanker. Most sources suggest it is one with a dead-weight capacity of 100,000 tonnes, although others suggest the capacity as 500,000 tons. Some sources suggest the defining capacity as 75,000 tonnes or tons.

surface tension A physical characteristic of liquids in which molecular attraction appears to create a thin elastic membrane across their surfaces. This tension is that which makes water droplets tend towards the spherical in the absence of other forces, and it is also the force that produces CAPILLARY ATTRACTION.

See also: MENISCUS

surface water Water derived directly from rainfall that has flowed across the surface of the ground to collect in streams, rivers and lakes. It is generally more likely to be contaminated with organic material and microflora than GROUNDWATER.

surfactant A material that reduces the interfacial tension of a liquid enabling it to be better able to 'wet' a surface. Surfactants are used in detergents to break down the surface tension of biofilms on surfaces to improve the action of the detergent in the cleaning process.

surveillance The ongoing, systematic collection, analysis and interpretation of health data essential for planning, implementing and evaluating public health practice. The information derived from the surveillance must be disseminated promptly to those who need to know in order to apply the findings to prevention and control strategies.

susceptibility A condition in which an individual has no pre-existing IMMUNITY or resistance to invasion by a PATHOGENIC organism and is therefore liable to produce the clinical symptoms of disease if infected.

sushi A dish of Japanese origin comprising variously raw (or cold) fish, cold rice and vegetables with a vinegar sauce.

suspended ceiling A false ceiling constructed below a true ceiling creating an intervening void space. Suspended ceilings are created for a number of different reasons including aesthetic, to use the void as a housing for services and ducting or to reduce the overall room volume and thereby reduce heating or air-conditioning costs.

suspended construction A structure in which the component units are held in place by cables fastened at their other end to a structural member positioned at a level above the main structure.

sustainable community A relatively ill-defined term, however, the concept describes a town, city or other defined centre of human population that has a mutually supportive and dynamic balance between environmental quality, economic opportunity and social health and well being. Sustainable communities should generally embrace the concepts of SUSTAINABLE DEVELOPMENT.

sustainable development Sustainable development has been widely and differently defined following its original introduction, but there is broad agreement as to the general understanding. Sustainable development can therefore be defined as that in which the needs of current generations can be met from the exploitation of the world's resources without prejudicing the ability to satisfy the needs of future generations. The definition that was adopted by the United Nations President's Council on Sustainable Development in 1993, following the report from the BRUNTLAND COMMISSION, was development that allowed people 'to meet the needs of the present without compromising the ability of future generations to meet their own needs'.

In purist theory any resource that is used should be replaced. Clearly this cannot, by definition, happen in relation to non-renewable resources such as fossil fuels, but the concept has been modified by some to incorporate elements of good husbandry and recycling to make best use of that which is available. The key elements of sustainable development are recognition of the long-term impacts and consequences of development; recognition of the interdependence of economic, environmental and social factors; decision making that is participatory and transparent; systems that promote equity within the community; and proactive prevention of problems.

As a concept sustainable development has been widely supported by world governments but translating belief into action has proved to be more problematic. This is especially so if one tries to implement scenarios in which each nation is pegged in resource usage to existing levels. Many developing nations, for example, disagree that their developmental needs should be frozen simply to maintain the higher standards of living of those in more developed countries.

See also: INTEGRATED PEST MANAGEMENT; ORGANIC FOOD; UNITED NATIONS CONFERENCE ON ENVIRONMENT AND DEVELOPMENT

sustainable waste management The application of the principles of SUSTAINABLE DEVELOPMENT to the treatment of waste material. Such management involves using raw materials effectively, reducing the amount of waste produced, recycling where possible and dealing with other waste to further the economic, environmental and social principles of sustainable development.

See also: ENERGY FROM WASTE

suture The series of stitches used to close a wound or surgical incision. The term is also used to describe the joints between the bones of the dome of the skull.

sward A type of environment typified by low-growing HERBACEOUS plants.

swine fever A highly infectious viral disease of pigs. In young animals the disease is often fatal but it can become chronic in older animals. It manifests with a range of symptoms including fever, malaise and diarrhoea; pregnant animals often abort. It is transmitted by inhalation of the virus, infected semen, contaminated feed or FOMITES. It is not considered to pose a threat to humans.

swine vesicular disease (SVD) A viral disease of pigs that causes symptoms which are clinically indistinguishable from Foot and Mouth Disease. The disease was first identified in Italy in 1966 and in subsequent years into the early 1970s the disease spread across Europe and the Far East. An outbreak occurred in the UK in December 1972. In more recent years the disease has largely been brought under control in developed countries. The disease in pigs is manifest by fever and by the production of vesicular lesions of the feet; these usually appear on the heel where the trotter joins the foot. In a small number of cases vesicles may appear elsewhere, such as for example on the snout or in the mouth. Lesions may also be generated on the upper leg or the torso.

SWOT analysis A systematic evaluation conducted through the respective categories of Strengths, Weaknesses, Opportunities and Threats. A SWOT analysis encompasses both internal and external considerations.

See also: PEST ANALYSIS

symbiotic Refers to a functional relationship between two living organisms of different species. In a mutual relationship (such as in relation to LICHEN) both organisms benefit, in a COMMENSAL relationship one organism benefits and the other is unharmed (such as between lichen and a tree), and in a parasitic (see PARASITE) relationship one organism benefits to the detriment of the other.

synchronous orbit A circular orbit around the earth at a distance of 6.6 times one EARTH RADIUS. At this point the orbital time equates to 24 hours, thereby keeping the orbiting object at a constant position over the earth's surface. This orbital position is used predominantly for communications and positioning satellites. A synchronous orbit is sometimes known as a geostationary orbit.

syncope A synonym for fainting or swooning.

synergy The phenomenon in which the combined effect of two or more entities or agents is greater than the sum of the individual entities of agents acting in isolation.

synthetic pyrethroids A group of insecticides that were introduced in the mid-1980s. They comprise a group of chemicals including flumethrin, cypermethrin and deltamethrin. They are potentially more toxic to wildlife (and especially to aquatic wildlife) even than the ORGANOPHOSPHOROUS INSECTICIDES.

See also: ORGANOCHLORINE INSECTICIDES

syphilis Colloquially known as the 'pox', a disease principally transmitted through sexual contact, although transmission by non-sexual contact with infected secretions and exudates is not unknown. Maternal transmission through the placenta or by blood transmission during delivery may also occur. The disease is caused by infection with the spirochaete *Treponema pallidum*. Primary infection usually presents as an infected lesion, with a secondary stage of more widespread lesions involving the skin or mucous membranes potentially occurring after around 4–6 weeks. This secondary stage usually resolves clinically within 12 months, but latent infectivity continues, potentially for life. The disease may develop unpredictably from the latency period at any time and result in lesions arising in internal organs, blood vessels, the central nervous system (CNS) and the skin. The development of these symptoms can potentially result in severely impaired health, serious disability or death.

system building A constructional technique in which modules are manufactured off-site and assembled on-site to create the final structure.

systematic error A consistent form of error in-built within an epidemiological study or clinical trial that will serve to produce consistent and repeatable results, even though those results are incorrect. Systematic errors are forms of BIAS.

See also: CONFOUNDING FACTOR; INFORMATION BIAS; SELECTION BIAS

systole The contraction phase of the cardiac cycle when arterial blood pressure is at its greatest.

See also: DIASTOLE

T

t *see* BIOLOGICAL HALF-LIFE

t-test A test of statistical significance. There are two types of t-test, the UNPAIRED T-TEST and the PAIRED T-TEST.

table d'hôte In a restaurant refers to a meal that consists of a set number of courses at a fixed price. There is usually a choice between dishes in each course.

tachycardia An increase in the heart and (therefore) pulse rate.

See also: BRADYCARDIA

tactic The term used to describe the process in which a strategy (see STRATEGIC) will (or is intended to be) achieved by identifying or specifying OPERATIONAL activities, potentially from a range of options.

tailing A non-specific term used to describe generically waste materials left over from mining or refining processes. Although much of the component of tailings is inert it is possible, dependent on source, for there to be high levels of contaminants, especially of heavy metals.

tapping The process of obtaining LATEX from certain plants by the action of making an incision into the outer bark and draining the sap into a collection vessel.

See also: RUBBER

taxa The term used to describe any of the classifications of TAXONOMY at whatever level (i.e. such as at levels of species or genera).

taxonomy The branch of biology devoted to the classification of animals and plants into different groupings based on their similarities of structure or characteristics.

TCC *see* AEROBIC COLONY COUNT

TCDD An acronym for the DIOXIN, the full name of which is 2,3,7,8-tetrachlorodibenzo-*p*-dioxin ($C_{12}H_4Cl_4O_2$). It is the most toxic of the dioxin group and was a contaminant of the defoliant known as Agent Orange used during the Vietnam War. It was also present as a major constituent in the release of chemicals following an industrial accident at Seveso, Italy, in 1976.

TDI *see* TOLERABLE DAILY INTAKE

tectonic plate One of a series of huge slabs or 'plates' of rock, potentially of continental size, which are postulated to comprise the earth's crust (known as the lithosphere). The theory holds that these plates move relative to each other causing upheaval when they move towards each other (leading to the creation of mountain ranges such as the Alps and the Himalayas). Alternatively they create ridges (such as on the floor of the Atlantic and Pacific oceans) when they move away from each other. The abrasion of one plate against another is thought to be responsible for SEISMIC shocks known as earthquakes.

See also: OROGENESIS

TEF *see* TOXIC EQUIVALENCE

temperature gradient The difference in temperature across a substance or structure whose outer surfaces are in contact with materials whose own temperatures are different from each other.

temperature inversion layer *see* INVERSION LAYER

tempering A process of strengthening or reducing brittleness of the surface of a construction material by heating it followed by a process of cooling. Tempering is often carried out in conjunction with QUENCHING, the combination of the processes varying with the material and the desired effect.

template A framework or outline used to guide and delineate the production of a finished product.

temporary threshold shift A temporary diminution in an individual's capacity to hear. It is usually occasioned by exposure to loud or high-energy sound. The effects are variable, sometimes reducing the ability to hear certain frequencies (especially in the higher registers, usually around 4,000–6,000 Hz (see HERTZ)) as well as previously or sometimes distorting the sound so as to make some characteristics indecipherable. The duration of the condition is relatively short term, possibly of some days, but the hearing gradually returns, although this may not necessarily be to previous capacity.

See also: PERMANENT THRESHOLD SHIFT

tender A written, priced bid, usually submitted in competition, to undertake specified work or a contract. Usually the bids are sealed and opened only on a specified date in the presence of designated individuals. This procedure is intended to prevent dishonesty through price fixing.

See also: OPEN TENDER; SELECTIVE TENDER

tenon The projecting end of a piece of wood, lock or similar that fits into a rectangular hole (the 'MORTISE') for the purpose of securing the two pieces together, either as a permanent joint (in woodwork) or as a security device (as in a lock).

tenon saw A short, fine-toothed saw with a reinforced back used in woodworking. It is used especially in cutting the TENON for a 'MORTISE and tenon' joint.

tension A state of stress derived from a force tending to pull or stretch that which is in tension.

See also: COMPRESSION

TEQ *see* TOXIC EQUIVALENCE

teratogen A substance that has the potential to be TERATOGENIC.

teratogenic Means capable of inducing deformity or malformation of the foetus during its development in the womb. The risk that a particular substance will cause abnormalities in the developing foetus is known as the teratogenic risk.

termination of pregnancy An ABORTION brought about by specific external intervention without which the pregnancy could be expected to proceed to birth.

terra firma Literally means 'mainland'; it is often incorrectly translated as 'firm land'. The term is usually applied to designate dry land as opposed to that which is the sea.

See also: MARITIME; TERRESTRIAL

terrace A level area of land sculpted either from a hillside or sloping site. The term is alternatively applied to a row of houses of uniform construction.

terracotta A burnt composite of clay and sand, usually of reddish-brown appearance, used to make building materials or statues, or the artefacts made therefrom.

terrazzo A type of decorative CONCRETE that uses marble as an AGGREGATE, the surface being polished to provide a hard, smooth and durable surface.

terrestrial Means of, or relating to, the land.

tessera The basic small piece of marble, glass, pottery or similar from which a MOSAIC is made.

tetanus An infection caused by the bacterium *Clostridium tetani*. The main animal reservoirs are various but most infections are contracted from horses. Transmission is through the contamination of puncture wounds, lacerations and burns with soil or faeces containing tetanus spores. Symptoms start as stiffness in the muscles near the infection site; stiffness of the jaws is common (the infection is commonly known as 'lockjaw'). The lips tend to pull back from the jaw in a condition known as *risus sardonicus*. Rigidity of other muscles can cause bent posture to develop and seizures or convulsions are common. Normal respiration may be affected. Tetanus is now rare in developed countries due to widespread immunisation (generally fewer than ten cases per year on average are notified in the UK). Those over the age of 50 are particularly at risk if they contract the infection and the case fatality rate is high, although the prognosis is good provided the disease is caught early enough.

tetraplegia *see* PARAPLEGIA

thalidomide A mild sedative drug that was prescribed widely in the late 1950s and early 1960s. It was subsequently found to cause congenital malformations in the foetus, notably adversely affecting the development of the limbs. Around 10,000–20,000 babies were affected worldwide of which around 5,000 remained alive at the turn of the millennium; the corresponding figures for the UK were around 800 and 450 respectively.

Following discovery of its adverse effects its use was initially discontinued. It is however now used in the treatment of certain cancers as it is an effective drug to prevent growth in certain tissues.

thatch A traditional roof covering comprising reeds, rushes or straw. The waterproofing is obtained as physical laws determine that rain runs along the individual downwards-sloping reeds before falling off onto a lower reed. Each reed therefore carries the water further downwards along the roof line until eventually it is discharged at the EAVES. In this way in a properly constructed whole only the upper surface of the thatch ever gets wet. As a consequence deterioration or rotting of the material occurs predominantly at the upper layer of thatch and it is this that is generally replaced during renovation.

Thatch is a very efficient thermal insulation material but is very susceptible to fire. Although fire-resisting agents can be used to soak the thatch before installation they can never completely fireproof the structure and some have been found adversely to affect water conductivity off the roof, allowing water to penetrate deeper into the thatch itself.

theodolite A surveying instrument used to measure both horizontal and vertical angles.

therapeutic The term relating to techniques or substances applied to a body with the aim of producing healing.

thermal cracking The process of breaking down large molecules (usually of hydrocarbons) into smaller molecules by the application of heat and pressure.

thermal resistance is the intrinsic resistance of a body or substance to the transmission of heat energy. It is usually quantified as the R-value and is the reciprocal of THERMAL TRANSMITTANCE, which is quantified as the U-value.

thermal transmittance A measure of the heat energy that can pass through a structure or body over time for each degree of temperature difference between its two sides expressed per unit area. It is usually quantified as the U-value.

See also: THERMAL RESISTANCE

thermistor A semiconductor in which electrical resistance is inversely correlated to temperature, i.e. the greater the temperature the less the electrical resistance. This capacity is employed in using thermistors as temperature sensors, their small size making them useful particularly in measuring microclimates in electronics and medical applications.

See also: THERMOCOUPLE

thermocline The layer of water in a large body of water, such as a lake, sea or ocean, which separates a warmer surface layer from a colder deeper layer. The thermal or TEMPERATURE GRADIENT across the thermocline usually exceeds the temperature differential within the layers that lie above and below it.

thermocouple is a temperature-measuring device comprising two wires of different metals connected to form a circuit. If one junction of the wires is maintained at a constant temperature and the other junction subjected to a change in temperature an electrical current is created, the strength of which is proportional to the temperature change.

See also: THERMISTOR

thermoduric count A numerical assessment of those micro-organisms that have survived a heating process.

thermodynamics The study of the relationship between heat and energy.

thermopause The layer of the earth's atmosphere that forms the junction between the THERMOSPHERE and the EXOSPHERE.

thermophilic Refers to the preference or necessity of a living organism to grow in the presence of higher temperatures (often but not exclusively) than others of the same species. In microbiology, for example, thermophilic campylobacter will grow between 37–42°C but not at 25°C.

thermopile An assemblage of thermoelectric sensors that are used to measure temperature.

thermoplastic A physical characteristic of a substance in which it becomes softer and more malleable upon the application of heat and harder or more brittle upon cooling.

thermosphere That layer of the earth's atmosphere extending between the MESOSPHERE and the EXOSPHERE. It extends to a height of around 400 km from the earth's surface at which height temperatures can reach 1,000°C.

See also: THERMOPAUSE

thermostat A device incorporated into heating or cooling system to maintain a constant temperature.

thinner A chemical substance that when added to a liquid lowers its VISCOSITY. Thinners are used notably in oil-based paints to lower viscosity and aid application. Common thinners used for oil-based paints are turpentine and white spirit.

thiol An alternative name for MERCAPTAN.

Third World country Nowadays, a somewhat misleading term used originally to describe a country belonging to the less developed areas of the world, particularly to those of Africa and Asia but also including to some extent those of South America. The term arose to provide a convenient shorthand to distinguish such countries from those of the First World (i.e. of industrialised capitalist states) and those of the Second World (i.e. the predominantly communist countries).

thixotropy The characteristic of a fluid in which VISCOSITY increases upon standing and decreases upon mechanical agitation.

thoron The name commonly used to describe RADON-220.

Three Mile Island An island located on the Susquehanna River in Pennsylvania, USA, where a major accident occurred at a nuclear power plant in March 1979. The accident occurred when the temperature in the reactor's core rose dramatically following loss of coolant. Half of the core melted and released radiation within the containment facilities, some of which subsequently escaped into the local environment. Opinion is divided as to whether human health effects have resulted from the leakage but it is certain that some of the local population were exposed to higher than normal levels of radioactivity. The accident turned popular opinion in the USA against the development of nuclear power.

See also: CHINA SYNDROME

threshold Loosely defined as an entry point or a position from which something starts. In architecture it is the bottom portion of a doorway (especially one for entering a building). In toxicology it is a point at which manifestation of a response by a living organism is first manifest.

The concept of a threshold for toxic effect is one that is intrinsic in our current assessment of health effects from various chemical substances in the knowledge that many substances exhibit a DOSE-RESPONSE EFFECT. The concept is based on the potential response from an 'average' recipient, allowing that there is no such being in reality. The threshold is determined as the point at which no discernible effect is manifest and safety is built in by determining a factorial dilution from this point as judged by expert opinion.

threshold limit value The term sometimes used to describe an OCCUPATIONAL EXPOSURE LIMIT.

thrombosis A complete or partial blockage by a bloodclot of one or more of the major blood vessels of the body or of the heart.

See also: CORONARY HEART DISEASE; EMBOLISM

tie A structural component holding two or more components together in a state of TENSION.

See also: STRUT

time series study One in which a defined population is studied over a specified period of time.

time-weighted average The average atmospheric concentration of a substance, usually described in parts per million by volume, measured over a specified time period. This latter is usually taken as an 8-hour work day and a 40-hour week.

tinnitus A perceived sensation of whistling, ringing or clicking in the ear, discernible by the sufferer even in the absence of any external stimulus.

tintometer Another word for a COLORIMETER.

titration A laboratory technique in which the amount of a particular substance contained within a solution or its concentration is calculated by measuring or observing a change in one characteristic (such as the colour change of a standard reagent or a change in electrical conductivity) as another substance is added to the original.

tolerable daily intake (TDI) The estimated amount of a particular substance per unit mass of bodyweight that it has been assessed can be ingested each day by a 'normal' individual over the period of a lifetime without such presenting a significant risk to health.

tomography A medical procedure in which X-RAYS or ultrasound are used to scan a section of the body in order to build up a focused image of tissues and organs at various depths. A collection of these sections can then in turn be used to build up a three-dimensional image.

ton A measure of weight. In English measure it is synonymous with the 'long ton' comprising 2,240 lbs. In USA measurement the term refers to the 'short ton' comprising 2,000 lbs. The metric TONNE is equivalent to 0.984 of a long ton or 1.102 of a short ton.

tongue and groove joint A method of connecting two pieces of material (especially timber) in which the projecting edge of one piece (the 'tongue') fits neatly into the hollowed section (the 'groove') of the other. Tongue and groove joints are not fixed by adhesive and consequently allow a degree of lateral movement. They are used notably in joining floorboards where such movement is unavoidable in response to changes in moisture content.

tonne A metric TON, equivalent to 1,000 kg or 0.984 of a ton.

topsoil The uppermost layer of the ground. It contains HUMUS and is vital in the support and nurture of natural vegetative growth.

See also: REGOLITH

Torrey Canyon An oil tanker that ran aground off the coast of Cornwall, UK, in March 1967 leading to an oil spillage of 117,000 tonnes of crude oil. It was a large-scale environmental disaster, not least because the measures that were taken to try to deal with the spillage were largely ineffective and contributed to the environmental damage sustained.

See also: SEA EMPRESS

torsion A synonym for twisting. Torsional strength is that exhibited by a structure or object in response to twisting. Torsional rigidity is the stiffness exhibited by a structure or object in response to twisting.

See also: SHEAR FORCE

tort A transgression or wrong that does not arise as a result of a contractual obligation but for which compensation or damages might legally be due. The term is also used to describe the violation of a duty (such as a duty of care) imposed by statute and in respect of which damages might be claimed.

total births Defined as the summation of all live births and STILLBIRTHS, usually reported as occurring within a defined period.

total colony count *see* AEROBIC COLONY COUNT

total plate count *see* AEROBIC COLONY COUNT

total power *see* COMBINED HEAT AND POWER

total suspended particles The term applied to the amount of PARTICULATE MATTER contained within a given volume of gas expressed by weight.

total viable count *see* AEROBIC COLONY COUNT

toxic equivalence (TEQ) A technique used to describe the toxic effects of a single chemical or group by equating the effects as a proportion of those of a reference substance. For example the effects of DIOXINS are often quoted using 2,3,7,8-tetrachlorodibenzo-*p*-dioxin (TCDD) as the reference substance.

toxic equivalency factor *see* TOXIC EQUIVALENCE

toxicokinetics The study of what happens to a chemical following its introduction into a living organism. A toxico-kinetic evaluation will include a description of how the chemical is absorbed into the body, how it is distributed around the body, the production of METABOLITES and its ultimate excretion or bioaccumulation.

toxicovigilance The systematic collection and evaluation of harmonised data related to the accidental or occupational exposure (normally of humans) to toxic substances or emissions. The purpose of characterising such events is for aiding future policy development, improving response activity and informing chemical risk assessment activities.

toxin A poisonous substance produced by plants or animals. It is noteworthy that many toxins are highly stable in the presence of heat. This is particularly important if toxins are released into food, as subsequent heating (e.g. cooking), which may kill pathogenic micro-organisms, is frequently ineffective against toxins.

See also: MYCOTOXINS

toxoplasmosis An infection caused by the protozoa *Toxoplasma gondii*. There are around 750 laboratory reports each year in the UK; many more probably go undiagnosed, as infections may be asymptomatic or mild and self-limiting. Immunocompromised patients and congenital infections are of particular concern as severe debilitating disease and death may result. Congenital infection can lead to HYDROCEPHALUS, mental retardation or loss of sight. The main animal reservoirs are cats or their kittens, rodents, birds, sheep, pigs, cattle and goats. Transmission is by the ingestion of cysts from soil or water that has been contaminated with infected cat faeces, ingestion of tissue cysts in raw goat's milk or in raw or undercooked meat, transplantation of organs or blood from infected persons or transplacental from mother to foetus.

TPC *see* AEROBIC COLONY COUNT

trace element One that is essential for the proper maintenance of the life of a particular organism (animal or plant), but which is required only in very small quantities. The list of trace elements includes cobalt, iron, lead, manganese, nickel, selenium and zinc. Trace elements are usually incorporated as minor constituents of some of both vitamins and enzymes, and are consequently natural constituents of living tissues. Although these elements are essential to the proper functioning of living organisms many are harmful or toxic in large quantities.

trachea The main air passage from the back of the throat to the lungs. It is also known as the 'windpipe'.

See also: BRONCHUS

transducer A device used to transform sound or vibration energy into electrical or magnetic energy or VICE VERSA.

transformation In molecular biology, the process whereby a normal cell becomes capable of neoplastic growth (i.e. it becomes capable of producing a TUMOUR).

transformer A device in an electrical system used to convert one level or type of voltage to another.

transgenerational effect One in which an exposure to a substance for an individual in one generation has a subsequent and causal effect on the offspring of those people. Ionising radiation is commonly cited as having potential in this area.

transgenic That living organism or tissue which has been modified to contain genetic material from another species.

See also: GENETICALLY MODIFIED ORGANISM; TRANSGENIC ANIMAL MODEL

transgenic animal model Any animal into whose GENOME fragments of DEOXYRIBONUCLEIC ACID (DNA) from another species have been incorporated to further research or experimentation, such animals acting as a MODEL for (usually) human experience.

transmissible spongiform encephalopathy One of a group of diseases, in which changes in the grey matter of the brain appear as holes or vacuoles, giving the brain a spongy appearance under the microscope. The disease is generally specific to one species of animal and variants for a number of different species are known. In 1996 it was recognised that the disease in cattle known as BOVINE SPONGIFORM ENCEPHALOPATHY (BSE) was capable of transmission to humans giving rise to a new variant of Creutzfeldt-Jacob disease (see CJD) (vCJD).

transmittance A measure of a particular medium's capacity to allow the passage of electromagnetic radiation. It is quoted per specified unit mass of the medium and is derived as being the ratio between the radiation leaving the medium to that entering it.

transom A horizontal dividing rail within the glazed portion of a window.

See also: MULLION

transpiration The process of evaporation of water from the leaves of plants thereby creating an imbalance in the osmotic pressure between the top of the plant and its roots. This leads to the passage of water up the plant, a process vital to the transportation of dissolved nutrients.

See also: OSMOSIS

transposon A sequence or fragment of DEOXYRIBONUCLEIC ACID (DNA) that has the capacity to be located variously along the length of a GENOME.

trap A self-contained water-filled dip or bend incorporated into a drainage system to prevent the upstream passage of odours or (potentially) of pests within the system.

trauma A condition brought about as a result of wound or injury.

tread The horizontal portion of a step in a staircase. The term is also applied to the horizontal distance travelled between the RISERS; a synonym for this distance is the 'going'.

See also: NOSING

trespass A wrongful act, the commission of which leads to some harm or deprivation to another party. The term is strictly applied to something that is done wilfully; it is not applied to acts of negligence or to the failure or omission to do something. The term is often used in relation to the wrongful incursion into property or onto land in respect of which the perpetrator has no rights, and in relation to which harm or damage is done as a consequence.

See also: TORT

trestle A self-supporting frame upon which a horizontal working platform can be placed.

triangulation A method of surveying in which the position of a third point or object is determined using angles from two points of a known distance apart.

tributyl compounds A group of substances that are especially highly toxic to aquatic life. They are used predominantly as constituents in marine ANTI-FOULING AGENTS.

Trichophyton verruscosum *see* RINGWORM

triglyceride An ESTER of fatty acids. FATS and OILS are defined as substances that contain one or more triglycerides.

tritium A rare naturally occurring and radioactive ISOTOPE of hydrogen. It has an atomic mass of 3 and a half-life of 12.5 years. It can be manufactured artificially in nuclear reactors and is used as a tracer in medical and laboratory research. It is also used as a constituent of hydrogen bombs. It is not thought to bioaccumulate to a significant degree in humans, although early research has indicated that some fish and shellfish species in estuaries receiving wastewater from establishments that routinely use tritium have been found to contain higher than expected tritium levels. It is possible that different species exhibit differing bioaccumulation capacity but further research is needed in this area.

See also: DEUTERIUM

tropopause That layer of the earth's atmosphere which forms the junction between the TROPOSPHERE and the STRATOSPHERE.

troposphere That layer of the atmosphere closest to the earth and lying immediately below the STRATOSPHERE. It is characterised by a decrease in temperature of around 6.5°C per kilometre increase in height rather than by definite depth, being around 18 km thick at the equator and around 4 km thick at the poles.

See also: MESOSPHERE; TROPOPAUSE

TSE *see* TRANSMISSIBLE SPONGIFORM ENCEPHALOPATHY

TSP *see* TOTAL SUSPENDED PARTICLES

tuber The swollen underground root of a plant (such as a potato) used for the storage of food and for the generation of new plants.

See also: BULB; CORM

tuberculosis The name given to a group of diseases caused, in humans, by three main types of bacteria, being *Mycobacterium tuberculosis*, *Mycobacterium africanum* and *Mycobacterium bovis* (the last found primarily in cattle). The tuberculosis bacilli act by forming fine granules or nodules known as 'tubercules' (from whence the name tuberculosis is derived) in the tissue where they are deposited. These tubercules start as small, scarcely visible granules but can multiply to the extent that they destroy the organ or tissue in which they are found.

The most important to human health is adult pulmonary tuberculosis caused by

Mycobacterium tuberculosis. Tuberculosis can affect most areas of the body, although generally it rarely affects muscle tissue, sinews or cartilage. It is usually transmitted around the body via the lymphatic system. Many manifestations of the disease only develop after a long period, often of many years (this is a stage of chronic but passive infection). Tuberculosis has been a serious disease of humans for over 2,400 years. Hippocrates (460–375 BC) knew the pulmonary form as 'phthisis' and in the Middle Ages the form that caused glandular swelling of the neck was known as 'scrofula' or the 'king's evil'. A disfiguring variety of skin tuberculosis is known as 'lupus vulgaris' (often shortened simply to 'lupus'). The term 'consumption' is a more modern term used to describe the pulmonary disease and the accompanying general wasting of the body.

Transmission of the infection can occur by three main routes, inoculation, inhalation or ingestion. In Western Europe and Northern America at the turn of the century it was estimated that one in ten of the population harboured the *Mycobacterium tuberculosis* bacillus. The high prevalence of the bacilli in the general population is a major factor in the association with tuberculosis infections developing in people with immunocompromising diseases such as AIDS. Globally there are around 16 million cases of active tuberculosis. Around 8 million new cases of tuberculosis will develop per year, of which 3.5 million of these will be of infectious pulmonary tuberculosis. Around 2 million of these per year will die (a rate of 5,000 deaths per day). The development of resistance to the antibiotics used in the traditional methods of treatment is a significant public health problem. The failure to complete treatment is considered to be a significant factor in the development of antibiotic resistance.

tumour Also known as a neoplasm, any abnormal mass of tissue generated by the growth of cells often independently of surrounding tissue. The term applies equally to both MALIGNANT and BENIGN growths.

turbine A device that translates the movement of a gas or liquid flowing through a cylinder or along a pipeline into the rotational force by impacting on the rotor and/or fixed blades of the turbine. The three principal substances that produce the force necessary to power a turbine are gases produced by combustion, steam or directly by moving water, the last usually acting under the force of gravity.

turpene An organic compound that is an intermediate in the formation of a number of secondary products formed in living organisms, the latter including STEROIDS, pigments and certain VITAMINS.

TVC *see* AEROBIC COLONY COUNT

Type I error A mistake in a clinical trial or similar in which a difference between two values is ascribed or apparently detected where one does not exist. It is also known as a 'false positive' result.

Type II error A mistake in a clinical trial or similar in which a difference between two values is not detected where one actually exists. It is also known as a 'false negative' result.

typing In the microbiological sense, the application of any one, or multiple, of a range of tests used to differentiate between members of the same species. The intention of the exercise is to be able more closely to identify organisms from a common source or to exclude organisms of the same species but of a different 'type'.

U

u-value A measure of the THERMAL TRANSMITTANCE of a structure expressed as the amount of heat energy per hour that passes through it for each degree of temperature difference between either side of the structure.

See also: K-VALUE

ulcer A breach in the surface of the skin or mucous membrane of the body. It is characterised by having resulted in the activation of the body healing mechanism (suppuration, redness, etc.) but in which the healing process is slow in concluding or fails to resolve.

ulceration The process in which the cells of a defined area of the outer body die in the development of an ULCER. The condition whereby larger areas of dead tissue develop on the outside of the body is known as GANGRENE.

See also: NECROSIS

ultra vires Means beyond the legal power or authority (e.g. of a person, assemblage or corporation). The term is from the Latin and literally means 'beyond strength'.

ultra-fine particle In relation to atmospheric contamination refers to particles with an AERODYNAMIC DIAMETER of less than 0.1 micrometres.

ultrasound Sound of a frequency above that which can be detected by the human ear. This is generally taken as above a frequency of around 20,000 Hz (see HERTZ). Ultrasound is widely used in medical diagnostics such as Ultrasound Tissue Characterisation and Ultrasound Imaging.

ultraviolet radiation A form of ELECTROMAGNETIC RADIATION with a short WAVELENGTH just outside the visible spectrum to humans of between that of X-RAYS and visible LIGHT. Ultraviolet radiation is capable of causing photochemical reactions. There are three types of UV radiation, designated UVA, UVB and UVC. UVA has a long wavelength and contributes mainly to ageing of the skin. UVB is blocked naturally by clouds and is the type that causes burning of the skin. UVC does not naturally come to earth, but can be generated artificially and is used as a bactericidal light. Contrary to popular opinion, all types of UV radiation are considered to be a risk factor in the generation of SKIN CANCER; none can be considered 'safe'.

UNCED The acronym for the UNITED NATIONS CONFERENCE ON ENVIRONMENT AND DEVELOPMENT.

UNCSD The acronym for the UNITED NATIONS COMMISSION ON SUSTAINABLE DEVELOPMENT.

under-ascertainment A situation in which either all the cases of a disease or all the data relating to such have not been collected. There are many reasons for this including lack of response from patients, non-reporting by medical practitioners and wide geographic distribution of cases making contact difficult. As a general rule the less serious a disease of condition the more under-ascertainment there is likely to be. Some studies try to give an indication of the level of under-ascertainment when they are reported.

underpin The provision of a new foundation for an existing structure without dismantling it. The process usually involves exposing the existing foundations, building a new structure underneath them and fixing the two together.

unitary authority A local authority in the UK that carries out the functions of both a County Council and a District Council.

unitary development plan One developed by local authorities to describe policies in relation to land use and planning, including policies for waste management.

United Nations (Organization) Originally an association formed by the Allies against the Axis powers during the Second World War (1939–45). It is now a supranational transglobal organisation dedicated to the promotion of peace and the improvement of the welfare of humanity.

United Nations Commission on Sustainable Development (UNCSD) An organisation of the United Nations that was set up in 1993 following the UNITED NATIONS CONFERENCE ON ENVIRONMENT AND DEVELOPMENT with the specific remit of monitoring progress in implementing the agreements made there.

United Nations Conference on Environment and Development (UNCED) A major conference organised by the United Nations and held in Rio de Janeiro in 1992. The conference is commonly referred to as the 'Earth Summit'.

See also: RIO DECLARATION

United Kingdom Originally established in 1801 with the Union between Great Britain and Ireland. It was established in its present form in 1922 with the partitioning of the Republic of Ireland (formerly the Irish Free State). The United Kingdom has no written constitution and, although the monarch is deemed to be the Head of State, the supreme authority in practice is exercised by the Government of the day. The United Kingdom is also known simply as 'Britain'. The UK joined the European Economic Community (now the European Community (EC)) in 1973.

unit price The installed cost of an identified constructional item or quantity.

unleaded petrol Petrol that uses high-OCTANE hydrocarbons or non-lead ANTIKNOCK COMPOUNDS to reduce KNOCK. The use of lead as a constituent of antiknock compounds has been phased out to improve environmental safety.

unpaired t-test A test of statistical significance used to compare the MEAN of two groups of individuals.

unsaturated hydrocarbon One that contains double or triple bonds between certain atoms in its structure. These bonds may be broken without disrupting the existing molecular skeleton to allow the molecule to combine with other atoms. An unsaturated hydrocarbon is therefore less stable than a SATURATED HYDROCARBON.

unscheduled DNA synthesis A process whereby DEOXYRIBONUCLEIC ACID (DNA) is manufactured by the cell at a time other than the normal scheduled intervals (known as the 'S period' – i.e. scheduled period). This unscheduled synthesis normally occurs following DNA damage.

untypeable In microbiology refers to the organisms whose response to laboratory examination is insufficiently characteristic of known organisms or strains to make identification possible.

upper-limb disorder Also known as work-related upper-limb disorder, another term used to describe REPETITIVE STRAIN INJURY.

upstream assessment An evaluation of the events that preceded a given phenomenon, essentially if it is an analysis of the cause.

See also: DOWNSTREAM ASSESSMENT

urban renewal The totality of activities implemented with the aim of reconstructing and redeveloping the social, economic and architectural environment of disadvantaged or socially deprived communities. It is important to address all of the components of deprivation and decay to ensure that the community becomes attractive and sustainable in its new form. Frequently it is necessary for initiatives to proceed on several fronts such as housing, recreation, transportation, employment, crime reduction and health and social services improvement or development for the totality to be successful. Such an approach requires considerable multi-agency working and the participation of the community itself is often seen as vital to success.

See also: BROWN FIELD SITE

urbanisation The process whereby a GREEN FIELD SITE is converted into use as a town, city or part thereof.

urine An excretion of the kidneys consisting of around 96 per cent water and 4 per cent bodily waste as dissolved solids. In a healthy individual in a temperate climate around half of the body loss of water is due to urine production; the remainder is shed as sweat. Both the composition and relative ratios of shed are variable dependent on climate (especially temperature and HUMIDITY) and state of health.

See also: DIURETIC

'User Pays' Principle A system in which the payment by an end user for a product or service reflects the full cost of all the (especially natural) resources used in their production. The term is used primarily in provision of goods for the wider community or society such as in relation to transport infrastructure and water supply.

See also: 'POLLUTER PAYS' PRINCIPLE

utility The value that an individual or group ascribes to goods or services. It is therefore determined subjectively rather than objectively.

See also: OBJECTIVE ASSESSMENT; SUBJECTIVE ASSESSMENT

Utopia Originally the title of a book by Sir Thomas More (1477–1535). The term is now used to describe an imaginary or hypothetical (especially impossible to achieve) realm or State in which everything is perfect.

V

vaccination The process whereby dead or attenuated PATHOGENIC organisms or toxins are introduced into the body in the form of a VACCINE with the aim of conferring increased resistance or immunity. The process of introduction of the vaccine to the body is often referred to as INOCULATION.

See also: IMMUNISATION

vaccine A substance containing dead or attenuated pathogenic organisms or toxins artificially introduced into the body with the aim of conferring increased resistance or immunity.

vacuum packing A process used in the food industry in which air is removed from a food container and sealed to prevent its re-entry. The object of the technique is to remove any respirable atmosphere that would sustain microbial growth.

validity An expression of the capability of a scientific study to be able to identify or ascertain that which is being sought. This will depend primarily on the design of the study, the SENSITIVITY and SPECIFICITY of ascertainment techniques, and the conduct of the study itself.

valve A device or structure incorporated into a passageway through which liquid or gas is conducted in order to regulate the flow or the direction of the flow.

Van Allen radiation belts Two streams of radioactive particles, probably of cosmic origin and from the sun, which are trapped at a height of between around 3,000–16,000 km (approximately 1,900–10,000 miles) by the earth's magnetic field.

vapour barrier An airtight membrane incorporated into the structure of a building with the aim of preventing moisture-laden air entering into the fabric where it could condense and cause damage.

See also: DAMP-PROOF COURSE

vapour lock A phenomenon in which a volatile liquid substance vaporises within a system designed to carry the liquid, causing impedance to the design flow.

vapour pressure The pressure exerted by the vapour in the presence of its own liquid when the two are confined together in equilibrium at a specified temperature. The vapour pressure provides an indication as to how VOLATILE a liquid is.

variable Any one of a number of factors in an epidemiological study, scientific investigation or similar that does not have a consistent value or which is capable of change sufficient to produce different results in the study should such variation occur.

vascular Refers to living organisms (either animal or plant) and means possessing vessels or ducts through which blood or sap can circulate.

vaulting An arched ceiling (especially of stone) over a space.

vCJD *see* CJD

vector An organism or entity that carries an infectious agent on or within its body or structure in such a way as to facilitate its transmission to another HOST.

See also: CARRIER

vegetative cell An expression in microbiology to describe the state of a (usually bacterial) cell that is growing and developing.

veins The major BLOOD vessels of the body that carry blood back to the heart after it has circulated through the body's tissues. The blood that they contain will be depleted of oxygen having delivered it to the tissues using the HAEMOGLOBIN component of the red blood cells (ERYTHROCYTES) as a transportation mechanism. The depleted oxygen levels make venous blood a much darker red than the oxygen-rich arterial blood. Veins are more numerous than arteries and generally run parallel to them. In addition smaller veins may be seen just under the skin.

See also: CARDIOVASCULAR SYSTEM; CAPILLARY

velocity Correctly defined as an expression of distance covered in a linear direction over a specified period of time. The term is often used commonly as a synonym for speed, but this is incorrect as the latter term does not incorporate the criterion of progress in a linear direction.

veneer An overlay or facing of decorative or protective material over a surface. It is often created to disguise an underlying constructional material that is considered to be of lower aesthetic value.

venturi tube A waisted restriction in a pipe inserted with the aim of impeding the flow of that which passes through the pipe. The restrictions can be used to create differential pressure at either end of the restriction for the purposes of measuring flow rates. Venturi tubes are also used to accelerate airflow to create suction, such as in some vacuum cleaners.

verdigris The greenish coloured coating, predominantly of copper carbonate, created on the surface of brass, bronze and copper on exposure to the atmosphere. Some consider this PATINA to be aesthetically pleasing and the process can be artificially speeded up or enhanced to produce desired decorative finishes.

verge board Another name for BARGE BOARD.

vernal Means pertaining to that which happens in the spring.

vernalisation The treatment of seeds or seedlings with either low or high temperatures in order to induce germination or to induce them to flower once the plants are mature.

verocytotoxin A toxic protein produced (predominantly) by certain types of the bacterium ESCHERICHIA COLI. The name derives because the protein is differentiated because it is toxic IN VITRO to the laboratory-cultured kidney cells of monkeys (Vero cells). There are several different forms of verocytotoxin.

See also: SHIGA-LIKE TOXIN; VTEC

verrucas Warts, of varying sizes, on the plantar area of the feet, (i.e. the fleshy, 'padded' areas on the sole); they are also known as 'plantar warts'. Verruca is the Latin name for wart. They are caused by type 1 of the human papillomavirus (HPV). In spite of the common nature of the infection (which generally ranges from around 5–50 cases per 1,000 population) there is very little detailed knowledge available and, consequently, varying opinion as to public health significance. Transmission is thought to be by inoculation from contaminated surfaces through an abrasion or cut, allowing the virus to attack the granulosum and keratin layers of the skin. The incubation period is thought to be between 1–20 months, usually 2–3 months, with a period of communicability that is uncertain but likely to be at least as long as the lesion is manifest.

The warts are most frequently seen in children and teenagers, and the traditional route of transmission is often thought to be changing rooms and communal shower facilities at swimming baths. People suffering from verucas are often excluded from such facilities although medical opinion and fashion varies as to the effectiveness of this control strategy. The warts themselves are frequently painful and medical advice on both the treatment of the individual and on exclusion policies should be sought.

vertebra One of the bones that comprise the backbone or spinal column.

See also: INVERTEBRATE

vertebrate An animal that possesses a backbone (also known as the spinal column). All the higher animals, including man, are vertebrates.

See also: ENDOSKELETON; INVERTEBRATE; VERTEBRA

vertical Means upright. It is applied in medical practice as vertical transmission and means the transmission of a disease usually between mother and child (i.e. from one generation to the next, particularly during pregnancy or childbirth). In legislature it is often applied to Directives from the EUROPEAN UNION (EU) as in the term vertical Directive. In this latter case it means a Directive applicable to a specific food or food type (e.g. meat, meat product, dairy product) at all or several levels in the food chain, (e.g. from primary production through to manufacture).

See also: HORIZONTAL

veterinary public health Has been defined by the World Health Organization in 1999 as 'the contribution to the complete physical, mental and social well-being of humans through an understanding and application of veterinary medical science'.

viable In the microbiological sense refers to a micro-organism that is capable of growth and development given suitable conditions.

viaduct A bridge, usually incorporating a number of arches, used to convey a road or railway over a chasm, void, space or similar.

See also: AQUEDUCT

vibrated concrete That to which an oscillating force has been applied with the purpose of removing air pockets and compacting the whole. Vibrated CONCRETE is therefore generally stronger than simple poured concrete, although there is a potential risk in relation to reinforced concrete that the reinforcing materials could possibly be displaced by the vibration.

vice chancellor In the UK is the chief executive or chief administrator of a university. The title of CHANCELLOR in the UK is honorary.

vice versa Means 'the other way around'.

virino A hypothetical fragment of viral DEOXYRIBONUCLEIC ACID (DNA) that has been suggested as being capable of infecting a cell such as to induce it to produce multiple copies of the fragment and, consequently, thereby to cause an associated disease.

virtual reality A proxy computer-generated or conceptualised environment that as closely as possible resembles real life. It enables exploration and alteration of potential scenarios for research or evaluation purposes without committing the observer to actions that would affect the real-life environment.

virulence A measure of the power of a PATHOGENIC micro-organism to cause disease. Virulence can also be used variously as an expression of the severity of the symptoms, the speed of spread of the infection or a measure of the ATTACK RATE of a particular organism.

See also: PATHOGENICITY

virulence factor Any phenomenon or circumstance that can act on or in a micro-organism in a way such as to affect either its ability to cause disease or the severity of the disease if initiated.

virulent factors Those characteristics of a pathogenic micro-organism that enable or give it the potential to cause disease.

virus An exceedingly small microscopic organism (around 300 nm) containing either single stranded RIBONUCLEIC ACID (RNA) or double stranded DEOXYRIBONUCLEIC ACID (DNA) surrounded by a protein coat and a glycoprotein/lipid envelope. As both RNA and DNA are required for reproduction, viruses cannot reproduce outside a living cell of another living organism. These they invade inducing the cell to reproduce more viruses, which are released when the cell ruptures. Antibiotics do not affect viruses and the injudicious use of these medicines to 'treat' viral infection is thought to contribute to ANTIBIOTIC RESISTANCE.

See also: ADENOVIRUS; ASTROVIRUS; EBOLA AND MARBURG VIRUSES; ROTAVIRUS; SMALLPOX; SRSV; VERRUCAS

vis-à-vis Literally means 'face-to-face' (with). It is usually used as a preposition meaning 'in relation to' or 'having regard to'.

viscera (singular viscus) The term to describe the organs of the abdominal and thoracic cavities of the body. The removal of these organs, such as for preparing food animals for cooking, is known as evisceration.

viscosity An expression of the physical property of a fluid that enables it to flow. The greater the viscosity the harder it is to pour; the less the viscosity the easier it is to pour. Viscosity can alter with temperature or by physical agitation. The instrument used to measure viscosity is known as a viscometer and is calibrated to measure the flow rate of the liquid passing through a capillary tube at a controlled temperature.

See also: FLUIDITY; THIXOTROPY; VISCOUS

viscous A term applied to a fluid meaning that it is resistant to flowing. The property derives from the forces acting between the molecules of the substance.

See also: VISCOSITY

visible radiation A synonym for LIGHT.

vital capacity That volume of air which an individual can expel in one rapid and forcible exhalation.

See also: FORCED EXPIRATORY VOLUME

vitamin One of a group of organic compounds that cannot, in general, be made by the organism itself, but which are required, in very small quantities, to ensure correct growth and function. They are supplied through the diet. Vitamin requirements can vary dependent on animal species. The term is thought to have arisen as a corruption of the phrase 'vital AMINE', although now it is known that not all vitamins are amines. Additionally, not all substances originally thought to be vitamins actually proved to be the case, and this accounts for the gaps in the nomenclature structure as non-vitamins were deleted from the list.

See also: CARBOHYDRATE; FAT; PROTEIN

vitreous enamel A hard protective finish formed on the surface of steel or aluminium by the process of fusing a thin coating of molten glass onto the surface of the metal.

vitreous humour The thick, transparent, jelly-like material that fills the posterior cavity of the eye.

See also: AQUEOUS HUMOUR

VOC The abbreviation for VOLATILE ORGANIC COMPOUND.

volatile Means capable of rapid evaporation, rapid change or explosive reaction.

volatile organic compound (VOC) Any ORGANIC compound that will react with oxides of nitrogen in the presence of sunlight to form ozone.

volt A unit measure of electrical potential. One volt is the potential difference against which one JOULE of work is done in the transference of one COULOMB.

See also: ELECTROMOTIVE FORCE

voussoir A wedge-shaped stone or brick predominantly used in the construction of an arch. Such arches are often semicircular. The voussoir at the apex or crown of the arch is known as the KEYSTONE.

VTEC The abbreviation for 'Verotoxin-producing ESCHERICHIA COLI'. The term refers to the ability of certain types of *E. coli* (notably, but not exclusively, type O157) to produce the protein known as VEROCYTOTOXIN. The abbreviation for this type of bacterium that is capable of producing the toxin is '*E. coli* O157 VTEC'.

vulcanisation The process of applying heat to convert a rubber compound into its final configuration in use. The application of heat produces chemical bonding between adjacent molecules within the rubber. The greater the heat, the more bonds between adjacent molecules and the harder the final product.

W

wainscot Panelling, usually of wood, installed on the lower portion of an internal wall for decorative or protective purposes. The material used to construct a wainscot is known as 'wainscoting'.

wall plate A horizontal timber fastened to the top of a wall and to which the RAFTERS are affixed.

wall tie A device incorporated into the structure of both sides and spanning the void of a CAVITY WALL in order to improve structural integrity by holding both sides together. The wall tie usually incorporates a feature such as a twist or dip that conveys any rainwater that penetrates the external skin to drip into the cavity and prevent it from crossing the void on the wall tie.

See also: TIE

warp In textiles is any yarn that runs lengthways through a woven fabric. The yarn that runs across and between the warp is known as the WEFT.

waste Generally described as that which is left over or surplus to requirements to the extent it is in need of disposal. In the UK waste has a defined meaning in law and includes scrap material, effluent or unwanted surplus substance or article that requires to be disposed of because it is broken, worn out, contaminated or otherwise spoiled. The legal definition excludes explosives and radioactive material that might otherwise be considered as waste as their disposal is specifically dealt with under the provisions of other legislation.

waste arisings The term used to describe the amount of waste generated within a defined locality over a given period of time.

waste hierarchy A system of classifying waste according to the way in which its generation or handling should be managed to further the principles of SUSTAINABLE DEVELOPMENT. The hierarchy is one of preference in which ideally the production of waste should be *reduced*. Where reduction is impractical the next option for consideration should be *reuse*. If this is not possible the recovery of the raw materials or part should be considered through such initiatives as *recycling*, *composting* or *energy recovery from waste*. Only where all other options have been determined as impractical should waste be sent for *disposal*.

waste management A systematic and applied plan for the control of waste material from its creation to ultimate disposal or recycling. Ideally waste management should also include consideration of the manufactured article and its packaging when it is made or created to ensure that the potential for waste is minimised and that the materials of construction used can either be satisfactorily recycled or come from renewable sources, embracing the principles of sustainable development.

waste management industry An all-embracing term used to describe any business or non-profit-making organisation involved in the collection, transportation, handling, management or disposal of WASTE.

waste management licensing The system in the UK whereby anyone who wishes to engage in the business of the deposit, recovery or disposal of WASTE requires a licence. The purpose of the licence is to regulate and control the operations through the imposition of conditions designed to protect human health, wildlife and the environment.

waste transfer station A facility to which waste is transported and collected prior to being transferred (usually as bulked loads) to another place for recycling, treatment or disposal. Waste transfer stations may process or sort the waste in some way prior to onwards transportation, the defining function of the site however being that such an establishment is essentially a holding place rather than being the ultimate destination.

water activity (Aw) A measure of the amount of water available within a substance to support the growth of micro-organisms. Pure water is assigned a value of 1.0, while other substances not being pure water will therefore contain water to a value of less than 1.0.

water hammer The banging noise within a water pipe generated by the sudden creation of an area of high pressure, often arising when water flow is suddenly curtailed or stopped. Similar symptoms can sometimes be generated due to the presence of air trapped within the pipes. Although not normally a cause for concern as regards the structural integrity of a sound plumbing system, the noise can be aesthetically displeasing and can place undue strain (possibly leading to damage) within an aged or weakened water distribution system.

water hardness *see* HARD WATER

water of crystallisation Water that is chemically combined in the structure of certain minerals upon crystallisation.

See also: ANHYDROUS

water softener A device used to reduce the undesirable effects of HARD WATER by improving the lathering of soaps, reducing tide marks on sanitary ware and reducing the potential for FURRING. Some softeners replace calcium and magnesium compounds with those of sodium. This can be inadvisable as an increase in the sodium intake is potentially dangerous for premature babies and those with heart problems. Unless they are maintained properly some water softeners can develop colonies of bacteria, which can ultimately adversely affect water quality.

water table The level below which the ground is saturated. A well will fill to the level of the water table. The level may be variable over time for a given location dependent on the nature of the ground, although some sub-strata can create water tables of surprising consistency.

water test An assessment to measure the integrity of a (usually new) drainage system. The system is filled with water and pressure is generated at the highest end of the system within a graduated transparent tube filled and held at the appropriate level. It is important when setting up the test to ensure that there is no air trapped within the system as this can lead to false results. A successful result is one in which the pressure is maintained within the system within defined limits for a specified period of time, these latter being specified by statute, guidance or by the equipment manufacturer.

See also: AIR TEST; COLOUR TEST; SMOKE TEST

wattle and daub A type of constructional infill between the structural timbers of buildings. The name derives from the insertion of 'wattles' (laths or twigs) between the timbers, these then being covered with 'daub' (clay plaster).

wavelength The linear distance between successive identical points (usually taken as the peak or trough) on a wave of energy as it is transmitted past the reference point. The wavelength has a direct relationship with the FREQUENCY and with the velocity of transmission.

See also: ELECTROMAGNETIC RADIATION

weatherboard A length of boarding (usually of timber) in which the thickness tapers from one edge to the other. It is used predominantly for cladding. In the USA it is known as 'clapboard'.

weather front *see* FRONT

weather struck *see* POINTING

weathering First, the topmost facing of a flat surface inclined or shaped so as to ensure that rainwater runs off away from the structure. Second, it is the process of abrasion and loss of the surface of, for example, rock or the facing material of buildings, including the breaking down of the surface by ice, chemicals (such as atmospheric pollution) or wind-blown particles, and their subsequent EROSION.

weft In textiles is any yarn that runs across the width of a woven fabric, from edge to edge. It is woven between the yarn that runs lengthways, this latter being known as the WARP.

welding The name given to a number of techniques employed whereby molten metal is generated at source to facilitate the process of joining two or more metals together either with or without the application of additional pressure. The most common welding techniques are gas welding (usually using a mixture of oxygen and acetylene) or electric arc welding (essentially the heat is generated by a controlled arc or spark between a hand-held electrode and the welding surface). There are three principal electric arc welding techniques.

Tungsten inert gas welding employs a fixed tungsten electrode surrounded by an introduced inert gas (such as argon or helium) to shield the welding. Manual arc welding produces a protective gas 'bubble' generated by the chemical coating on a consumable hand-held electrode (i.e. the electrode is continually melted to provide the metal weld). The final technique uses a consumable electrode but piped gas is introduced to provide the controlled atmosphere. The gas in this latter technique may be inert (called metal inert gas welding) or be an active component (such as carbon dioxide) in the welding process (called metal active gas welding). Other welding techniques use electron beams, friction or ultrasound as the heat generation source but these latter are generally employed in specialist situations.

Welsbach mantle *see* GAS MANTLE

western approaches The term used to refer to the fishing area within the EC's 200 mile fishing zone of the Atlantic Ocean located to the south and west of the UK. It is defined broadly as that area lying to the south-west of Land's End in the UK, to the south of the Republic of Ireland (Eire) and to the west of Brittany, France.

Westminster The area of London, UK, within which the main administrative base of the UNITED KINGDOM Government is located. The name is also used (post-DEVOLUTION) as a shorthand for the administrative base for the principal Civil Service Departments in England.

wet (and dry) bulb thermometer Used to measure RELATIVE HUMIDITY. It comprises two thermometers, one that measures the AMBIENT air temperature (the 'dry bulb') and another whose bulb is wrapped in gauze or absorbent material that is wet by water (the 'wet bulb'). The reading of this latter thermometer is affected by the rate of evaporation of the water from the gauze, which itself is determined by the relative humidity. The drier the air (another way of saying the lower the relative humidity), the greater the rate of evaporation and hence the greater the loss of LATENT HEAT. This is manifest as a lower temperature reading on the wet-bulb thermometer when compared to the dry-bulb thermometer reading. A chart is then used to translate the recorded difference into an equivalent relative humidity. At 100 per cent relative humidity there is no difference in the recorded temperatures between the two thermometers because no water can be lost by evaporation.

Simple wet-bulb thermometers rely on natural evaporation from the gauze, while more sophisticated and accurate instruments are equipped with a small fan that supplies a known air movement across the wet bulb. Static wet-bulb thermometers are not useful to measure rapid changes in relative humidity as they require time to stabilise; they should only be used therefore as an indication of the average relative humidity over time.

See also: WHIRLING HYGROMETER

wet rot A condition of decay in wood or timber primarily brought about by fungal attack arising as a result of a high degree of dampness. The term wet rot is used in building science essentially to distinguish the condition from 'DRY ROT', which, in spite of its name, is a similar decay process brought about by fungal attack but in circumstances in which the wood is less damp than that needed to bring about wet rot.

whet Means to sharpen (usually a blade) by rubbing the edge against an abrasive material such as a whetstone (a 'sharpening stone').

whirling hygrometer Also known as the whirling psychrometer, an instrument containing a pair of WET AND DRY BULB THERMOMETERS in a housing held at 90° to a handle, the housing and thermometers being rotated manually in the fashion of an old-style football rattle. The instrument is used to measure RELATIVE HUMIDITY.

white blood cell *see* LEUCOCYTE

white noise Sound of a broad-band frequency in which both phase and intensity vary randomly but in which the overall energy per proportional frequency bandwidth is equal. White noise is used to measure the frequency response of noise monitors across all frequencies.

See also: NOISE WEIGHTING

Whitehall A collective term used to describe the major Government departments in either England or (if national) the UK. The name derives from the street in London where many (but not all) such departments are located or based.

WHO *see* WORLD HEALTH ORGANIZATION

widdershins *see* WITHERSHINS

Wilcoxon test A test of statistical significance derived by using the MEDIANS of two different groups of individuals.

wind chill factor A measurement of the reduction in temperature perceived by an individual in response to air movement across the skin.

wind rose A schematic diagram of wind speed and/or direction over a given period of time for a given fixed location point. The diagram uses lines or arcs radiating out from a central point, the length of the lines or area covered by the arc indicating wind speed or the duration that the wind was in that particular direction.

wind shadow The area of low pressure to the LEEWARD side of an object created by the wind's passage. It can be an important consideration in predicting the rate and pathway of discharge of sub-soil gases such as RADON or METHANE.

windpipe Another name for the TRACHEA.

windward The side exposed towards the wind or that located in the direction to which the wind blows. It is the opposite of LEEWARD.

wireless A telecommunications system in which some or all of the communications path is travelled using electromagnetic waves rather than by cable. The first wireless transmitters used Morse code. Nowadays cellular telephones, pagers, cordless computer peripherals, remote control devices and satellite communication such as the Global Positioning System are all examples of wireless equipment.

withershins A synonym for ANTICLOCKWISE. It is sometimes rendered (especially in Scotland) as 'widdershins'. The term is also used to describe a movement in the opposite direction to that of the sun or to that which is 'normal'.

wood alcohol *see* METHANOL

World Bank The popular name for the INTERNATIONAL BANK FOR RECONSTRUCTION AND DEVELOPMENT.

World Health Organization (WHO) An agency of the United Nations established in 1948 whose aim broadly is to improve human health by addressing global health issues.

Worldwide Fund for Nature (WWF) The name given to the international organisation, originally known as the World Wildlife Fund, dedicated especially to the preservation of endangered species and conservation.

wrought iron That produced by one (i.e. forging or rolling) of two traditional processes of making iron; the other produces CAST IRON. Wrought iron is very pure, containing generally less than 0.1 per cent carbon. It is highly malleable and resistant to rust.

X

xenobiology The (currently entirely theoretical) study of potential or possible life from elsewhere in the universe.

xenobiotic An adjective used to describe a chemical that is alien to a particular biological system.

xenoestrogen A chemical from outside a living organism with the potential to promote oestrogenic activity if introduced into that organism.

xenotransplant The transplantation of the tissues or organs of one species or genus to another. Although some scientists argue that xenotransplantation could resolve many shortages in human transplantation, others argue that there are unknown health threats, especially from the potential mutation of (as yet unknown) viruses from the host organism. A common example of xenotransplantation in practice is the replacement of defective human heart valves with those from pigs.

xerophyte A plant that has become adapted through evolution to be suited to life in ARID conditions.

xerosphere The term used to describe collectively those areas of the earth that are covered by hot or cold DESERTS.

X-rays A high-energy electromagnetic IONISING RADIATION generated when electrons collide with atomic nuclei. X-rays are described in some scientific literature as a having a wavelength shorter than ultraviolet (UV) radiation (i.e. below 1×10^{-8} metres) but in practice electromagnetic radiation that has a wavelength shorter than 1×10^{-11} metres is generally known as GAMMA RADIATION. X-rays were originally known as Röntgen rays in recognition of their discovery by Wilhelm Konrad Röntgen (1845–1923).

X-rays have neither mass nor electric charge and are less able to penetrate matter than is gamma radiation. X-rays have differential penetration dependent on the density of the material through which they are passing. This feature is used in X-ray photography of bone as it is denser than surrounding tissue and shows up as a darker image on the photographic plate.

Y

yard An imperial measure of length comprising 36 inches. It is equivalent to 0.9144 of a METRE.

yeast A class of spore-forming, unicellular fungi. They are widely found in nature, being active in the soil and contributing to the decay process of vegetable material. Certain yeasts may cause FERMENTATION under anaerobic conditions, a trait that has been exploited commercially in both brewing and baking industries (in particular with the yeast *Saccharomyces cerevisiae*). Some yeasts can colonise and cause infections or irritations of the skin in humans.

See also: ALCOHOL; BREAD

Z

Z *see* ATOMIC NUMBER

z-value The number of degrees of temperature (Celsius or Fahrenheit) required to reduce the number of a specified micro-organism within a specified substrate to one-tenth of the original organisms present.

See also: D-VALUE

zenith The highest point or, alternatively, that portion of the sky which is directly overhead, at 90° to the horizontal, of an observer.

zero population growth (ZPG) That state in a community or country when the crude birth rate is equal to the death rate and the number of people therefore remains stable. Some populations have achieved this as an unplanned state, while others (notably China) are pursuing this as the aim of legislative processes that limits the number of children a couple are allowed to have. In theory a community with zero population growth consumes fewer resources and becomes a self-sustaining unit. In practice technological advancement and population expectation largely determine the rate of resource usage, and imposed birth quotas distort population distributions and have reportedly resulted in parents killing female babies in order to protect their opportunity to produce males.

See also: EUGENICS; SUSTAINABLE DEVELOPMENT; UTOPIA

zoonosis A disease caused in humans by the passage of a pathogenic organism from an animal. Classically the definition of a zoonosis was qualified to exclude infections in man that also produced disease in animals (or vice versa). More modern usage suggests that any disease passed from an animal to man can be defined as a zoonosis.

zooplankton Microscopic animals living in salt or freshwater. They are the animal component of PLANKTON and often comprise the larval forms of crustaceans and shellfish as well as adult microscopic animal life and krill.

See also: PHYTOPLANKTON

zootoxin Any toxin (such as the venom of certain species of snakes and spiders) produced by an animal.

zygospore A thick-walled spore formed by the ZYGOTE of some species of fungi and algae.

zygote Generally taken as being the fertilised ovum (i.e. the union of ovum and spermatozoon) prior to division.

Some sources also ascribe the term to the cells of such a union immediately following division.

zymosis/zymolysis Two terms meaning FERMENTATION.

APPENDIX I
UNITS AND MEASUREMENTS

Weight

Symbol	*Unit*
g	Gram
kg	Kilogram (1,000 grams)
mg	Milligram (one-thousandth of a gram)
μg	Microgram (one-millionth of a gram)
ng	Nanogram (one-thousand millionth of a gram)
pg	Picogram (1 million-millionth of a gram)

Conversion

To convert	*Multiply by*
Ounces to grams	28.3495
Pounds to kilograms	0.4536
Stones to kilograms	6.3503
Hundredweights to kilograms	50.8023
Tons to tonnes	1.016

To convert	*Divide by*
Grams to ounces	28.3495
Kilograms to pounds	0.4536
Kilograms to stones	6.3503
Kilograms to hundredweights	50.8023
Tonnes to tons	1.016

Apothecary weight

Apothecary weight	*Grams* (approximate equivalent)
1 grain	0.06
60 grains (1dram)	4
2.5 drams	10
8 drams (1 ounce)	30

Length

Symbol	*SI units*	*Equivalent*
m	Metre	39.37 inches
km	Kilometre (1,000 metres)	0.621 mile
mm	Millimetre (one-thousandth of a metre)	0.03937 inch
μm	Micrometre (1 thousand-thousandth of a metre)	
nm	Nanometre (1 thousand-millionth of a metre)	
pm	Picometre (1 million-millionth of a metre)	
Å	Ångström unit (= 10^{-10} metre)	

To convert	*Multiply by*
Inches to centimetres	2.54
Feet to metres	0.3048
Yards to metres	0.9144
Miles to kilometres	1.6093

To convert	*Divide by*
Centimetres to inches	2.54
Metres to feet	0.3048
Metres to yards	0.9144
Kilometres to miles	1.6093

Temperature

To convert Fahrenheit to Celsius (Centigrade):

(°F − 32) × 5/9

To convert Celsius (Centigrade) to Fahrenheit:

9/5 × °C (+ 32)

	Temperature in degrees		
Reference point	*Kelvin*	*Centigrade*	*Fahrenheit*
Absolute zero	0	−273.15	−459.67
Freezing point of mercury	234.28	−38.87	−37.966
Ice point (i.e. the intermediate between ice and water)	73.15	0	32
Steam point (i.e. the intermediate between water and steam)	373.15	100	212
Melting point of cadmium	594.05	320.9	609.62
Melting point of lead	600.45	327.3	621.14
Boiling point of mercury	629.73	356.58	673.844
Melting point of zinc	692.65	419.5	787.1
Sulphur intermediate between liquid and vapour	717.75	444.6	823.28
Melting point of aluminium	933.25	660.1	1220.18
Melting point of silver	1,233.95	960.8	1,761.44
Melting point of gold	1,336.15	1,063	1,945.4
Melting point of copper	1,356.15	1,083	1,981.4
Melting point of platinum	2,042.15	1,769	3,216.2
Melting point of tungsten	3,653.15	3,380	6,116

Speed

1 knot is defined as a speed equivalent to 1 nautical mile per hour.
1 knot is equivalent to 1.15 statute miles per hour or 1.852 km per hour.

Nautical measures

Nautical linear measure		
6 feet	1 fathom	1.829 metres
1 cable length	100 fathoms	182.0 metres
10 cable lengths	1 nautical mile	1.852 kilometres
10 cable lengths	1 nautical mile	1.150779 statute miles
10 cable lengths	1 nautical mile	1,852.0 metres

Volume

Metric system		
1,000 cubic millimetres	1 cubic centimetre	0.6102 cubic inch
1,000 cubic centimetres	1 cubic decimetre (1 litre)	61.023 cubic inches
1,000 cubic decilitres	1 cubic metre	35.314 cubic feet

Imperial system		
1 cubic inch	16.387 cubic centimetres	
1,728 cubic inches	1 cubic foot	0.0283 cubic metre

Liquid and dry measure

Imperial	*Imperial*	*Cubic inches*	*Litres*
1 gill	5 fluid ounces	9.0235	0.1480
4 gills	1 pint	34.68	0.568
2 pints	1 quart	69.36	1.136
4 quarts	1 gallon	277.4	4.546
2 gallons	1 peck	554.8	9.092
4 pecks	1 bushel	2219.2	36.37

APPENDIX II
ABBREVIATIONS AND ACRONYMS

A&E	Accident and Emergency
ABM	Assured British Meat
ABMP	Association of British Meat Processors
ACC	Aerobic Colony Count
ACCPE	Advisory Committee on Consumer Products and the Environment
ACDP	Advisory Committee on Dangerous Pathogens
ACE	Association for the Conservation of Energy
ACFDP	Advisory Committee for Food and Dairy Products (*PHLS*)
ACGM	Advisory Committee on Genetic Modification
ACMSF	Advisory Committee on the Microbiological Safety of Food
ACNFP	Advisory Committee on Novel Foods and Processes
ACP	Advisory Committee on Pesticides
ACRE	Advisory Committee on Releases to the Environment
ADAS	Agricultural Development Advisory Service
ADI	Acceptable Daily Intake
ADP	Adenosine Diphosphate
AEA	Atomic Energy Authority
AFD	Accelerated Freeze Drying
AGVR	Advisory Group on Veterinary Residues
AIDS	Acquired Immune Deficiency Syndrome
ALARA	As Low As Reasonably Achievable
APC	Aerobic Plate Count
AQMA	Air Quality Management Area
AQS	Air Quality Standards
ASC	Assured Safe Catering
ASH	Action on Smoking and Health
ASMR	Age-Standardised Mortality Rate
ASP	Amnesiac Shellfish Poisoning
ATP	Adenosine Triphosphate
ATSDR	Agency for Toxic Substances and Disease Registry (*USA*)
Aw	Available Water
BAP	Biologically Active Principle
BAT	Best Available Techniques
BATNEEC	Best Available Technology Not Entailing Excessive Cost
BBSRC	Biotechnological and Biological Sciences Research Council

BCG	Bacillus Calmette–Guérin
BDA	British Dental Association
BGS	British Geological Survey
BHF	British Heart Foundation
BIP	Border Inspection Post
BIV	Bovine Immunodeficiency-like Virus
BMA	British Medical Association
BMJ	*British Medical Journal*
BNFL	British Nuclear Fuels plc
BOD	Biochemical Oxygen Demand *or* Biological Oxygen Demand
BPCA	British Pest Control Association
BPEO	Best Practical Environmental Option
BPM	Best Practicable Means
BRE	Building Research Establishment
BS	British Standard
BSC	British Safety Council
BSE	Bovine Spongiform Encephalopathy
BSI	British Standards Institute
BST	Bovine Somatotrophin
BVA	British Veterinary Association
BWEA	British Wind Energy Association
CA	Consumers Association
CAC	Codex Alimentarius Commission
CAP	Common Agricultural Policy (*EU*)
CAPIC	Centre for Accident Prevention and Injury Control
CBI	Confederation of British Industry
CCDC	Consultant in Communicable Disease Control
CCFRA	Campden and Chorleywood Food Research Association
CCP	Critical Control Point
CCW	Countryside Council for Wales
CDC	Centers for Disease Control and Prevention (*Atlanta, Georgia, USA*)
CDSC	Communicable Disease Surveillance Centre
CE	Comité Europa (*European Kite Mark Accreditation System*)
CEFAS	Centre for Environment, Fisheries and Aquaculture Science
CFC	Chlorofluorocarbon
CFS	Chronic Fatigue Syndrome
CFU	Colony-Forming Unit
CHD	Coronary Heart Disease
CHP	Combined Heat and Power (*Plant*)
CHPA	Combined Heat and Power Association
CIEH	Chartered Institute of Environmental Health
CIMAH	Control of Industrial Major Accident Hazards (*Regulations 1984*)
CIMSU	Chemical Incident Management Support Unit
CIP	Clean In Place
CJD	Creutzfeld-Jacob Disease
CJDSU	Creutzfeld-Jacob Disease Surveillance Unit
CMO	Chief Medical Officer
CNS	Central Nervous System
COC	Committee on Carcinogenicity of Chemicals in Food, Consumer Products and the Environment

COD	Chemical Oxygen Demand
COHb	Carboxyhaemoglobin
COM	Committee on Mutagenicity of Chemicals in Food, Consumer Products and in the Environment
COMA	Committee on the Medical Aspects of Food and Nutrition Policy
COMAH	Control of Major Accident Hazards (*Regulations 1999*)
COMARE	Committee on Medical Aspects of Radiation in the Environment
COMEAP	Committee on the Medical Effects of Air Pollutants
CONCAWE	Group for the Conservation of Clean Air and Water In Europe
CoP	Code of Practice
COSHH	Control of Substances Hazardous to Health (*Regulations 1988*)
COSLA	Convention of Scottish Local Authorities
COT	Committee on the Toxicity of Chemicals in Food, Consumer Products and the Environment
CPD	Continuing (or Continuous) Professional Development *or* Central Postcode Directory
CPHL	Central Public Health Laboratory
CPRE	Council for the Protection of Rural England
CPS	Crown Prosecution Service
CSF	Cerebrospinal Fluid
CTB	Council Tax Benefit
CVL	Central Veterinary Laboratory
CVO	Chief Veterinary Officer
DALY	Disability-Adjusted Life Year
DAT	Drug and Alcohol Team
DH (*or* DoH)	Department of Health
DNA	Deoxyribonucleic Acid
DSP	Diarrhetic Shellfish Poisoning
DVLA	Driver and Vehicle Licensing Agency
EA	The Environment Agency
EALs	Environmental Action Levels
EC	European Community *or* European Commission
EEB	European Environmental Bureau
EfW	Energy from Waste
EHCS	English House Condition Survey
EHO	Environmental Health Officer
EIA	Environmental Impact Assessment
ELISA	Enzyme-Linked Immune Sorbent (*or* Immunosorbent) Assay
EMAS	Employment Medical Advisory Service *or* (*European*) Eco-Management and Audit Scheme
EMF	Electromagnetic (*or*) Electromotive Fields
EPAQS	Expert Panel on Air Quality Standards
FAC	Food Advisory Committee
FAO	Food and Agriculture Organisation (*UN*)
FDA	Food and Drugs Administration (*USA*)
FDF	Food and Drink Federation
FEPA	Food and Environmental Protection Act 1985
FMD	Foot and Mouth Disease
FOE	Friends of the Earth

FPHM	Faculty of Public Health Medicine (*of the UK Royal Colleges of Physicians*)
FSA	Food Standards Agency *or possibly* Food Safety Act (1990)
GATT	The General Agreement on Tariffs and Trade
GHQ	General Health Questionnaire
GIS	Geographical Information System
GM	Genetic Modification *or* Genetically Modified
GMC	General Medical Council
GMO	Genetically Modified Organism
GMP	Good Manufacturing Practice
GP	General Practitioner
HACCP	Hazard Analysis Critical Control Point
HAGCCI	Health Advisory Group for Chemical Contamination Incidents
H&S	Health and Safety
HBV	Hepatitis B Virus
HCV	Hepatitis C Virus
HELA	Health and Safety Executive/Local Authorities Enforcement Liaison Committee
HIA	Health Impact Assessment
HIV	Human Immunodeficiency Virus
HMIP	Her Majesty's Inspectorate of Pollution
HMO	House in Multiple Occupation
HMSO	Her Majesty's Stationery Office (*now* The Stationery Office)
HSC	Health and Safety Commission
HSE	Health and Safety Executive
HTST	High Temperature Short Time
HUS	Haemolytic Uraemic Syndrome
ICD	International Classification of Disease (*especially ICD codes*)
IEH	Institute for Environment and Health (*MRC – University of Leicester, UK*)
IFOAM	International Federation of Organic Agricultural Movements
IFST	Institute of Food Science and Technology
ILGRA	Interdepartmental Liaison Group on Risk Assessment
IOSH	Institution of Occupational Safety and Health
IPC	Integrated Pollution Control
IPCC	Inter-governmental Panel on Climate Change
IPCS	International Programme on Chemical Safety (*WHO*)
IPPC	Integrated Pollution Prevention and Control
ITSA	Institute of Trading Standards Administration
KAP	Knowledge/Attitude/Power (*social study evaluation technique*)
LAAPC	Local Authority Air Pollution Control
LACORS	Local Authorities Co-ordinating Office on Regulatory Services
LACOTS	Local Authorities Co-ordinating Body on Food and Trading Standards (now LACORS)
LAPC	Local Air Pollution Control
LAPPC	Local Air Pollution Prevention and Control *or* Local Authority Pollution Prevention and Control
LAQM	Local Air Quality Management
LAWDC	Local Authority Waste Disposal Company
LOAEL	Lowest Observed Adverse Effect Level

MBM	Meat and Bone Meal
MEP	Member of the European Parliament
MMR	Measles, Mumps and Rubella
MRC	Medical Research Council
MRL	Maximum Residue Level/Limit
MRM	Mechanically Recovered Meat
MRSA	Methycillin-Resistant *Staphylococcus aureus*
MSG	Monosodium Glutamate
NCC	National Consumer Council
NELH	National Electronic Library for Health
NFU	National Farmers' Union
NGO	Non-Governmental Organisation
NHBC	National House-Building Council
NHS	National Health Service
NHSiS	National Health Service in Scotland
NICE	National Institute for Clinical Excellence
NIOSH	National Institute for Occupational Safety and Health
NOAEL	No Observable Adverse Effect Level
NRPB	National Radiological Protection Board
NRT	Nicotine Replacement Therapy
NSCA	National Society for Clean Air and Environmental Protection
OECD	Organisation for Economic Co-operation and Development
OEL	Occupational Exposure Limit
OES	Occupational Exposure Standard
OF&G	Organic Farmers and Growers
OFF	Organic Food Federation
OFT	Office of Fair Trading
OTMS	Over Thirty Months Scheme (*BSE control for cattle*)
OVS	Official Veterinary Surgeon
PACE	Police and Criminal Evidence Act 1984
PAH	Polycyclic (or Polynuclear) Aromatic Hydrocarbons
PHA	Port Health Authority *or* Pollution Hazard Appraisal
PHLS	Public Health Laboratory Service (*England and Wales*)
PPE	Personal Protective Equipment
PSP	Paralytic Shellfish Poisoning
PT	Phage Type
QALY	Quality Adjusted Life Year
R&D	Research and Development
RAPEX	The Rapid Alert System for Food (*an EU system*)
RCEP	Royal Commission on Environmental Pollution
RDF	Refuse Derived Fuel
REHIS	Royal Environmental Health Institute of Scotland
RIDDOR	Reporting of Injuries, Diseases and Dangerous Occurrences (*Regulations*)
RIFE	Radioactivity in Food and the Environment
RIMNET	Radioactive Incident Monitoring Network
RIPHH	Royal Institute of Public Health and Hygiene
RNA	Ribonucleic Acid
ROSPA	Royal Society for the Prevention of Accidents
RSH	Royal Society of Health

SAHSU	Small Area Health Statistics Unit
SBS	Sick Building Syndrome
SCF	Scientific Committee for Food
SCIEH	Scottish Centre for Infection and Environmental Health
SEAC	Spongiform Encephalopathy Advisory Committee
SEPA	Scottish Environmental Protection Agency
SI	Statutory Instrument
SMR	Standardised Mortality Ratio
SOLACE	Society of Local Authority Chief Executives
SRM	Specified Risk Material
SRSV	Small Round Structured Virus(es)
SSSI	Site of Special Scientific Interest
STD	Sexually Transmitted Disease
STEL	Short-Term Exposure Limit
SVD	Swine Vesicular Disease
SVS	State Veterinary Service
TB	Tuberculosis
TBE	Tick-borne Encephalitis
TDI	Tolerable Daily Intake
TEF	Toxic Equivalency Factor
TEQ	Toxic Equivalent
TLV	Threshold Limit Value
TSE	Transmissible Spongiform Encephalopathy
TSO	Trading Standards Officer *or* The Stationery Office
TSS	Toxic Shock Syndrome
TVC	Total Viable Count
TWA	Time-Weighted Average
UKCIP	United Kingdom Climate Impacts Programme
UKROFS	United Kingdom Register of Organic Food Standards
UKWRAP	United Kingdom Waste Resources Action Programme
UNCEP	United Nations Conference on Environment and Development
UNEP	United Nations Environment Programme
UNICEF	United Nations Children's Fund
vCJD	(New) Variant Creutzfeld-Jacob Disease
VLA	Veterinary Laboratories Agency
VOC	Volatile Organic Compound
WHO	The World Health Organization
WTO	The World Trade Organization
ZD	Zero Defects
ZT	Zero Tolerance

BIBLIOGRAPHY

Advisory Group on the Medical Aspects of Air Pollution Episodes (1995) *Health Effects of Exposures to Mixtures of Air Pollutants*, London: HMSO.

Anderson, R.M. and May, R.M. (1991) *Infectious Diseases of Humans*, Oxford: Oxford University Press.

Ayres, D.C. and Hellier, D.G. (1998) *Dictionary of Environmentally Important Chemicals*, London: Blackie Academic and Professional.

Bassett, W.H. (ed.) (1999) *Clay's Handbook of Environmental Health*, London: E. & F.N. Spon.

Beaghole, R. Bonita, R., and Kjellstrom, T. (1993) *Basic Epidemiology*, Geneva: World Health Organization.

Bennet, J.C. and Plum, F. (eds) (1996) *Cecil Textbook of Medicine*, W.B. Philadelphia: Saunders and Co.

Berry R.W. *et al.* (1996) *Indoor Air Quality in Homes – Parts 1 and 2*, Watford: Building Research Establishment.

Bowman R.C. and Emmett E.D. (1998) *A Dictionary of Food Hygiene*, London: Chadwick House Group Limited.

Campden and Chorleywood Food Research Association (1997) *HACCP: A Practical Guide*, 2nd edn, Technical Manual No. 38, Chipping Campden, Gloucestershire.

Committee on the Medical Effects of Air Pollution, Department of Health (1997) *Handbook on Air Pollution and Health*, London: The Stationery Office.

Communicable Disease and Public Health (journal) (various), Colindale, London: PHLS.

Detels, R., Holland, W.W., McEwen, J. and Omenn, G.S. (1997) *Oxford Textbook of Public Health*, Oxford: Oxford University Press.

Environmental Health Today (2002), CD-ROM version, London: Chadwick House Group Limited.

Frazier, W.C. and Westhoff, D.C. (1988) *Food Microbiology*, London: McGraw-Hill.

Genetically Modified Foods: Facts Worries, Policies and Public Confidence (1999), London: Department of Trade and Industry.

Guidelines for the Production of Heat Preserved Foods (1994), London: HMSO.

Hobbs, B.C. and Roberts, D. (1993) *Food Poisoning and Food Hygiene*, London: Edward Arnold.

Holland, W.W. and Stewart, S. (1997) *Public Health, the Vision and the Challenge*, London: The Nuffield Trust.

Kemp D.D. (1998) *The Environment Dictionary*, London: Routledge.

The Impact of Genetic Modification on Agriculture, Food and Health, (1999), London: The British Medical Association.

Institute for Environment and Health (1996) *IEH Assessment on Indoor Air Quality in the Home*, Leicester: IEH.

Institute of Hydrology (1994) *The Implications of Climate Change for the National Rivers Authority*, London: HMSO.

The Interdepartmental Working Group on Tuberculosis (1998) *The Prevention and Control of Tuberculosis in the United Kingdom*, London: Department of Health.

Living with Radiation (1998), Oxford: National Radiological Protection Board.

MacPherson, G. (ed.) (1999) *Black's Medical Dictionary*, 39th edn, London: A. & C. Black.

The Microbiological Safety of Food; Part 1 (1990), Report of the Committee on the Microbiological Safety of Food, Chairman Sir Mark Richmond, London: HMSO.

The Microbiological Safety of Food; Part 2, (1991), Report of the Committee on the Microbiological Safety of Food, Chairman Sir Mark Richmond, London: HMSO.

Palmer, S.R., Lord Soulsby and Simpson, D.I.H. (1998) *Zoonoses*, Oxford: Oxford Medical Publications.

Salisbury, D.M. and Begg, N.T. (eds) (1996) *1996 Immunisation against Infectious Disease*, London: HMSO.

Sustainable Development – The U.K. Strategy (1994), London: HMSO.

Sutherland, J.P., Varnam, A.H. and Evans, M.G. (1986) *A Colour Atlas of Food Quality Control*, London: Wolfe Publishing Ltd.

Todd E.C.D. and MacKenzie J.M. (eds) (1993) *Escherichia coli O157:H7 and other Verotoxigenic E.coli in Foods*, Ottawa: Bureau of Microbial Hazards.

West, G.P. (2001) *Black's Veterinary Dictionary*, 20th edn, London: A. & C. Black.